Mechanical Response of Engineering Materials

Second Edition

Richard A. Queeney
Albert E. Segall

Penn State University

Cover images courtesy of Albert E. Segall.

www.kendallhunt.com
Send all inquiries to:
4050 Westmark Drive
Dubuque, IA 52004-1840

ISBN 978-1-4652-5193-0

Printed in the United States of America
10 9 8 7 6 5 4

DEDICATION

This book is dedicated to the memory of Richard A. Queeney who has always been, and always will be the "heart and soul," of this book, regardless of how many changes are made down the road. Rick was a good friend who taught me as much about engineering as he did about life and unfortunately death. His great sense of humor and an innate ability to see the ironic (and yes, moronic) things in life, even when very ill, will be dearly missed. My only hope is that this book does Rick justice by continuing to be a good guide for engineering students, while still capturing his wit, knowledge, and love for the profession and all it entails!

A significant share of the royalties from this book is donated to support cancer research, and in particular, Glioblastoma Multiforme.

ACKNOWLEDGEMENTS

Few things are possible without a cast of thousands, and this book is no exception. I would therefore like to acknowledge and thank the following who have helped me in so many ways that space and time will not allow a full rendering.

I would first like to thank Donna Queeney for her guidance and support that was a crucial mix of inspirational, legal, and financial.

I would also like to thank Judith Todd of Engineering Science and Mechanics for her patience and assistance on so many levels I do not know where to begin.

Thanks also go to my wife Beth, daughter Sarah, and son Jake Segall who helped with typing, editing, and all survived countless blank stares over the past few years.

I would also like to thank Barbra Shaw, Cliff Lissenden, Ivi Smid, Dinesh Agrawal, Don Heaney, and Chantal Binet for their suggestions, edits, and help.

Liz Liadis is gratefully thanked for retyping the original manuscript.

Last, but certainly not least, I would like to thank Lucas Passmore for creating the new figures used in this version and for technically editing the manuscript with many great suggestions.

CONTENTS

CHAPTER ONE: THE ANALYSIS OF STRAIN 1

1.1 DEFINITION OF STRAIN COMPONENTS 1
1.2 STRAINS AND DISPLACEMENT FIELDS 2
1.3 STRAIN TRANSFORMATIONS 5
1.4 TECHNIQUES OF STRAIN MEASUREMENT 12
1.5 DETERMINATION OF STRAIN IN A PLANE ON A FREE SURFACE 12

CHAPTER TWO: THE ANALYSIS OF STRESS 17

2.1 STRESS AT A POINT 17
2.2 UNITS OF STRESS 20
2.3 STRESS SIGN 20
2.4 PLANE AND THREE-DIMENSIONAL STRESS STATES 21
2.5 EQUILIBRIUM OF STRESSES 22
2.6 AXIS ROTATIONS AND STRESS TRANSFORMATIONS 24
2.7 THREE-DIMENSIONAL STRESS TRANSFORMATIONS 32
2.8 COMPLETE THREE-DIMENSIONAL ANALYSIS 37
2.9 STRUCTURAL STRESS ANALYSIS 39
2.10 PRINCIPLE OF SAINT VENANT 42
2.11 STRESS CONCENTRATIONS 42

CHAPTER THREE: ELASTIC BEHAVIOR OF MATERIALS 51

3.1 ELASTIC RESPONSE 51
3.2 HOOKE'S LAW FOR ISOTROPIC SOLIDS 53
3.3 THERMAL STRAINS IN ISOTROPIC SOLIDS 55
3.4 CONSTRAINED THERMAL STRESSES AND STRAINS 56
3.5 HOOKE'S LAW FOR ANISOTROPIC MATERIALS 63
3.6 ORTHOTROPIC COMPOSITES 66
3.7 ELASTIC CONSTANT TRANSFORMATIONS 69

CHAPTER FOUR: TENSILE AND COMPRESSIVE RESPONSE 75

4.1 INTRODUCTION 75
4.2 TENSILE TEST VARIABLES 75
4.3 TEST PROCEDURES 76
4.4 TEST MEASUREMENTS 76
4.5 ELASTIC RESPONSE MEASURES 77
4.6 INELASTIC (PLASTIC) TENSILE RESPONSE 81
4.7 STRAIN HARDENING 84
4.8 YIELD POINTS 85
4.9 INHOMOGENEOUS STRAINS AND YIELDING 85
4.10 FRACTURE IN TENSION 86
4.11 BRITTLE STRENGTH 88
4.12 BENDING STRENGTH MEASURES 89
4.13 HARDNESS MEASURES AND METHODS 90
4.14 ENERGY MEASURES IN TENSION 92

CHAPTER FIVE: YIELDING FOR MULTIAXIAL STRESS STATES 101

5.1 INTRODUCTION 101
5.2 YIELDING IN PURE SHEAR 101
5.3 YIELDING FOR ARBITRARY STRESS STATES 104
5.4 MAXIMUM SHEAR STRESS THEORY 105
5.5 DISTORTION ENERGY THEORY 109
5.6 INTERNAL FRICTION THEORY 113
5.7 THEORY VALIDITY 117

CHAPTER SIX: PLASTIC DEFORMATION AND DUCTILE FAILURE 121

6.1 INTRODUCTION 121
6.2 INSTANTANEOUS (TRUE) STRAIN 121
6.3 INSTANTANEOUS OR TRUE STRESS 123
6.4 STRAIN HARDENING IN TENSION 123
6.5 LOAD INSTABILITY IN TENSION 125

CHAPTER SEVEN: BRITTLE AND LOW DUCTILITY FAILURE ANALYSIS 129

7.1 INTRODUCTION 129
7.2 FRACTURE OF PERFECTLY BRITTLE MATERIALS 129
7.3 THEORETICAL COHESIVE STRENGTH 132
7.4 BRITTLE FAILURE UNDER COMBINED STRESS 133
7.5 MODES OF CRACK EXTENSION 135
7.6 MATERIAL FRACTURE RESISTANCE 139
7.7 HIGH-STRENGTH ALLOYS AND PLASTICITY ASPECTS 140
7.8 PLANE STRAIN FRACTURE TOUGHNESS TESTING 144
7.9 FRACTURE MECHANICS DESIGN CONCEPTS 145
7.10 FAILURE ANALYSIS AND FRACTOGRAPHY 147

CHAPTER EIGHT: FATIGUE FAILURE 157

8.1 INTRODUCTION 157
8.2 FATIGUE ENDURANCE 157
8.3 THE STATISTICAL NATURE OF S-N DIAGRAMS 158
8.4 MEAN STRESS EFFECTS 160
8.5 CUMULATIVE DAMAGE CONCEPTS 162
8.6 NOTCH EFFECTS 165
8.7 THE PHYSICAL NATURE OF FATIGUE FAILURE 166
8.8 FATIGUE CRACK PROPAGATION 166
8.9 FATIGUE FRACTOGRAPHY 170

CHAPTER NINE: VISCOELASTICITY 177

9.1 INTRODUCTION 177
9.2 VISCOELASTIC MODEL ELEMENTS 177
9.3 MAXWELL VISCOELASTIC MODEL 178
9.4 VISCOELASTIC TESTING 179
9.5 MAXWELL MATERIAL RESPONSE 180
9.6 VOIGHT-KELVIN AND COMPOUND VISCOELASTIC MODELS 183

9.7 GENERAL LINEAR VISCOELASTIC RESPONSE 186
9.8 REAL POLYMER VISCOELASTICITY 186
9.9 EXPERIMENTAL RESPONSE DETERMINATION 187

CHAPTER TEN: CREEP 199

10.1 INTRODUCTION 199
10.2 CREEP BEHAVIOR 199
10.3 CREEP STRAIN PREDICTION 200
10.4 CREEP STRAIN PREDICTION 201
10.5 CREEP RUPTURE 207
10.6 CREEP DAMAGE MECHANISMS 211
10.7 HIGH TEMPERATURE MATERIALS 211

NUMERICAL ANSWERS TO PROBLEM ASSIGNMENTS 217

INDEX 223

CHAPTER ONE

THE ANALYSIS OF STRAIN

1.1 DEFINITION OF STRAIN COMPONENTS

It should hopefully come as no great surprise that except for kryptonite under Superman's sway, all solids deform under the action of loads such as forces and moments. In fact, the reader is probably familiar with the large visible deformations associated with rubbery materials (elastomers) such as rubber bands, especially their ability to induce pleasure in someone else when "snapped." In addition to these great qualities, changes in field variables may lead to deformations in selected solids; temperature fluctuations change the dimensions of all solids while nuclear radiation or variations in water content can induce swelling. In fact, anyone with an old wooden boat knows that it must be allowed to "swell" after placement in the water, lest the boat (and unfortunate boater) goes straight to Davy Jones's locker. Offshore pleasures aside, there are numerous reasons why these size and shape changes, or deformations in engineering parlance are of concern.

For example, a precision machine tool can not be so flexible that it is unable to maintain the required dimensions of the work piece it processes. Of a less obvious nature is the need to understand deformations in the solution of statically indeterminate structural problems. Interestingly, many if not most structures are of this type, and their analysis cannot be completed without consideration of deformations. Finally, as shall be seen in subsequent chapters, experimental determination of internally transmitted forces usually involves measuring deformations, and relating the latter to the former via constitutive laws and algebraic expressions relevant to the service application.

To arrive at the deformation measures most commonly employed in engineering practice, one must first consider a small, rectangular plate in the xy-plane, as simplistically (but still eloquently) depicted in Figure 1.1.

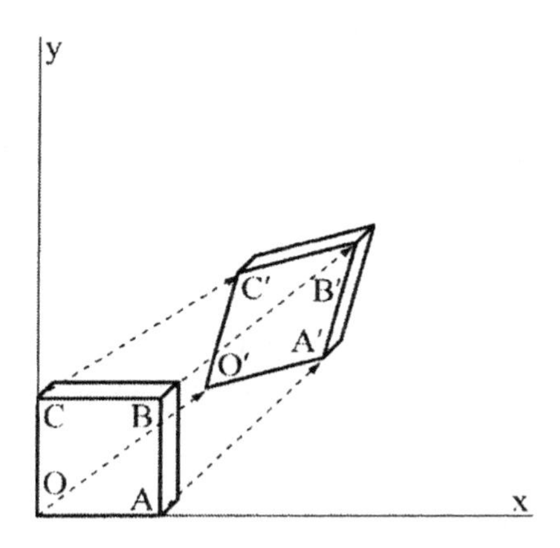

Figure 1.1 Deformation of a plane element.

Originally, the plate, or material element, is bounded by points OABC in the undeformed state. After being subjected to a system of forces and moments (unseen here), the points OABC are translated to O'A'B'C', as shown in the figure. In addition to being translated and rotated, the element has also been deformed by being stretched (or shrunken) in the x- and y-directions; as such, it has also been distorted from a rectangle to a parallelogram. To describe the deformation of a plane element (in the plane), three measures are needed: two for stretches and shrinks, called *normal* deformations, and one for distortion, or *shear* deformation.

Normal Strains equal fractional changes in length per unit-length of a predetermined line (sometimes referred to as a gage line or mark). The symbol chosen for normal strain is ε, and it is doubly subscripted with the coordinate sense of interest. Thus:

$$\varepsilon_{xx} = \lim_{OA\to 0} \frac{O'A' - OA}{OA} \tag{1.1a}$$

$$\varepsilon_{yy} = \lim_{OC\to 0} \frac{O'C' - OC}{OC} \tag{1.1b}$$

It should be noted that the limit is specified because the deformations under the applied load(s) may not be uniform and change from point to point. Moreover, the normal strains are unit-less (length/length), so the strain retains the same numerical value whether measured as inches/inch, meters/meter, or Light-Years/Light-Years. Clearly, the sign of the normal strain is fixed by the sign of the change in length or numerator: Positive size changes lead to positive strains, referred to as *tensile*, whereas negative changes lead to *compressive*, or negatively signed strains.

Shear Strains equal the tangent of the change in angle of what was originally a right angle as determined by the coordinate axis directions. As shown by Equation 1.2, the symbol chosen for shear strain is γ and is also doubly subscripted with the axis directions defining the originally right angle.

$$\gamma_{xy} = \lim_{OA,OC\to 0} \tan\left(\angle AOC - \angle A'O'C'\right) \tag{1.2}$$

As before, the limit is taken to accommodate potential changes in distortion or shear from point to point. For many of the materials of interest to the structural engineer or designer, the ensuing distortions and deformations will be small (less that 10^{-2}) because the materials are relatively stiff. Therefore, the tangent of the angular change will be indistinguishable from the angle change itself (in radians), and one may simply use:

$$\gamma_{xy} = \gamma_{yx} \cong \lim_{OA,OC\to 0} \left(\angle AOC - \angle A'O'C'\right) = \lim_{OA,OC\to 0} \left(\frac{\pi}{2} - \angle A'O'C'\right) \tag{1.3}$$

The sign of the shear strain depends upon whether A′O′C′ is greater than $\pi/2$ (negative shear strain) or less than $\pi/2$ (positive shear strain), with the angle change measured from x-axis to y-axis, counterclockwise positive, for γ_{xy}, and vice versa for γ_{yx}.

1.2 STRAINS AND DISPLACEMENT FIELDS

Strains, and the deformations from which they arise, typically vary from point to point within a structure; recall the case of a beam subjected to a pure moment where the axial flexural strains are zero on the neutral axis and increase linearly to their maxima on the outer fibers of the beam. Even for an apparently uniform strain state such as a thin strip subjected to tension, strains may vary on a local and near-microscopic scale, as real materials contain pores, inclusions, and/or reinforcements. In order to analyze this potential variation, it is prudent to refer back to Figure 1.1. In this case, let the displacement of point O to point O′ be described vectorially by *r*, where:

$$r = u\hat{i} + v\hat{j} \tag{1.4}$$

Using this definition, the x- and y-components of the vector, *r* become *u* and *v*, respectively. What about the displacement of point A to A′? If all of the point displacements were *r*, there would be no relative displacement or deformation, and, hence, no strain for the system.

Thus, for the case of interest where there strain exists, point A must move more (or less) than point, O in either the x- or y-sense. Consider Figure 1.2, which depicts the motion of the element OABC to final state O′A′B′C′ (note that the original element is Δx by Δy in extent). If point O moves by an amount *u* in the x-sense as it becomes positioned at O′, then A must move an amount $u + \delta u$ while still reaching A′; δu might be either positive or negative.

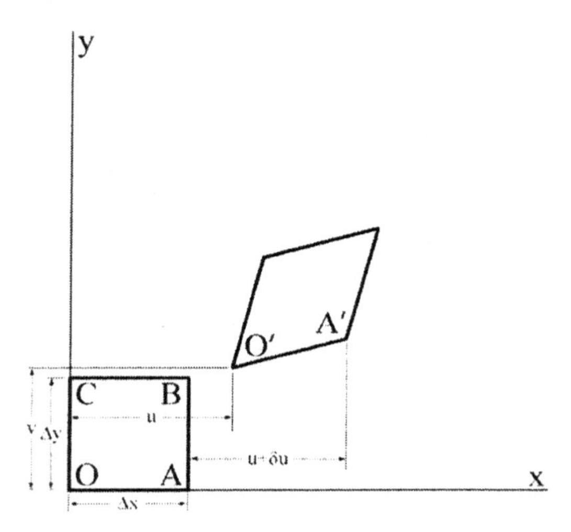

Figure 1.2 Deformation and relative displacements of a plane element.

Under this scenario, the variation quantity δu may be expressed as:

$$\delta u = \frac{\partial u}{\partial x}\Delta x \tag{1.5}$$

Equation (1.5) assumes that the displacement, *u* is a continuous function of position *x* and is differentiable; this assumption is equivalent to denying the development of holes or cracks during displacement. In actual practice, holes and/or cracks may be a consequence of service, but the reader must deal with them as already existing, not appearing during deformation; more on such discontinuities and the techniques to handle them will follow in subsequent chapters. Similar variations must take place, from point to point, in both the horizontal and vertical senses. The resulting total displacement field is shown in Figure 1.3.

Using the now complete strain definitions, Equations (1.1) and (1.3), can finally be written as:

$$\varepsilon_{xx} = \lim_{OA \to 0} \frac{O'A' - OA}{OA} = \lim_{\Delta x \to 0} \frac{\left(\Delta x + \frac{\partial u}{\partial x}\Delta x\right) - \Delta x}{\Delta x} = \frac{\partial u}{\partial x} \tag{1.6a}$$

$$\varepsilon_{yy} = \lim_{OC\to 0} \frac{O'C' - OC}{OC} = \lim_{\Delta y \to 0} \frac{\left(\Delta y + \frac{\partial v}{\partial y}\Delta y\right) - \Delta y}{\Delta y} = \frac{\partial v}{\partial y} \tag{1.6b}$$

$$\gamma_{xy} = \lim_{OA,OC\to 0} (\angle AOC - \angle A'O'C') =$$

$$\lim_{\Delta x, \Delta y \to 0} \left(\frac{\left(v + \frac{\partial v}{\partial x}\Delta x\right) - v}{\Delta x} + \frac{\left(u + \frac{\partial u}{\partial y}\Delta y\right) - u}{\Delta y} \right) = \frac{\partial v}{\partial x} + \frac{\partial u}{\partial y} \tag{1.6c}$$

Equations (1.6) now define the three in-plane strain components in terms of continuous displacement fields.

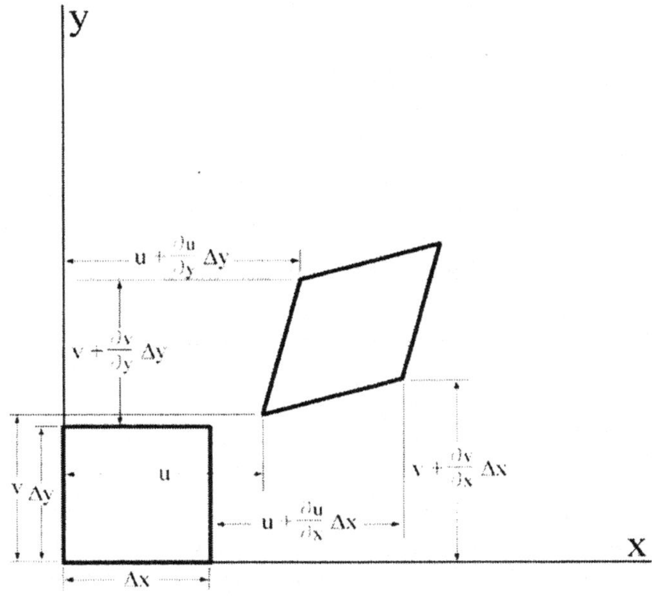

Figure 1.3 Complete displacement fields for a plane material element.

More generally, strain might be expected to be a three-dimensional parameter, reflecting all stretches (or shrinks) and distortions in and out of the xy-plane. If the displacement vector, *r* is three-dimensional, then it may be rewritten as:

$$r = u\hat{i} + v\hat{j} + w\hat{k} \tag{1.7}$$

As before, each displacement component *u, v,* and/or *w* is a continuous differentiable function of *x, y,* and *z*. By inference from Equation (1.6), there are six independent strain components, or:

$$\varepsilon_{xx} = \frac{\partial u}{\partial x} \tag{1.8a}$$

$$\varepsilon_{yy} = \frac{\partial v}{\partial y} \qquad (1.8b)$$

$$\varepsilon_{zz} = \frac{\partial w}{\partial z} \qquad (1.8c)$$

$$\gamma_{xy} = \gamma_{yx} = \frac{\partial v}{\partial x} + \frac{\partial u}{\partial y} \qquad (1.8d)$$

$$\gamma_{yz} = \gamma_{zy} = \frac{\partial v}{\partial z} + \frac{\partial w}{\partial y} \qquad (1.8e)$$

$$\gamma_{zx} = \gamma_{xz} = \frac{\partial w}{\partial x} + \frac{\partial u}{\partial z} \qquad (1.8f)$$

For editorial simplicity, the symbol ε_{ij} will be understood to represent a three-by-three array consisting of all strain components at any point of interest:

$$\varepsilon_{ij} = \begin{vmatrix} \varepsilon_{xx} & \frac{\gamma_{xy}}{2} & \frac{\gamma_{xz}}{2} \\ \frac{\gamma_{yx}}{2} & \varepsilon_{yy} & \frac{\gamma_{yz}}{2} \\ \frac{\gamma_{zx}}{2} & \frac{\gamma_{zy}}{2} & \varepsilon_{zz} \end{vmatrix} \qquad (1.9)$$

It can be easily shown that $\gamma_{xy} = \gamma_{yx}$, $\gamma_{yz} = \gamma_{zy}$, and $\gamma_{zx} = \gamma_{xz}$. Moreover, the factor of ½ in the shear terms has been arbitrarily (at least at this point) inserted into the array. As it turns out, the inclusion of the ½ factor makes the array a square matrix, subject in its manipulation to the rules of matrix algebra, an endeavor we shall not pursue formally at this point.

1.3 STRAIN TRANSFORMATIONS

Strain measurement devices are often constrained to determine values in specific directions, perhaps due to the geometry of the structural member and the problems of attaching a device as detailed in section 1.4. As is usually (and unfortunately) the case, the desired measurement direction(s) may not turn out to be the ones of most intense design interest. In fact, these directions may be dependent on the magnitudes of various loadings, and not at all obvious to the observer. Given these concerns, a brief analysis of the effect of rotating coordinate axes about their origin on the strain magnitudes will in turn serve to develop the equations that lead to the strain measures of most interest (usually maximum values). Not unexpectedly, changed strain values obtained by rotating axes are also referred to as *strain transformations*.

For the analysis, two sets of coordinate axes will be employed as shown in Figure 1.4. The "original" axes are x and y, and the "new," or rotated (by angle θ) axes are now primed, or x' and y'. Displacements measured in the x-, y-axis frame are denoted u and v as before, with displacements in the new $x'y'$-frame denoted as u' and v', respectively. The positions of points x and y (or x', y'), as well as the displacements u

and v (or u' and v'), are all determined by using the appropriate coordinate axis set (original or new), respectively. From the geometry shown in Figure 1.5, the following transformational equivalences may be written:

$$x = x' \cos\theta - y' \sin\theta \tag{1.10a}$$

$$y = x' \sin\theta + y' \cos\theta \tag{1.10b}$$

$$u' = u \cos\theta + v \sin\theta \tag{1.10c}$$

$$v' = u \sin\theta + v \cos\theta \tag{1.10d}$$

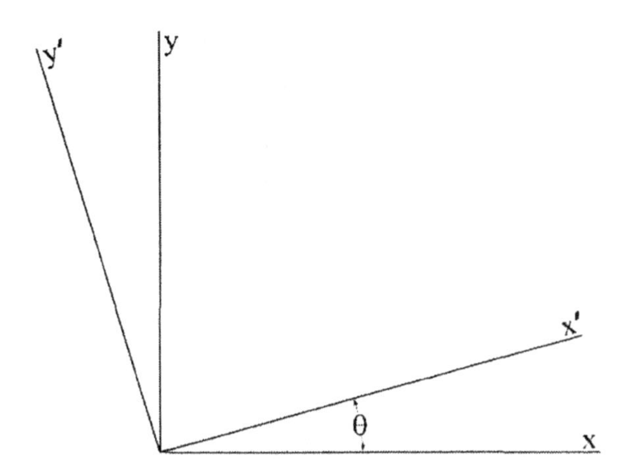

Figure 1.4 Original and rotated (primed) coordinate axes in the xy-plane.

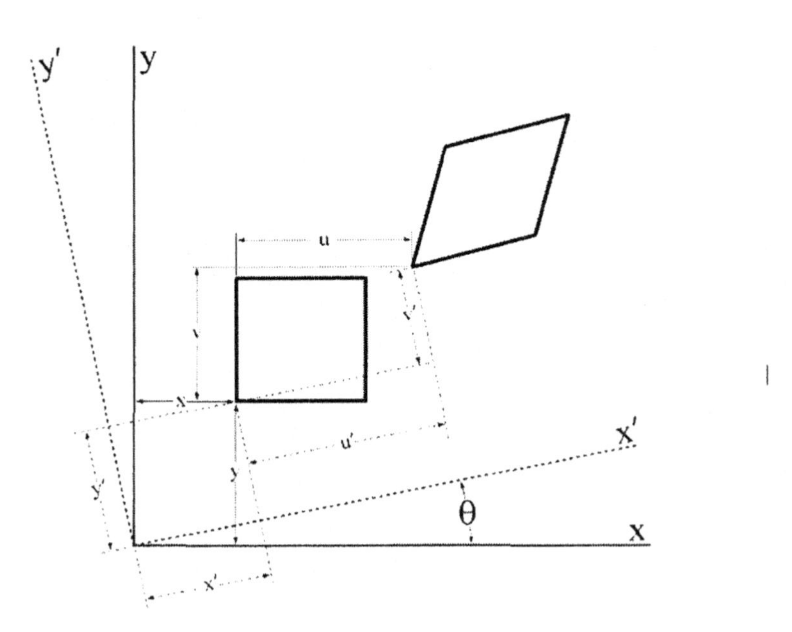

Figure 1.5 Measures in original and rotated x- and y-axes.

The strain components, ε_{ij} in the new axis set would still be resolved via equations (1.5), with displacements u' and v' determined in the x' and y' directions. On the other hand, the relationships between original and new, or rotated measures are found by recalling the much loved chain rule for partial derivatives:

$$\varepsilon'_{xx} \equiv \frac{\partial u'}{\partial x'} = \frac{\partial u'}{\partial x}\frac{\partial x}{\partial x'} + \frac{\partial u'}{\partial y}\frac{\partial y}{\partial x'} \tag{1.11a}$$

$$\varepsilon'_{yy} \equiv \frac{\partial v'}{\partial y'} = \frac{\partial v'}{\partial x}\frac{\partial x}{\partial y'} + \frac{\partial v'}{\partial y}\frac{\partial y}{\partial y'} \tag{1.11b}$$

$$\gamma'_{xy} \equiv \frac{\partial v'}{\partial x'} + \frac{\partial u'}{\partial y'} = \left(\frac{\partial v'}{\partial x}\frac{\partial x}{\partial x'} + \frac{\partial v'}{\partial y}\frac{\partial y}{\partial x'}\right) + \left(\frac{\partial u'}{\partial y}\frac{\partial y}{\partial y'} + \frac{\partial u'}{\partial x}\frac{\partial x}{\partial y'}\right) \tag{1.11c}$$

Terms such as $\partial x/\partial x'$ or $\partial x/\partial y'$ may be easily found by partially differentiating Equations (1.10) and utilizing Equation (1.5) where appropriate. After differentiating and substitution, not to mention some tedious algebra (the devil is always in the details), one finally obtains:

$$\varepsilon'_{xx} = \varepsilon_{xx}\cos^2\theta + \varepsilon_{yy}\sin^2\theta + \gamma_{xy}\sin\theta\cos\theta \tag{1.12a}$$

$$\varepsilon'_{yy} = \varepsilon_{xx}\sin^2\theta + \varepsilon_{yy}\cos^2\theta - \gamma_{xy}\sin\theta\cos\theta \tag{1.12b}$$

$$\gamma'_{xy} = 2\left(\varepsilon_{yy} - \varepsilon_{xx}\right)\sin\theta\cos\theta + \gamma_{xy}\left(\cos^2\theta - \sin^2\theta\right) \tag{1.12c}$$

Equation (1.12) relates strains measured in one coordinate frame to the strains in another for the same deformed body. Interestingly, these expressions are analogous to the cosine law of addition for vectors, except that the resulting Transformation Equations involve higher order trigonometric functions among other things.

As mentioned earlier, knowing the maximum values of strain is usually of prime importance for design. However, before searching out angles, θ for which either normal or shear strains are maximum, Equation (1.12) will be algebraically manipulated into a more telling and useful form. Using the following trigonometric identities as substitutes in Equation (1.12):

$$\sin 2\theta = 2\sin\theta\cos\theta \tag{1.13a}$$

$$\cos 2\theta = \cos^2\theta - \sin^2\theta = 2\cos^2\theta - 1 = 1 - 2\sin^2\theta \tag{1.13b}$$

The strain transformation equations become:

$$\varepsilon'_{xx} = \frac{\varepsilon_{xx} + \varepsilon_{yy}}{2} + \frac{\varepsilon_{xx} - \varepsilon_{yy}}{2}\cos 2\theta + \frac{\gamma_{xy}}{2}\sin 2\theta \tag{1.14a}$$

$$\varepsilon'_{yy} = \frac{\varepsilon_{xx} + \varepsilon_{yy}}{2} - \frac{\varepsilon_{xx} - \varepsilon_{yy}}{2}\cos 2\theta - \frac{\gamma_{xy}}{2}\sin 2\theta \qquad (1.14b)$$

$$\frac{\gamma'_{xy}}{2} = \frac{\varepsilon_{xx} - \varepsilon_{yy}}{2}\sin 2\theta + \frac{\gamma_{xy}}{2}\cos 2\theta \qquad (1.14c)$$

Although it may not be overly obvious to the reader at this point, especially if their eyes are beginning to glaze over due to the tedious math involved, Equation (1.14) is equivalent to Equation (1.12). The utility (and beauty) of the former over the latter is that Equation (1.14) can be graphically represented as a circle in a strain space as discovered by Otto Mohr in 1853. *Poor Otto; while his discovery has an inherent beauty and usefulness for engineering, it will nonetheless go down in history as the ultimate confusion factor for many engineering students.*

Confusion factor notwithstanding (and hopefully eliminated in this and subsequent chapters), the not-so-revered circular representation is constructed in the following manner via steps A-E:

A. The strain space has axes as depicted below, with the normal strain in the horizontal direction and one-half the shear in the vertical direction.

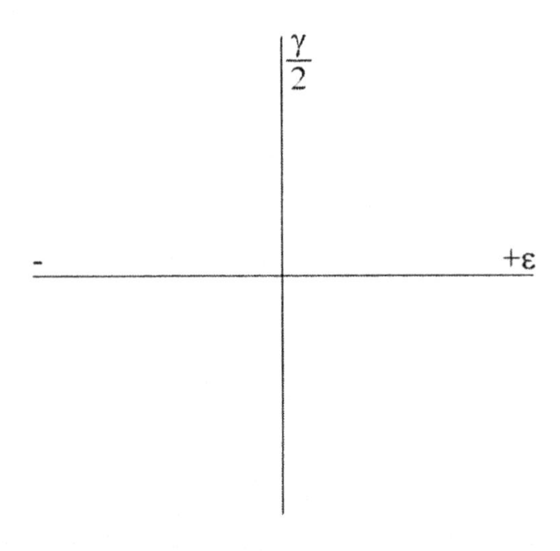

Figure 1.6 Normal and shear strain axes.

B. Normal strains are now plotted; their sign is the factor governing placement to the right (positive) or left (negative) of the origin.

C. A state of shear strain is represented by two points, one above the axis an amount $\gamma/2$, the other below by the same value. Two cases are possible as indicated in Figure 1.7

D. The two strain-space points of ε_{xx}, $\gamma_{xy}/2$ and ε_{yy}, $\gamma_{yx}/2$ when connected by a straight line, define the circle diameter. The circle center, C is then:

$$C = \frac{\varepsilon_{xx} + \varepsilon_{yy}}{2} \qquad (1.15a)$$

The circle radius, *R* about that circle center is:

$$R = \sqrt{\left(\frac{\varepsilon_{xx} - \varepsilon_{yy}}{2}\right)^2 + \left(\frac{\gamma_{xy}}{2}\right)^2} \qquad (1.15b)$$

E. All other states of strain, for different choices of θ, as given by equations (1.14) and (1.12), are points on the circle. Angles, θ in the actual physical space translate to rotations of twice that amount or 2θ, on the circle; the rotational sense, clockwise or counterclockwise, is the same.

a: Shear strain positive:

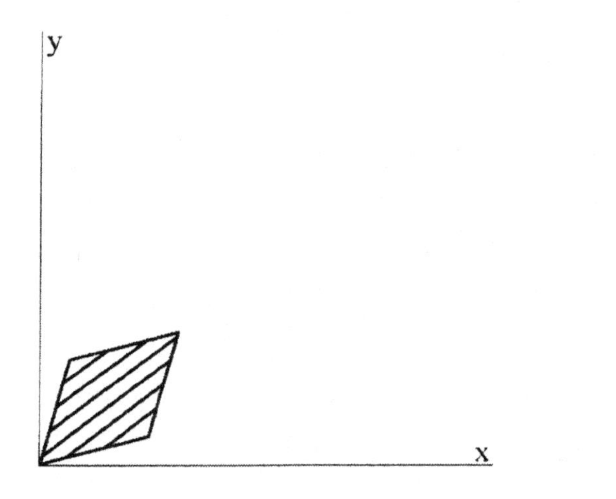

$\gamma_{xy}/2$ is plotted downward
$\gamma_{yx}/2$ is plotted upward

b: Shear strain negative:

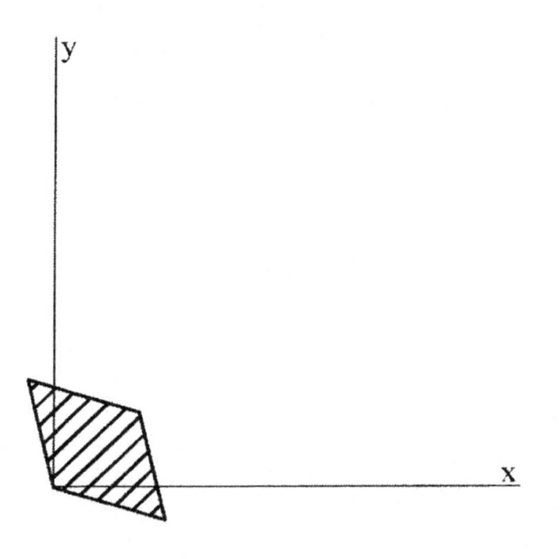

$\gamma_{xy}/2$ is plotted upward
$\gamma_{yx}/2$ is plotted downward

Figure 1.7 Positive (a) and negative (b) shear strain orientations.

Example Problem 1-1: A plane state of strain at a point on an aircraft wing has been determined to be:

$$\varepsilon_{ij} = \begin{vmatrix} 800 & 600 & 0 \\ 600 & -400 & 0 \\ 0 & 0 & 0 \end{vmatrix} \text{x}10^{-6}\ \text{(mm/mm)}$$

Construct Mohr's Circle for Strain and determine: (a) the maximum and minimum normal strains (hereafter referred to as *principal strains*) and the directions in which they act, and (b), the maximum shear strains and their directions of action. From rules B and C, above, the circle appears as shown in Figure 1.8e.

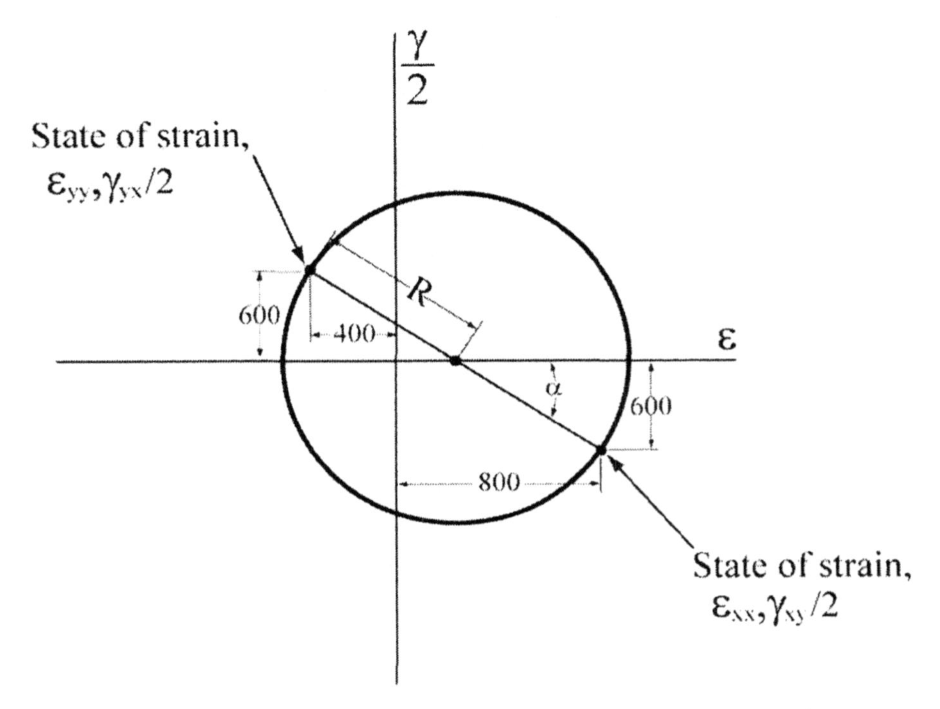

Figure 1.8 Strain-space points and corresponding Mohr's circle.

The circle center, *C* is:

$$C = \frac{\varepsilon_{xx} + \varepsilon_{yy}}{2} = \frac{800 + (-400)}{2} \text{x}10^{-6} = 200\ \text{x}10^{-6}$$

The circle radius, *R* is found to be:

$$R = \sqrt{\left(\frac{\varepsilon_{xx} - \varepsilon_{yy}}{2}\right)^2 + \left(\frac{\gamma_{xy}}{2}\right)^2} = \sqrt{\left(\frac{800 - (-400)}{2}\right)^2 + (600)^2}\ \text{x}10^{-6} = 848\ \text{x}10^{-6}$$

The rotational angle, α is found from:

$$|\tan\alpha| = \left| \frac{\frac{\gamma_{xy}}{2}}{\frac{\varepsilon_{xx} - \varepsilon_{yy}}{2}} \right| = \frac{600 \times 10^{-6}}{600 \times 10^{-6}} = 1 \qquad \rightarrow \qquad \alpha = 45^0$$

It is very important to note that the principal strain values are *always* located on the normal strain axis (where $\gamma = 0$). The principal values are 180 degrees apart on the circle, corresponding to 90 degrees apart in real space.

Calling the plane principal values ε_I and ε_{II}, then:

$$\varepsilon_I = C+R = (200 + 848) \times 10^{-6} = 1048 \times 10^{-6} \text{ (mm/mm)}$$

$$\varepsilon_{II} = C-R = (200 - 848) \times 10^{-6} = -648 \times 10^{-6} \text{ (mm/mm)}$$

The principle direction on the wing associated with the principal strain, ε_I is found by rotating from the state-of-strain point (ε_{xx}, $\gamma_{xy}/2$) to ε_I by 45 degrees in the counterclockwise sense around the circle periphery. In real space, rotating $\alpha/2$, or 22.5 degrees, in the counterclockwise sense from the x-axis brings one to the direction corresponding to the direction of ε_I or the x_I-direction. The x_{II}-direction is 90 degrees (or half of 180 degrees) from the x_I axis, all of which is graphically represented in Figure 1.9.

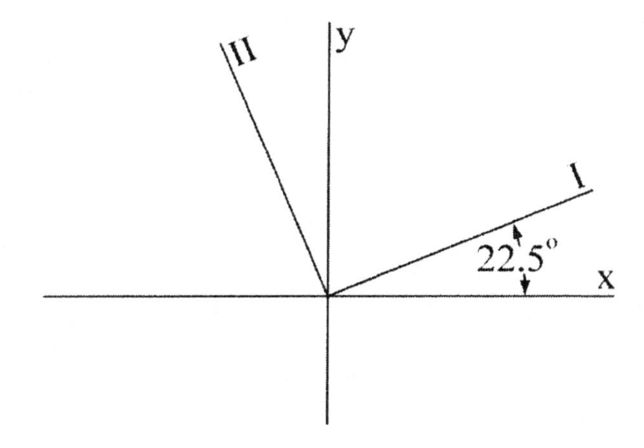

Figure 1.9 Orientation of the principal axes relative to the x- and y-axis.

The maximum shear strain in the xy-plane is the circle radius, *R*

$$\frac{\gamma_{max}}{2} \equiv R = 848 \times 10^{-6} \text{ (mm/mm)}$$

It is important to note that the maximum shear strain is 90 degrees from the principal strain states on the circle, or 45 degrees in real space. Its sense including how the angles become either larger or smaller depends on which side (top or bottom, respectively) of the circle is considered. Finally, strain transformations can be algebraically defined in three-dimensional axis rotation cases. However, strain measurements are almost exclusively carried out experimentally on actual surfaces, so that plane rotations are of the most practical interest.

1.4 TECHNIQUES OF STRAIN MEASUREMENT

Strain measurement has been a critical factor throughout the history of material and structural testing. In principle, rulers and protractors should suffice to determine *normal* and *shear* strains, respectively. Unfortunately, life is never easy (Mohr's circle anyone?), so in practice, strain determination is rather challenging as they may be less than 10^{-3} in magnitude or smaller. Nonetheless, tensile-strains were determined to one part in a million (10^{-6}) during the late Middle Ages using relatively crude microscopes with graduated eyepieces. The high degree of accuracy was possible due to the large "L" values in $\Delta L/L$ with long filaments being tested by hanging them from cathedral ceilings. In more recent times, various mechanical devices were invented that utilized smaller gage lengths; these read ΔL-values by magnifying the actual deformations through a complicated systems of gears or levers, or by ingeniously projected light beams that were reflected from mirrors affixed to the deforming body.

For the most part, modern structural strain measurements techniques trace their origins to the observed change in the electrical conductivity of a metallic wire as it is stretched. Among his many achievements, Lord Kelvin found that as long as the wire was elastically stretched, the fractional change in electrical resistance, $\Delta R/R$, was proportional to the fractional change in length, $\Delta L/L$, or:

$$\frac{\Delta R}{R} = f\,\frac{\Delta L}{L} = f\,\varepsilon \qquad (1.16)$$

The factor, f in Equation (1.16) is the *strain sensitivity* of the metal and is unique to the alloy composition and its thermo-mechanical history; f is usually between 2.0 and 4.0 for those alloys commonly employed as strain sensors. Resolutions of the order of 10^{-6} are more easily obtained in electrical resistance measurements than in deformation determinations by using Wheatstone bridge techniques in conjunction with precision (formerly wire-wound) resistors.

It was not until the latter half of the 1930s that capitalism prevailed and the Kelvin effect was successfully commercialized. In order to obtain sufficiently large values of resistance while still maintaining practically small gage lengths, *L* the wire was wound back and forth in the gage direction. A schematic of the resulting wire pattern is illustrated in Figure 1.10. The first commercial strain gages (dubbed "SR-4" gages, after their co-inventors Simmons and Ruge, who collaborated with a lawyer and an entrepreneurial engineer named Tatnall, for "4") were wire-wound and fixed to a piece of tissue paper, which could then be glued to the structure in question. Nowadays, most electrical resistance strain gages are punched from thin foils by precision dies or cut via lithography, encapsulated in epoxy or other polymeric materials, and conveniently mounted with quick-setting epoxy adhesives. Gage lengths vary from as small as 1/64 inch to 2 or 3 inches long. For the latest in strain gage technology (many other types of sensing elements have been developed) or any of the alternative means of deformation measurement, consult any recent reference on experimental stress analysis.

1.5 DETERMINATION OF STRAIN IN A PLANE ON A FREE SURFACE

At one time, stress and strain analysis was a field of endeavor typified by great mathematical complexity (to the average engineer), and real design geometries and loadings could only be solved in approximate ways. Unfortunately, these approximate solutions cried out for validation, which was dutifully provided by the experimentalist using strain gages or other appropriate (and unfortunately dying) techniques such as Photoelasticty and Moiré Interferometry. Indeed, often no analytical solutions could be provided, and experimental examination of prototype models was the only avenue open to the designer. At the present time, numerical calculation codes, such as "finite element" programs allow stress and strain analyses to be performed by ordinary mortals. However, a complete confidence in calculated results, particularly for new and difficult structural geometries, can only be established through the application of experimental techniques. *A very simple rule to live and engineer by should be "Trust, but verify!"*

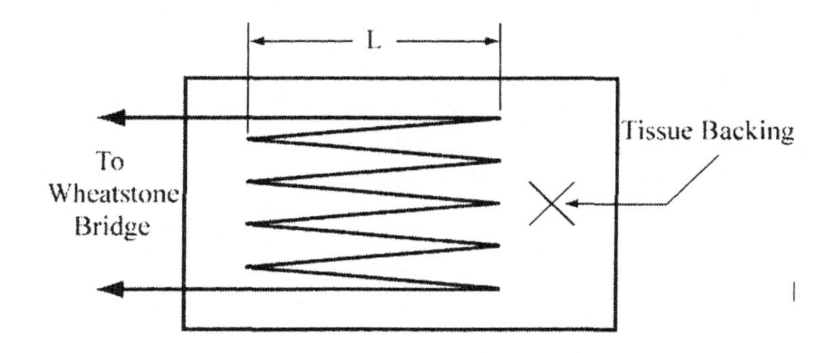

Figure 1.10 Tissue-mounted, wound wire strain gages such as the "SR-4" variety.

When the engineer finally succumbs to the inevitable, the appropriate experiments are usually, though not necessarily always, carried out on the surface of a structure. Thus, strains such as ε_{xx}, ε_{yy}, and γ_{xy} must, in general, be determined. Unfortunately, a strain sensor that directly determines values of shear strain, γ_{xy} for small gage sections has never been invented. However, electrical resistance strain gages that only determine normal strains can in-fact be employed to indirectly determine the state of plane shear, γ_{xy} based on three independent strain quantities. For this to work, the experimental technique must employ three resistance gages mounted in known and non-colinear directions; since capitalists always seem to come to the rescue, a readymade device already exists to perform this task as shown by the 45-degree strain-rosette in Figure 1.11.

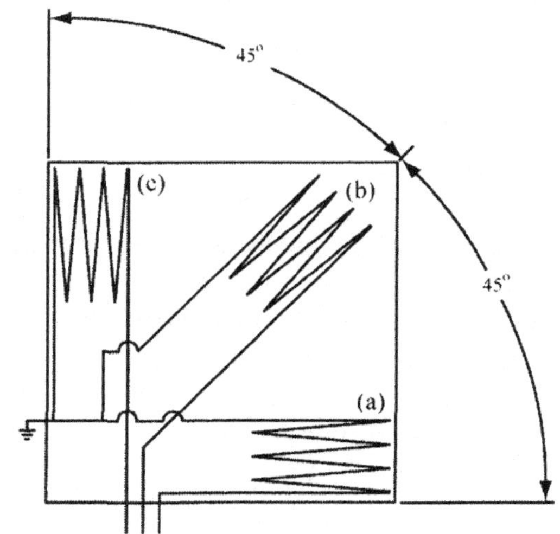

Figure 1.11 Schematic of a 45-degree strain-rosette.

Having affixed the rosette to the surface and the structure subsequently loaded, the strain in gages (a), (b), and (c) are determined. If the (a) and (c) gages are aligned in the *x*- and *y*-directions, respectively, then

$\varepsilon_{xx} = \varepsilon_a$ and $\varepsilon_{yy} = \varepsilon_c$. To find γ_{xy}, Mohr's Strain Circle must be utilized (I did tell you it would be useful) in the data analysis.

Note that gages (a) and (c) are 90 degrees apart so that strains ε_a and ε_c will be diametrically opposed (180 degrees apart) on the strain circle. As such, the circle center becomes $C = \frac{1}{2}(\varepsilon_a + \varepsilon_c)$. On the other hand, the circle radius cannot be directly calculated as that would presuppose knowledge of γ_{xy}. What the experimentalist does know are the three strain values, the circle center, and that the strain ε_b must lie on the circle midway between the points ε_a and ε_c, as sketched in Figure 1.12.

Note that in Figure 1.12, the analysis is equivalent to fitting a rigid "T" of equal arm length (equal to *R*), such that it is inclined at the proper angle, α to allow it to "touch" all the normal strain lines simultaneously. Given this arrangement, there is indeed a unique *R* that will fit the three normal-strains as measured.

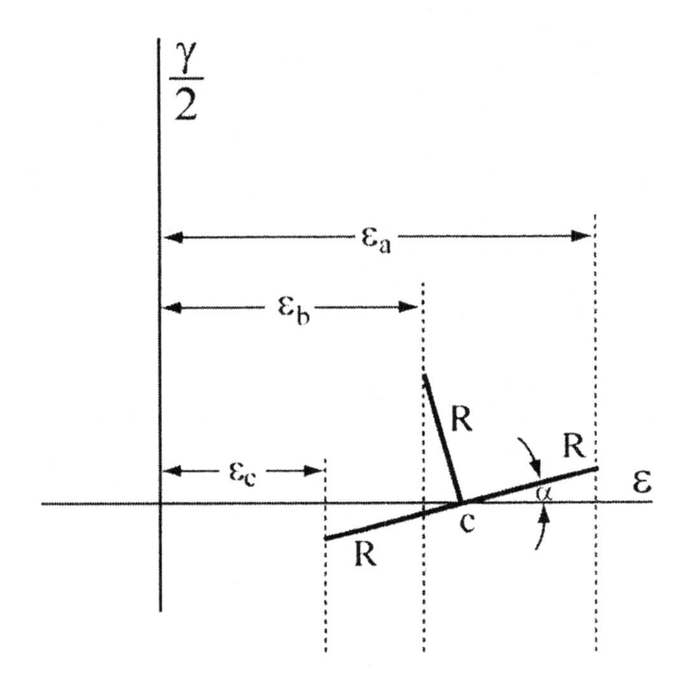

Figure 1.12 Alignment and orientation of the three strains measured by the strain-rosette.

Based on the geometry shown in Figure 1.12, the resulting two independent equations are therefore $\varepsilon_a = C + R\cos\alpha$ and $\varepsilon_b = C - R\sin\alpha$. Solving these equations simultaneously:

$$\cos\alpha = \frac{(\varepsilon_a - C)\sin\alpha}{(C - \varepsilon_b)} \tag{1.17a}$$

then:

$$R\cos\alpha = R\left[\frac{(\varepsilon_a - C)\sin\alpha}{(C - \varepsilon_b)}\right] = \varepsilon_a - C \tag{1.17b}$$

-or-

$$R \sin\alpha = C - \varepsilon_b = \frac{\varepsilon_a + \varepsilon_c}{2} - \varepsilon_b = -\frac{\gamma_{xy}}{2} \quad (1.17c)$$

And finally, the desired result:

$$\gamma_{xy} = 2\varepsilon_b - (\varepsilon_a + \varepsilon_c) = 2\varepsilon_b - 2C \quad (1.18)$$

Hence, the state of surface strain is now completely determined because the components in the x-, y-plane are specified.

$$\varepsilon_{ij} = \begin{vmatrix} \varepsilon_{xx}(=\varepsilon_a) & \frac{\gamma_{xy}}{2} & 0 \\ \frac{\gamma_{yx}}{2} & \varepsilon_{yy}(=\varepsilon_c) & 0 \\ 0 & 0 & 0 \end{vmatrix} \quad (1.19)$$

Note that the magnitude of ε_b relative to that of the circle center, C determines the sign of the shear strain. Finally, although the strain component ε_{zz} is set at zero in the above array, it could be anything for the rosette experiment because the gages cannot sense it.

PRACTICE EXERCISES

1-4. The strain states in Problems 1-4 were determined from measurements using electrical resistance strain gages. on (1) an automotive frame, (2) an aircraft wing, (3) reactor vessel, and (4) a large football stadium super-structure Determine the following for each state of strain (units of micro-strain: μ in./in. or μ m/m):

a. The magnitude and sign of the principal strains.

b. The magnitude of the maximum shear strain.

c. The coordinate directions of the principal strains relative to the original x, y axes.

1.
$$\varepsilon_{ij} = \begin{vmatrix} 900 & 150 & 0 \\ 150 & -100 & 0 \\ 0 & 0 & 0 \end{vmatrix} \text{x}10^{-6}$$

2.
$$\varepsilon_{ij} = \begin{vmatrix} 0 & -200 & 0 \\ -200 & 800 & 0 \\ 0 & 0 & 0 \end{vmatrix} \text{x}10^{-6}$$

3.
$$\varepsilon_{ij} = \begin{vmatrix} 800 & 0 & -500 \\ 0 & 200 & 0 \\ -500 & 0 & -100 \end{vmatrix} \text{x}10^{-6}$$

4.
$$\varepsilon_{ij} = \begin{vmatrix} 1000 & -200 & 0 \\ -200 & 200 & 0 \\ 0 & 0 & 0 \end{vmatrix} \text{x}10^{-6}$$

5. For a 45-degree strain gage rosette mounted to Darth Vader's prosthetic arm as he deftly "handles "rebel scum," the following observations were made for gages a, b, and c (see Figure 1.11). Determine the principle strains and the orientation of the principle axes for each case. All strains are in units of micro-strain and therefore have orders of magnitude of 10^{-6}.

Case	ε_a	ε_b	ε_c
i	600	200	-200
ii	-200	400	100
iii	-700	-320	200

CHAPTER TWO

THE ANALYSIS OF STRESS

2.1 STRESS AT A POINT

Except for the few remaining flat earth believers, it is certainly accepted by most that forces and moments applied to a structural member are transmitted through and along the member. In fact, at any point within the member, *equilibrium of forces and moments must be observed, lest the component collapse into itself or fly off into space or who knows where.* Since there are not a lot of mini black holes or projectiles about, it is safe to say that the forces and moments are internally reacted and more importantly (at least for the engineer), that these reactions do have implications for design. In terms of determining how the member reacts to these loads and the practical use of this knowledge for design, the most common method is to look at the resulting stresses.

Perhaps the best way of examining how the stresses are generated within a body is to examine the infamous "Stress Potato" as indicated below in Figure 2.1. More than just bad, albeit tasty "Carbs" or the remains of Mr. Potato Head after meeting a Vegamatic, the section as shown below is assumed to (and in fact, must) be in static equilibrium under the action of all externally applied forces, F and moments, M.

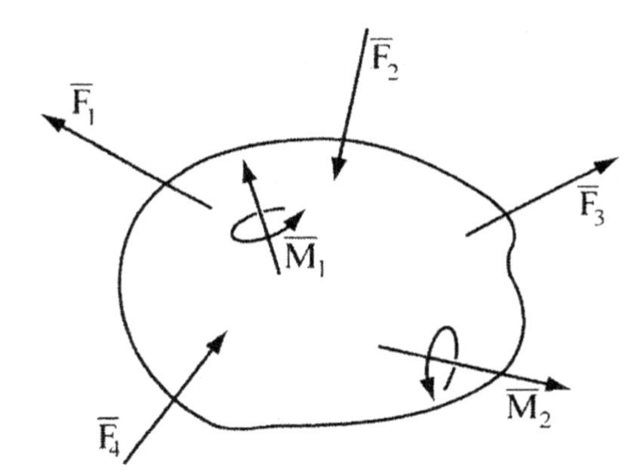

Figure 2.1 Behold the stress potato; no ordinary spud, this is an arbitrary structural member subjected to a combination of forces F_1, F_2 ...and moments M_1, M_2, ... etc.

From a design standpoint, the engineer must ultimately be concerned with the ability of the structural member to carry the prescribed loading. This in turn requires one to determine the very nature of the transmitted forces within the member. Accordingly, a cut is made such that a cross-sectional area "A" is exposed as shown in Figure 2.2. Because the right-side segment of the member has been eliminated by sectioning, the transmitted forces may be represented by a resultant force vector, denoted herein as $\overline{R}$.

The magnitude and inclination of $\overline{R}$ including its line of action are determined by the demands of force and moment equilibrium, respectively. In vector form, one may write $\overline{R}$ in the following way:

$$\overline{R} = R_x\hat{i} + R_y\hat{j} + R_z\hat{k} \tag{2.1}$$

where the resultant has been decomposed into components in the coordinate directions already indicated in Figure 2.2.

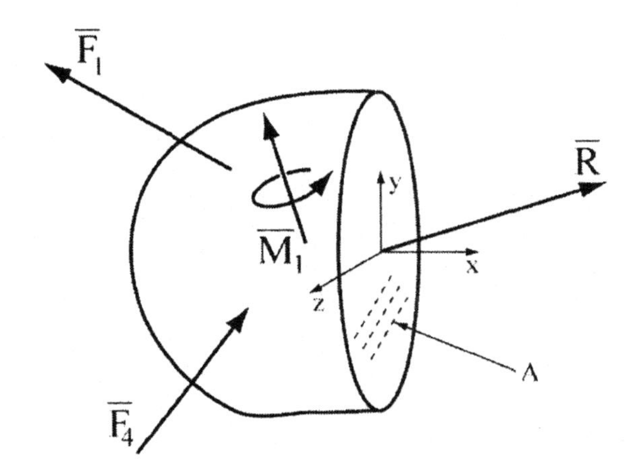

Figure 2.2 Free-body diagram of a sectioned structural member in equilibrium.

Because few things in life and universe are simple, it is not particularly realistic to assume that the resultant transmitted force physically acts at the point (0,0,0) as indicated in the figure. More generally, it might prove worthwhile to assume that the resultant is actually distributed in some way over the entire cross-section. Thus, the concept of stress as the distribution of a force over a cross-section was born.

In its most basic sense, stress and pressure can be considered analogous since both are a measure of a distributed force and in fact, share the same units (force/area). However, pressures are always normal to a surface and do not regularly vary with location for most (but not all) situations, so there are some distinct differences. In order to determine the more complex stress state, first consider a very small segment dA_x of the total area A, subscripted with an "x" to denote the coordinate sense of its *outwardly directed normal*. Acting on that area segment is an increment of force, dR, the whole of which is illustrated in Figure 2.3.

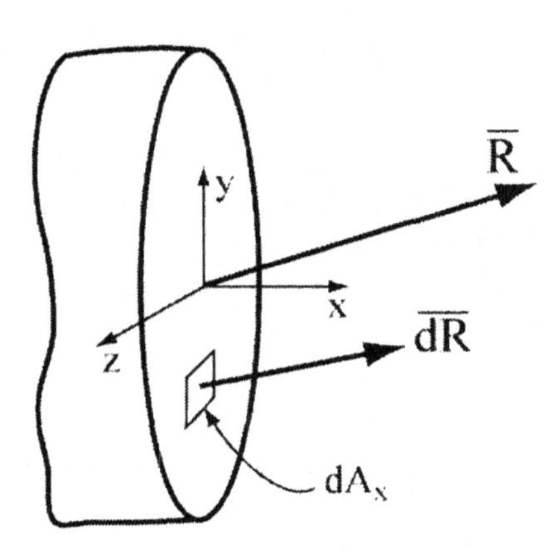

Figure 2.3 Transmitted force increment $d\overline{R}$ acting on area element dA_x.

Equilibrium still rules the day, so one can safely assume:

$$\sum_{\substack{\text{all} \\ dA_x}} d\bar{R} = \bar{R} \tag{2.2}$$

with

$$d\bar{R} = d\bar{R}_x\hat{i} + d\bar{R}_y\hat{j} + d\bar{R}_z\hat{k} \tag{2.3}$$

Although the force resultants and their increments serve to describe the overall equilibrium of the above structural member, the concept of *stress*, or a distributed force is certainly more useful and convenient in determining exactly what forces act at any point in the structure. Given the potential orientations of the forces with respect to our sectional cut, two types of stress must therefore be defined.

Normal Stresses: designated herein by the symbol, σ are associated with force components acting normal or perpendicular to the cross-section plane of interest. While not necessarily obvious from the name, *Normal Stresses* are generally associated with stretching (tension) and shrinkage (compression).

Shear Stresses: denoted by the symbol, τ relate to the internal force components parallel to the section plane. Although there is nothing "abnormal" about *Shear Stresses*, they are most easily and commonly associated with warping or twisting.

Since both types of stresses may vary from point to point on any given cross-section, it is necessary to precisely define them at each point. However, force increments and cross-section planes both possess directionality, so the stress components must ultimately reflect this dual directional sense; each component will therefore be assigned two subscripts as indicated below:

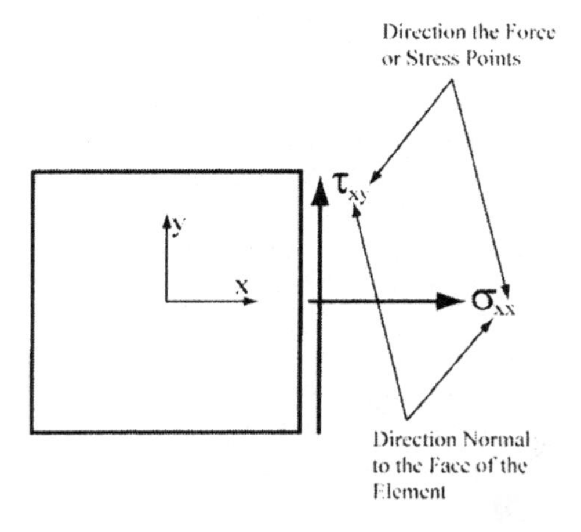

Figure 2.4 Dual directionality of stress; the first subscript reflects the direction of the outward normal while the second shows the force/stress direction.

For the internal force arrangement shown in Figure 2.3, the resulting three stress components may be defined as follows:

$$\sigma_{xx} = \lim_{dA_x \to 0} \frac{d\bar{R}_x}{dA_x} \tag{2.4a}$$

$$\tau_{xy} \equiv \lim_{dA_x \to 0} \frac{d\bar{R}_y}{dA_x} \tag{2.4b}$$

$$\tau_{xz} \equiv \lim_{dA_x \to 0} \frac{d\bar{R}_z}{dA_x} \tag{2.4c}$$

where the limit operations shown in Equations (2.4) are used to ensure that the stresses are defined on a point-by-point basis.

While we may have spared our friend Spud with only one cut, sections could also be made such that surface normals are created in the y- and z-directions, respectively. Repeating the entire discussion on the previous page that resulted in Equations (2.4), one would obtain one normal and two shear stress components at the physical point in question on each cross-section, or σ_{yy}, σ_{zz}, τ_{yx}, τ_{yz}, τ_{zx}, and τ_{zy}. Practically speaking, specific structural examples may lead to component values of zero for many of these nine components.

2.2 UNITS OF STRESS

Despite the futile attempts of many engineers, politicians, and even the very costly Martian probe fiasco (wrong units programmed into the controls), two systems of stress units coexist in the United States. The older traditional units, once referred to as "English" units, are pounds per square inch or *psi*. Because these units are almost solely used within the United States, they are called *U.S. Customary Units*. The International system of units (SI) related to metric measures, denotes stresses in Newtons per square meter (N/m^2) or *Pascals*. Not surprisingly, many segments of American industry have stubbornly resisted the adoption of SI units due to the costs involved in conversion of tooling, blueprints, and so on; while these may be valid excuses, plain, old human cantankerousness has led to some foot dragging in conversion efforts. Eventually, the need to compete in the international market place, or to communicate with more advanced aliens (such as the designers of the ill-fated Martian probe), will force the metric conversion issue and *Newtons* and *Pascals* will no longer conjure up images of fig cookies or celery.

Most materials employed in load-bearing structural or machine designs are capable of carrying thousands of *psi*, so that stress magnitudes are often given in units of *ksi*, where one *ksi* equals 1,000 psi. In the SI system, thousands of *psi*, or *ksi*, correspond to millions of *Pa*; accordingly, stress magnitudes are quoted in units of *MPa*, where one *MPa* is on the order of 10^6 *Pa*. The conversion factors between the two units of stress measure are: 1 *psi* = 6.895×10^3 *Pa*.

2.3 STRESS SIGN

As discussed earlier, each stress component is computed by dividing the force increment by an area element, with both measures having directionality; the direction of the area element is defined by the sense of its outwardly directed normal. Force components acting in the positive coordinate sense are positively signed, and vice versa. The same sign convention is applied to the outwardly directed normal vector for each plane section. Once this convention has been established, it is very easy to then determine the sign of the associated stress component by using a very simple rule that was probably learned in elementary school: a positive times another positive or negative times a negative yields a positive and any mixed products always produce a negative.

Hence:

> If the force increment and the outwardly directed area normal are of the same sign (even if both are negative), the resulting stress is therefore positive (+).
>
> If the force increment and the outwardly directed area normal are of differing signs, regardless of the order, then the resulting stress is negative (-).

Of course, directionality and sign are dependent upon the choice of coordinate system imposed by the engineer; several examples of potential stress and their associated signs are indicated in Figure 2.5.

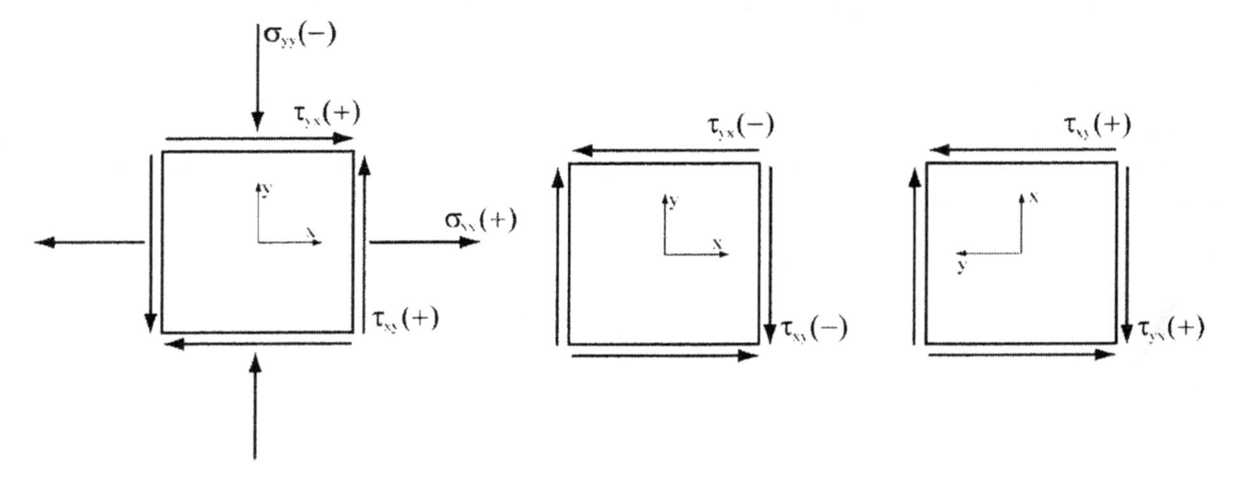

Figure 2.5 Two-dimensional stress states of various signs.

Note that the two elements on the right are physically identical, but the signs of the stresses differ as they are dependent on the arbitrary choice of positive axis direction. *Shear* stress signs in general are dependent on axis choice, but *normal* stresses are positive (tensile) or negative (compressive) regardless of the axis choice.

2.4 PLANE AND THREE-DIMENSIONAL STRESS STATES

Unfortunately, life is seldom simple and stress analysis is certainly no exception. Hence, in a general sense, there will actually be nine components of stress present at all points in a structure. These components may be indicated on a stress element as shown in Figure 2.6A. However, in many practical instances no doubt appreciated by "Stress Guessers" across the globe, the stress system may turn out to be two-dimensional in nature, with all stresses acting in one plane. Such often encountered two-dimensional stress states are appropriately referred to as *plane stress* as illustrated in Figure 2.6B where the xy-plane is assumed to be the plane of stress in question.

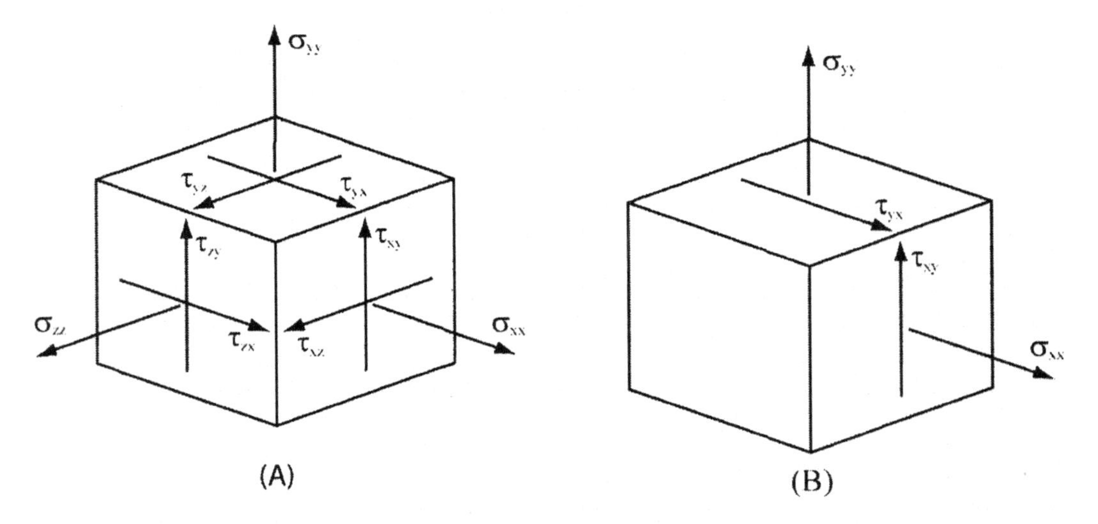

Figure 2.6 States of stress for (A) fully three-dimensional loading, and (B) plane stress in the xy-plane.

As with strain in Chapter One, stresses can be shown to possess the requisite algebraic properties that allow them to be treated as matrices, or alternatively, tensors of rank two. Although this text will mercifully not apply matrix or tensor analysis to any significant degree, it will often display stress states at a point in array form for editorial ease. In particular, if σ_{ij} is understood to stand for the three-by-three array of all nine stress components, then:

$$\sigma_{ij} = \begin{vmatrix} \sigma_{xx} & \tau_{xy} & \tau_{xz} \\ \tau_{yx} & \sigma_{yy} & \tau_{yz} \\ \tau_{zx} & \tau_{zy} & \sigma_{zz} \end{vmatrix} \qquad (2.5a)$$

For a state of *plane stress*, such as shown in Figure 2.6b, the matrix simplifies to:

$$\sigma_{ij} = \begin{vmatrix} \sigma_{xx} & \tau_{xy} & 0 \\ \tau_{yx} & \sigma_{yy} & 0 \\ 0 & 0 & 0 \end{vmatrix} \qquad (2.5b)$$

Obviously, a plane stress state can be defined by any two combinations of the x-, y-, or z-planes and not just the xy plane shown in Figure 2.6. Hence, do not always assume that xy plane is the only plane stress game in town.

2.5 EQUILIBRIUM OF STRESSES

Usefulness notwithstanding, force and moment equilibrium does indeed place constraints upon the manner in which stresses can vary from point to point within a structural member. For instance, consider the plane stress state shown in Figure 2.7; it is assumed that variations of stress in the element of thickness t, can be accurately described by the partial derivative of the stress in question with respect to a given coordinate direction times a small incremental distance moved in that direction. For example, the change in σ_{xx}, from $x = 0$ to $x = \Delta x$ is given by $(\partial\sigma_{xx}/\partial x)\ \Delta x$ and so on.

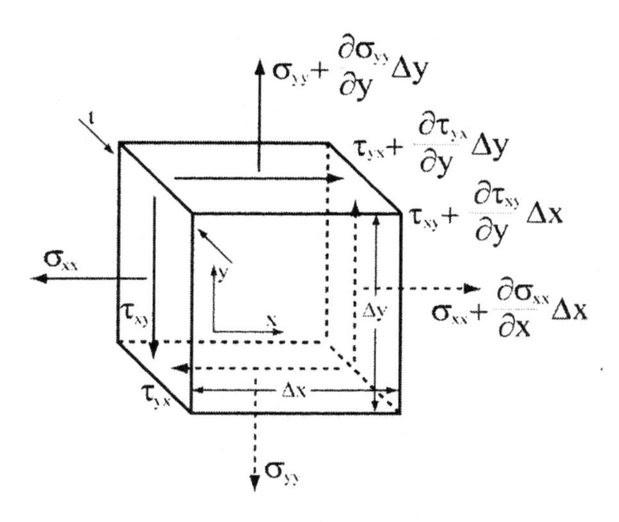

Figure 2.7 Stress variations in plane stress.

Considering moment equilibrium about point "o", at the lower left corner of the element in Figure 2.7:

$$\sum M_2 \equiv 0 = -(\sigma_{yy}\Delta x * t)\frac{\Delta x}{2} + \left[\left(\sigma_{yy} + \frac{\partial \sigma_{yy}}{\partial_y}\Delta y\right)\Delta x * t\right]\frac{\Delta x}{2} - \left[\left(\tau_{yx} + \frac{\partial \tau_{yx}}{\partial y}\Delta y\right)\Delta x * t\right]\Delta y + \ldots$$

$$\ldots + \left[\left(\tau_{xy} + \frac{\partial \tau_{xy}}{\partial x}\Delta x\right)\Delta y * t\right]\Delta x + (\sigma_{xx}\Delta y * t)\frac{\Delta y}{2} - \left[\left(\sigma_{xx} + \frac{\partial \sigma_{xx}}{\partial x}\Delta x\right)\Delta y * t\right]\frac{\Delta y}{2}$$

$$= \frac{\partial \sigma_{yy}}{\partial y}\left(\frac{\Delta x}{2}\right) - \frac{\partial \sigma_{xx}}{\partial x}\left(\frac{\Delta y}{2}\right) - \tau_{yx} + \tau_{xy} - \frac{\partial \tau_{yx}}{\partial y}(\Delta y) + \frac{\partial \tau_{xy}}{\partial x}(\Delta x) \tag{2.6}$$

In the limit as the stress element shrinks to a point and Δx and Δy approach zero, Equation (2.6) simply yields $\tau_{yx} = \tau_{xy}$. As such, shear stress components with permuted or interchanged indices, are equivalent; this result does not depend on the assumption of plane stress, so $\tau_{xz} = \tau_{zx}$ and $\tau_{yz} = \tau_{zy}$ as well. Thus, there are in reality, only three *independent* shear stress values (τ_{xy}, τ_{yz}, and τ_{xz}). More importantly, life is in fact, relatively simple, at least in this instance and one can rewrite the matrix array of stress components Equation (2.5a) as:

$$\sigma_{ij} = \begin{vmatrix} \sigma_{xx} & \tau_{xy} & \tau_{xz} \\ \tau_{xy} & \sigma_{yy} & \tau_{yz} \\ \tau_{xz} & \tau_{yz} & \sigma_{zz} \end{vmatrix} \tag{2.7}$$

Consider now force equilibrium in the x-direction, or:

$$\sum F_x \equiv 0 = \sigma_{xx}\Delta y * t + \left(\sigma_{xx} + \frac{\partial \sigma_{xx}}{\partial x}\Delta x\right)\Delta y * t - \tau_{yx}\Delta x * t + \left(\tau_{yx} + \frac{\partial \tau_{yx}}{\partial y}\Delta y\right)\Delta x * t \tag{2.8a}$$

-or-

$$\sum F_x = 0 = \frac{\partial \sigma_{xx}}{\partial x} + \frac{\partial \tau_{yx}}{\partial y} = \frac{\partial \sigma_{xx}}{\partial x} + \frac{\partial \tau_{xy}}{\partial y} \tag{2.8b}$$

Simply put, Equation (2.8) indicates that a gradient in one stress component must be compensated for by a related gradient in another apparently independent component. Although derived for fully three-dimensional stress states, Equation (2.8) remains the end result of moment equilibrium. On the other hand, the final form of the force equilibrium equations such as Equation (2.8) are somewhat altered. In this case, the three-dimensional form of the stress-equilibrium equations become:

$$\frac{\partial \sigma_{xx}}{\partial x} + \frac{\partial \tau_{xy}}{\partial y} + \frac{\partial \tau_{xz}}{\partial z} = 0 \tag{2.9a}$$

$$\frac{\partial \tau_{yx}}{\partial x} + \frac{\partial \sigma_{yy}}{\partial y} + \frac{\partial \tau_{yz}}{\partial z} = 0 \tag{2.9b}$$

$$\frac{\partial \tau_{zx}}{\partial x} + \frac{\partial \tau_{zy}}{\partial y} + \frac{\partial \sigma_{zz}}{\partial z} = 0 \tag{2.9c}$$

2.6 AXIS ROTATIONS AND STRESS TRANSFORMATIONS

When the stresses in a structural or machine member are analyzed as part of the design procedure, its natural geometry usually suggest an arrangement of coordinate axes. For example, it is "natural" when analyzing a beam to assign one cardinal direction to be coincident with the longitudinal axis (passing through the centroid of the beam's cross-section), and to align another axis with the direction of transverse loading (the true vertical in the case of gravity loading). The right-hand rule is then utilized to assign the third axis direction after choosing the two initial directions. Such "natural" choices usually lead to the easiest stress analysis and are thus, highly desirable. However, from a strength/design perspective, these axes and the ensuing analysis may not necessarily yield the extreme values of normal or shear stress that are relevant to setting upper limits on allowable loads. Moreover, these directions may not coincide with some structurally significant aspect or defect such as welds, cracks, or openings

As luck would have it (or lack thereof, to be more precise), the desired or maximum values of normal and/or shear stress will not necessarily be coincident with the designated axes of the structure. To illustrate this point, first consider a typical plane stress ($\sigma_{zz} = 0$) element as in Figure 2.8. An imaginary cut is made through the element and new axes x' and y' are assigned normal and parallel to the cut face, respectively. The axis x' is rotated an amount θ from the x-axis, as is the y' from the y-axis. Ultimately, the point of this maneuver is to answer the question: "is there an angle θ for which normal (or shear) stresses have extreme values?" The free body diagram for the cut stress element is given in Figure 2.9. Note that a normal stress σ'_{xx} and a shear stress τ'_{xy} have been assigned to the cut face to maintain equilibrium.

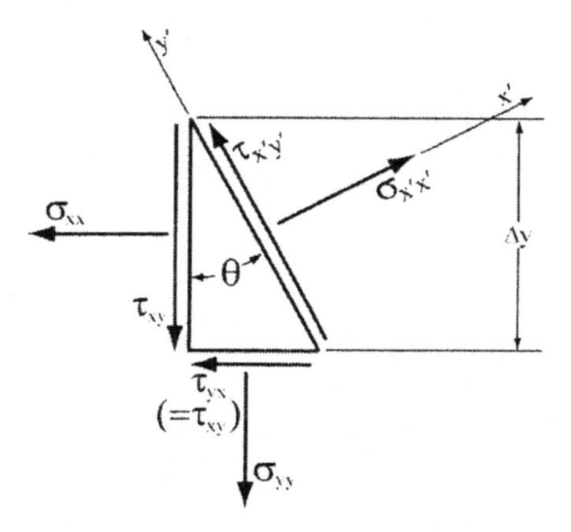

Figure 2.8 Cutting a plane stress element on a defined oblique plane.

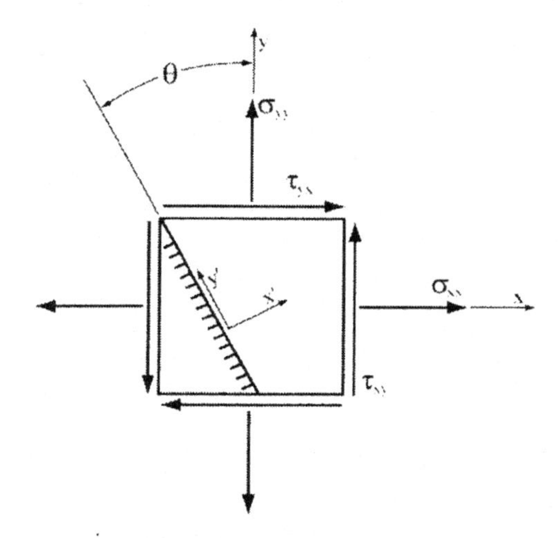

Figure 2.9 Equilibrium of stresses on a sectioned stress element.

Demanding moment equilibrium of the sectioned element of Figure 2.9 only leads to Equation (2.7), so not much gained here. Force equilibrium in the x'-direction is another matter however, resulting in:

$$\sum F_{x'} \equiv 0 = \sigma_{x'x'}\left(t * \frac{\Delta y}{\cos\theta}\right) - \sigma_{xx}(\Delta y * t)\cos\theta - \tau_{xy}(\Delta y * t)\sin\theta + \ldots$$
$$\ldots - \tau_{yx}(t * \Delta y \tan\theta)\cos\theta - \sigma_{yy}(t * \Delta y \tan\theta)\sin\theta \qquad (2.10a)$$

Similarly, in the y' direction:

$$\sum F_{y'} \equiv 0 = \tau_{x'y'}\left(t * \frac{\Delta y}{\cos\theta}\right) + \sigma_{xx}(\Delta y * t)\sin\theta - \tau_{xy}(\Delta y * t)\cos\theta + \ldots$$
$$\ldots + \tau_{yx}(t * \Delta y \tan\theta)\sin\theta - \sigma_{yy}(t * \Delta y \tan\theta)\cos\theta \qquad (2.10b)$$

In this form, Equations (2.10a and 2.10b) can be readily solved for both $\sigma_{x'x'}$ and $\tau_{x'y'}$, respectively. In addition, $\sigma_{y'y'}$ can be found without examining equilibrium further by noting that $\sigma_{y'y'} = \sigma_{x'x'}$ when $\theta \rightarrow \theta + 90^0$. Thus:

$$\sigma_{x'x'} = \sigma_{xx}\cos^2\theta + \sigma_{yy}\sin^2\theta + 2\tau_{xy}\sin\theta\cos\theta \tag{2.11a}$$

$$\sigma_{y'y'} = \sigma_{xx}\sin^2\theta + \sigma_{yy}\cos^2\theta - 2\tau_{xy}\sin\theta\cos\theta \tag{2.11b}$$

$$\tau_{x'y'} = \left(\sigma_{yy} - \sigma_{xx}\right)\sin\theta\cos\theta + \tau_{xy}\left(\cos^2\theta - \sin^2\theta\right) \tag{2.11c}$$

Equations (2.11 a-c.) are usually referred to as the *Stress Transformation Equations* for states of plane stress, and express (force) equilibrium among stresses at different coordinate axis positions. However, before searching for specific θ-values in Equation (2.11) that correspond to extreme normal and shear stress values, it is prudent to substitute the common trigonometric identities such that:

$$\sigma_{x'x'} = \left(\frac{\sigma_{xx} + \sigma_{yy}}{2}\right) + \left(\frac{\sigma_{xx} - \sigma_{yy}}{2}\right)\cos 2\theta + \tau_{xy}\sin 2\theta \tag{2.12a}$$

$$\sigma_{y'y'} = \left(\frac{\sigma_{xx} + \sigma_{yy}}{2}\right) - \left(\frac{\sigma_{xx} - \sigma_{yy}}{2}\right)\cos 2\theta - \tau_{xy}\sin 2\theta \tag{2.12b}$$

$$\tau_{x'y'} = \left(\frac{\sigma_{yy} - \sigma_{xx}}{2}\right)\sin 2\theta + \tau_{xy}\cos 2\theta \tag{2.12c}$$

As you may have already surmised, the preceding relationships serve a far greater purpose than just adding more equations to your lexicon. In fact and as mentioned earlier in Chapter One, Otto Mohr noted in 1853 that Equations (2.12) are those of a circle drawn in a space whose axes correspond to normal and shear stress. Equations (2.12) merely state that the stresses for any angle θ exist on this circle, including the extreme value cases of potentially the greatest interest for design. In the present discussion, Mohr's circle will be employed to find extreme values, rather than through the more tedious algebraic manipulation of Equations (2.11 and 2.12).

As shown by the space in question by Figure 2.10, the abscissa or normal-stress axis has positive (tensile) values plotted to the right and negative (compressive) values to the left. On the other hand, the vertical axis or ordinate represents shear stresses and features unusual labels: *cw* (clockwise or ↻) and *ccw* (counterclockwise or ↺) for up and down, respectively; if such directions are new or confusing to you, please do not despair as they have become increasingly confounding in the age of digital clocks. Digital downfalls notwithstanding, the designation *cw* refers to the twisting sense of the shear stress acting on one plane only, with *ccw* denoting the opposite sense. The examples of plane shear stress in Figure 2.10 show that if τ_{xy} has a *cw* sense, τ_{yx} will be *ccw*, and vice versa so that moment equilibrium is observed. All of these finer points, the construction of Mohr's circle, and much more are examined in the following example.

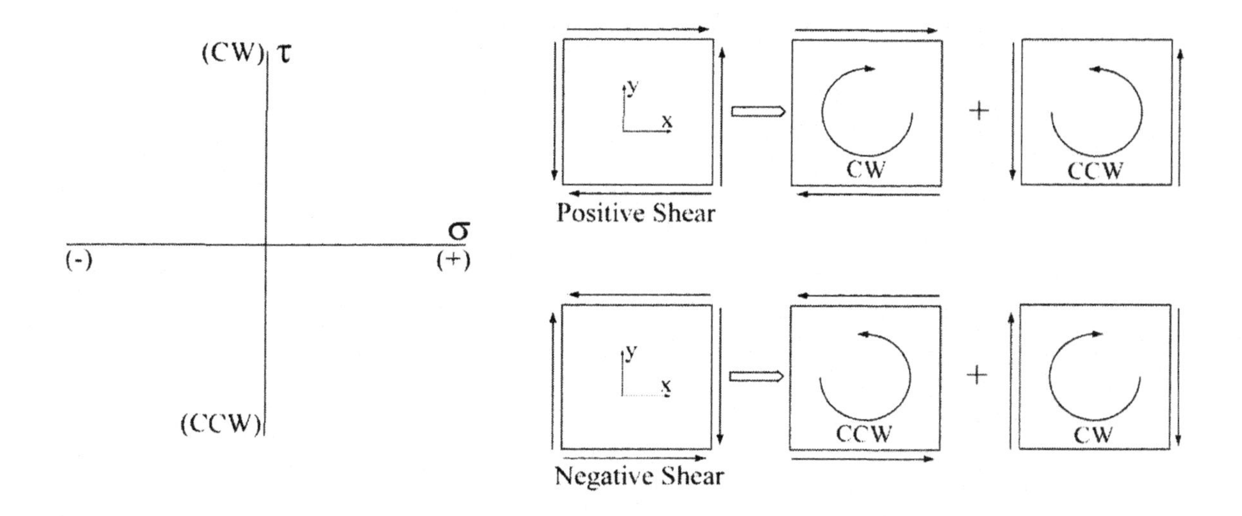

Figure 2.10 Mohr's stress space and twisting sense examples.

However, before the fun begins, it should be noted that poor Otto Mohr's circle is sometimes (alright, always) viewed by students as an archaic reminder of the "Dark Ages" when graphical methods reigned supreme and calculators and computers simply did not exist. Believe it or not, this is not the case as the circle (even when drawn by a computer) is a truly powerful and enlightening way to view a stress-state, its orientation, and most importantly, the maximum values required for design.

Example Problem 2-1: Your first assignment as a "Stress Guesser" at a local helicopter company is to analyze a Dunsel bracket of their new and improved "Death Copter ☠." A plane-stress analysis has been completed on the structure with the point of most concern found to have the following stress state.

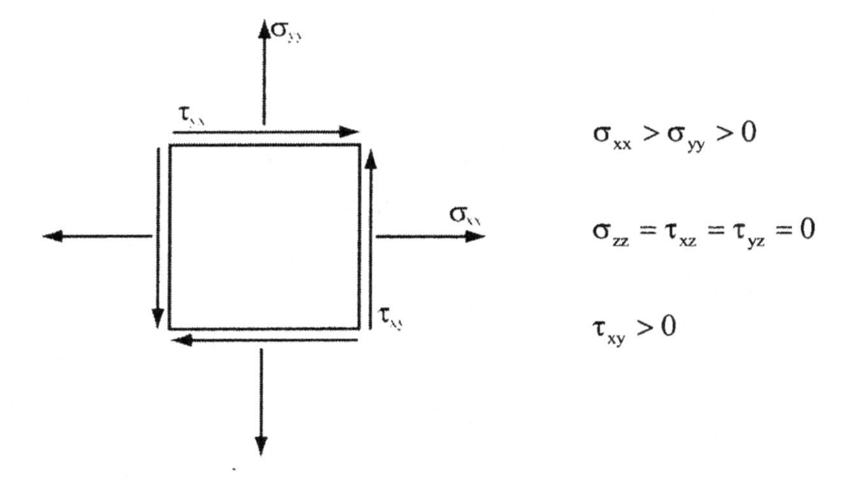

Figure 2.11 Stress element with normal and shear stresses.

On the σ–τ space below, vertical dashed lines are plotted that intersect the σ–axis at values of σ_{xx} and σ_{yy}. The shear values are plotted on those dashed lines, τ_{xy} on the σ_{xx}-line (it has a ccw sense) and τ_{yx} on σ_{yy} (it has cw sense).

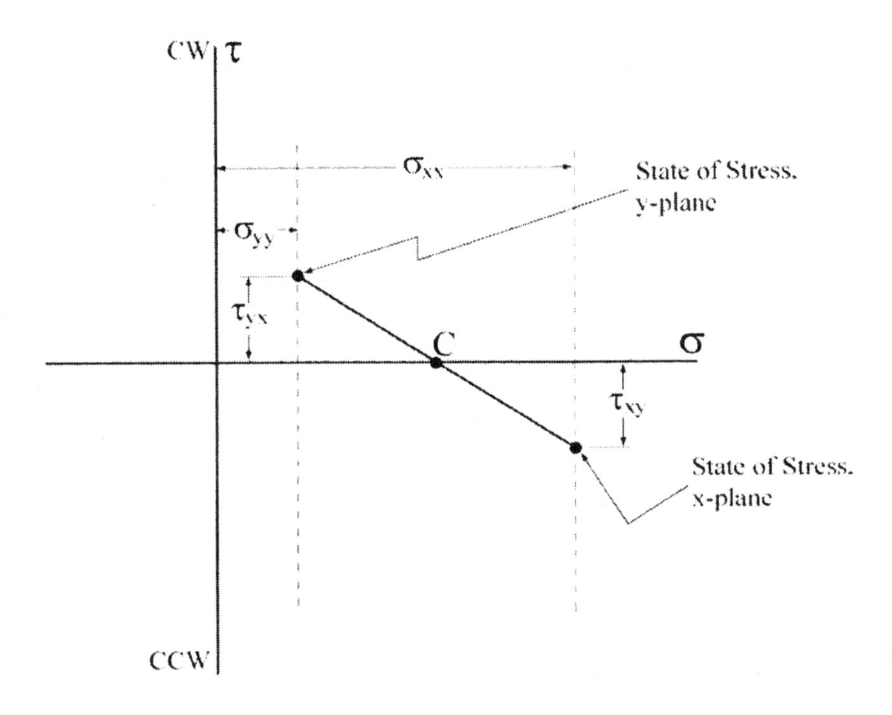

Figure 2.12 Normal and shear points.

Once these points are plotted and connected, the resulting line becomes the diameter of the circle with a center at point C with the following magnitude and the full circle as sketched below in Figure 2.13:

$$C = \frac{\sigma_{xx} + \sigma_{yy}}{2}$$

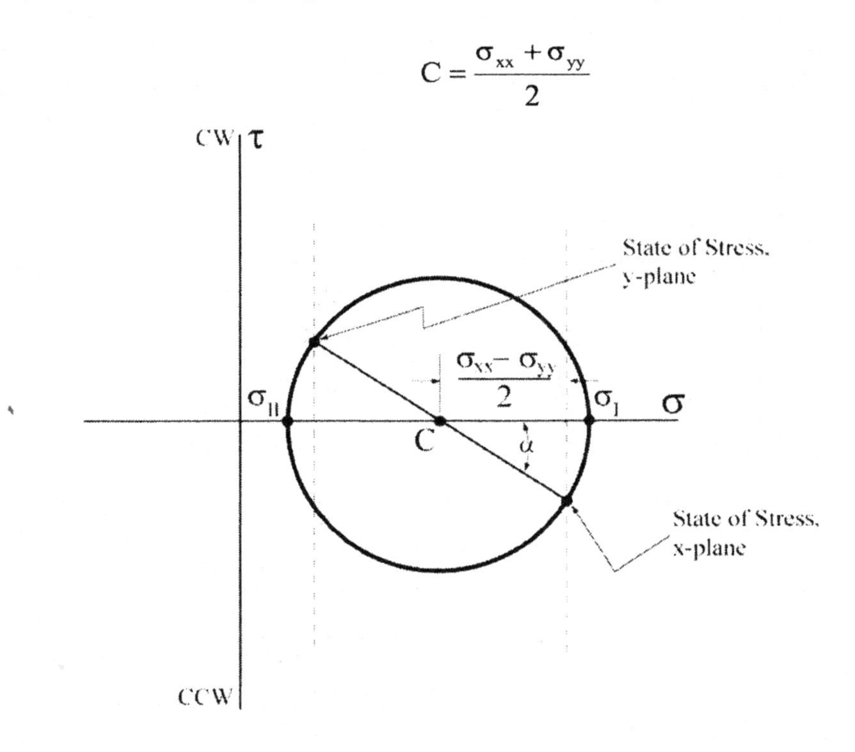

Figure 2.13 The circle begins to emerge after plotting the normal/shear pairs.

Although it will not be proven here (just trust us as the politicians always say), Equation (2.12) represents the circle above with the stress states for all possible values of θ; all conscientious readers are actively encouraged to verify the equivalence (and truthfulness of politicians for that matter) if desired. Given the appearance of an angular rotation in 2θ in Equation (2.12), it is not surprising that axis rotations of θ in real space are 2θ rotations on the circle. However, the rotational sense (not to be confused with the twisting sense of the shear stresses) is cw or ccw on the circle just as in the real space. Also note that stress-states on the x-plane are 180^0 apart from those on the y-plane when envisioned on the Mohr's circle, or twice the 90-degree separation of the planes in real space. With the stress states now elegantly represented by the circle, the next step is the determination of the extreme or maximum values, and planes they act upon.

The extreme values of normal stress (maximum or minimum values) have been labeled there as σ_I and σ_{II} and are referred to herein as *Principal Stresses*. Note that they act on perpendicular planes in real space, and that these planes correspond to zero shear stress. **Hence, the defining characteristics of the Principal Normal Stresses are that they are maximum values that reside on planes where there is no shear!**

If *R* is defined as the circle radius, the theorem of Pythagoras gives us:

$$R = \sqrt{\left(\frac{\sigma_{xx} - \sigma_{yy}}{2}\right)^2 + \tau^2_{xy}}$$

Hence, the principal values may be simply computed from:

$$\sigma_I = C + R$$

$$\sigma_{II} = C - R$$

Notice that the plane upon which σ_I acts is located by traversing the circle an amount "α" in the *ccw* rotational sense. Hence, in the real physical sense, we rotate by α/2, or:

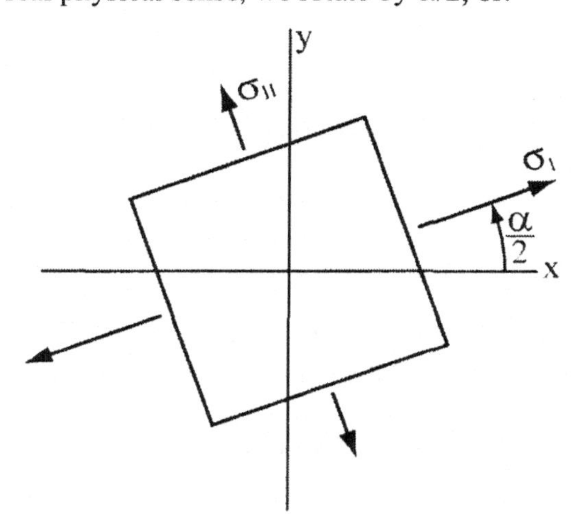

Figure 2.14. Principal axes and their orientation.

Where the angle, α is defined by:

$$\tan\alpha = \frac{|\tau_{xy}|}{\left|\frac{\sigma_{xx} - \sigma_{yy}}{2}\right|}$$

The extreme value of shear stress (*maximum shear stress*) is found at both the top and bottom of the circle and can be mainly described by the difference of the principal stresses:

$$\tau_{max} \equiv R = \frac{\sigma_I - \sigma_{II}}{2}$$

It should be noted that there may be normal stresses, σ_n on the planes of maximum shear stress simply equal to the circle center, or $\sigma_n = C$. Moreover, the planes of maximum shear stress are always rotated 45 degrees (in real space) from principal stress directions. Finally, the twisting sense of the maximum shear stress is determined by the location of the plane in question; the top of the circle has a *cw* twist while the bottom of the circle sees a *ccw* twist.

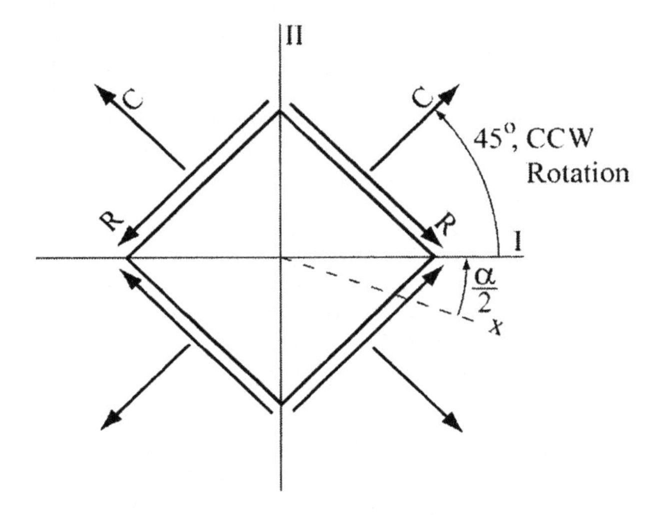

Figure 2.15 Maximum shear planes and their orientation.

It is in this manner that the principal and maximum shear stresses can be determined. Given their obvious importance to durability and failure, behavioral and failure theories using them will be presented in later chapters.

Example Problem 2-2: The following stress state at a point has been determined to be the "worst" in a mountain bike gooseneck. What are the principal stresses, the maximum shear stress, and the planes upon which they act?

$$\sigma_{ij} = \begin{vmatrix} 150 & -65 & 0 \\ -65 & -80 & 0 \\ 0 & 0 & 0 \end{vmatrix} \text{ (MPa)}$$

Although the creation of Mohr's circle has induced many a case of heartburn, there is no need for medication (at least not yet) as its construction is actually quite simple. In fact, all one has to do is plot the normal and shear stresses as if they are simple x-y points on a graph. Hence, the x-plane is defined by the points (150, -65) and the y-plane is (-80, -65) with the normal stress defining the plane. Using this approach and noting the appropriate direction from the sign of the stress value, *voila,* a circle emerges as shown below:

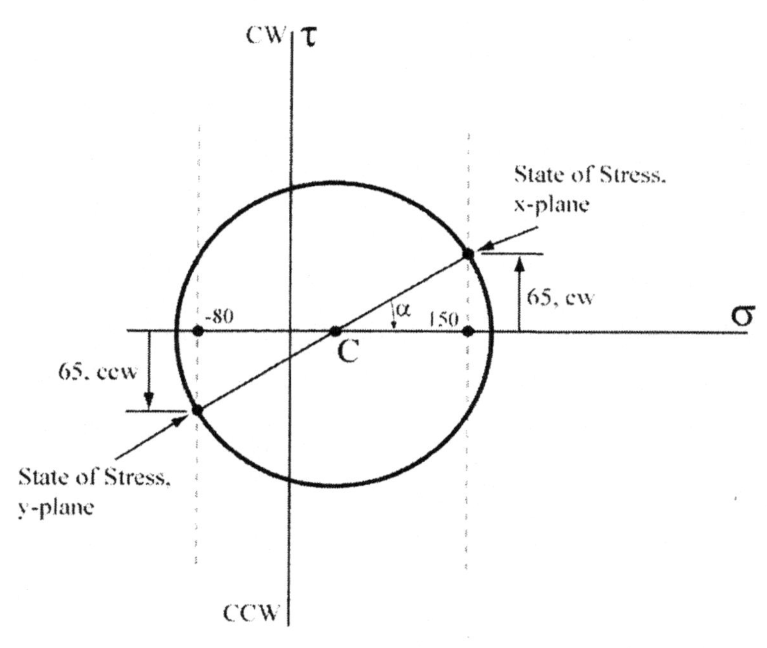

Figure 2.16 Mohr's circle showing stress orientations.

Solving for the center, *C* and the radius of the circle, *R*, one finds:

$$C = \frac{\sigma_{xx} + \sigma_{yy}}{2} = \frac{150 + (-80)}{2} = 35 \text{ MPa}$$

$$R = \sqrt{\left(\frac{\sigma_{xx} - \sigma_{yy}}{2}\right)^2 + \left(\tau_{xy}\right)^2} = \sqrt{\left(\frac{150 - (-80)}{2}\right)^2 + (-65)^2} = 132 \text{ MPa}$$

The principal stresses are subsequently calculated as:

$$\sigma_I = C + R = 167 \text{ MPa}$$

$$\sigma_{II} = C - R = -97 \text{ MPa}$$

Noting that the maximum shear stress in the x, y-plane is defined by the circle to be the radius of the circle, R:

$\tau_{max} = R = 132 \text{ MPa}$

The angle between the principal axes and the original coordinate axis system is:

$$\tan\alpha = \frac{|\tau_{xy}|}{\left|\frac{\sigma_{xx} - \sigma_{yy}}{2}\right|} = \frac{65}{115} = 0.565 \qquad \alpha = \tan^{-1}(0.565) = 29.5^\circ; \; \frac{\alpha}{2} \approx 14.8^\circ$$

with the resulting principal and maximum shear stress magnitudes and orientations shown in Figure 2.17.

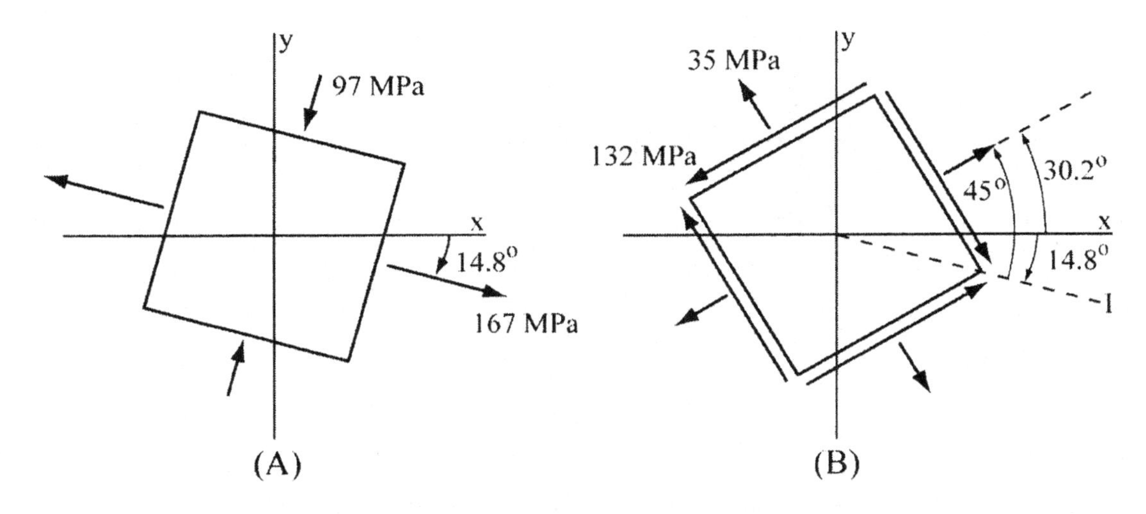

Figure 2.17 State of (A) principal and (B) maximum shear stresses.

2.7 THREE-DIMENSIONAL STRESS TRANSFORMATIONS

As mentioned earlier, designers often need to determine principal and maximum shear values for three-dimensional states of stress. In reality, this is important since all stress-states are truly three-dimensional in nature. However, situations often (and mercifully) arise where some may be neglected because of their low magnitude and/or sign that render them as little to no threat. In the cases where we cannot ignore the third plane, then Mohr's circle can be used as long as, and only if one of the planes is devoid of shear stresses.

Consider the stress state shown in Figure 2.18. A normal stress component σ_{zz} has been super-imposed upon a state of plane stress in the x, y-plane; *note that because there are no shear stresses acting on the z-plane, σ_{zz} is by definition, principal.* If a section were made parallel to the z-axis as was done in Figures 2.8 and 2.9, the existence of the stress σ_{zz} and its generated force $F_z = \sigma_{zz} (\Delta y) (\Delta y \tan\theta)$, would not alter any of the equilibrium equations in the x', y'-directions. Therefore, stresses may still be transformed in the xy-plane as before, and Mohr's circle constructed in the presence of an out-of-plane normal stress. *However, and this must be heavily stressed (all puns intended), the usage of Mohr's circle for three-dimensional stress states is only valid if, and only if there is no shear on at least one of the planes!*

If one can operate a Mohr's circle in the I, II-plane, performing rotations about the z-, or III-, axis, why not create a circle in the II, III-plane? Just as before, the force equilibrium equations in the II, III-plane would prove equivalent to the equations of Mohr's circle since the out-of-plane stress is strictly normal.

Similar arguments may be applied to the existence of a transformation circle for the III, I-plane. *As such, there exists not one transformation circle, but three that must be considered for the determination of the maximum shear stress.* However, these circles are special as they can only be constructed when one coordinate direction is a principal direction, as was the case for the z-axis in the above discussion as illustrated in Figure 2.20. The transformations implied by the circles are thus restrictive so that coordinate rotations may only be made about a principal axis; of course this axis may correspond to a zero normal stress as already explored in the plane stress state.

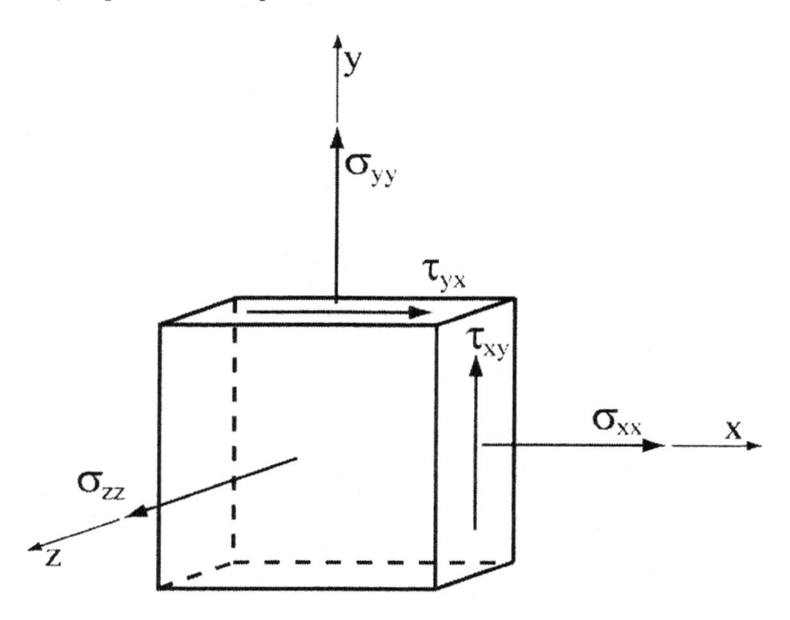

Figure 2.18 Plane-stress element in the x-y plane with superimposed normal stress in the z-direction.

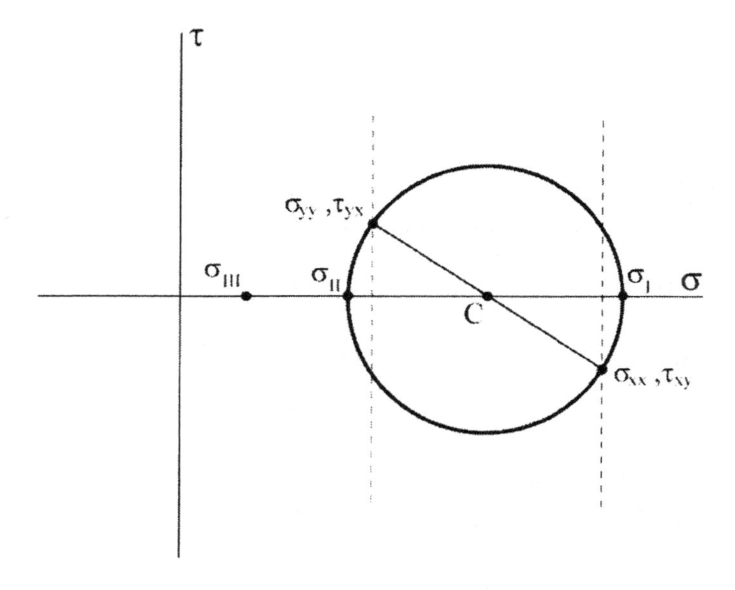

Figure 2.19 Principal stresses σ_I, σ_{II}, and σ_{III} and Mohr's circle in the I-II or x-y plane.

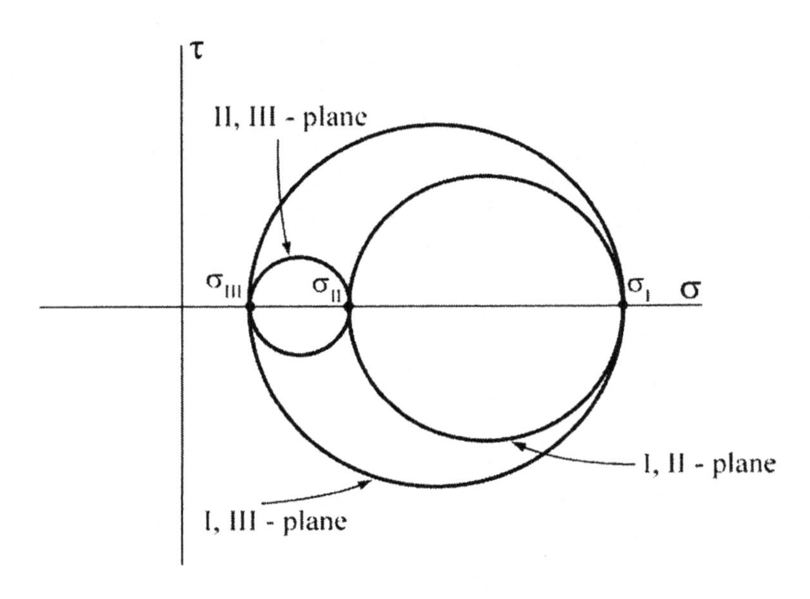

Figure 2.20 Three-Dimensional Mohr's Circles.

The restriction is not a true shortcoming in the only rotational context of interest, namely, the seeking of planes of maximum shear stress. The maximum shear stress states are always found directly from the principal planes, since their normals are at 45 degrees to the principal directions. The three maximum shear stresses corresponding to each principal plane are therefore:

$$\tau_{max} = \frac{\sigma_I - \sigma_{II}}{2} \tag{2.13a}$$

-or-

$$\tau_{max} = \frac{\sigma_{II} - \sigma_{III}}{2} \tag{2.13b}$$

-or-

$$\tau_{max} = \frac{\sigma_I - \sigma_{III}}{2} \tag{2.13c}$$

However, by always sticking to the convention that $\sigma_I > \sigma_{II} > \sigma_{III} > 0$, then one has only to worry about the I-III plane as it will always produce the largest, and therefore, true maximum shear stress.

Example Problem 2-3: The following stress state has been determined to be the one of most intense interest for a shield designed to counter the much-feared "Death Copter" mentioned earlier. What are the principal stresses and the maximum shear stress and upon which planes do they act?

$$\sigma_{ij} = \begin{vmatrix} 40 & 15 & 0 \\ 15 & 20 & 0 \\ 0 & 0 & 0 \end{vmatrix} \text{ (ksi)}$$

The Mohr's circles construction takes on the form shown in Figure 2.21. Note that there are now (and forever after) three circles, even though $\sigma_{zz} = 0$ and the convenient convention of $\sigma_I > \sigma_{II} > \sigma_{III}$ will be used. For the usual computations:

$$C = \frac{40+20}{2} = 30 \text{ ksi} \qquad R = \sqrt{\left(\frac{40-20}{2}\right)^2 + (15)^2} = 18 \text{ ksi}$$

$$\tan\alpha = \frac{15}{10} = 1.5; \ \alpha = 56.3^\circ; \ \frac{\alpha}{2} = 28.2^\circ$$

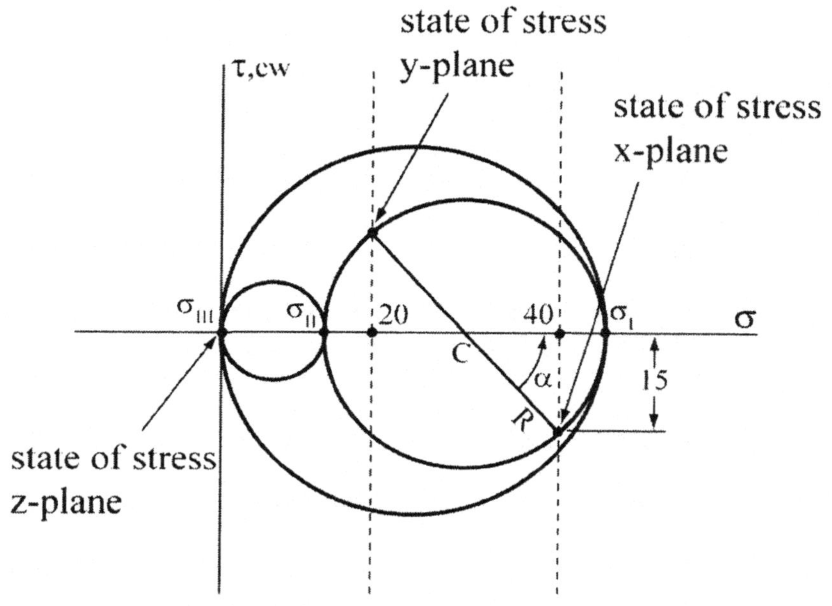

Figure 2.21 Three-dimensional Mohr's circles.

The principal values are computed as:

$\sigma_I = C + R = 48$ ksi

$\sigma_{II} = C - R = 12$ ksi

$\sigma_{III} = 0$

The orientation of the principal element is shown in Figure 2.22. From the three Mohr's circles, it can be seen that the maximum shear stress is on the I-III-plane, not the I, II, or x, y. Computing τ_{max}, we have:

$$\tau_{max} = \frac{\sigma_I - \sigma_{III}}{2} = \frac{48-0}{2} = 24 \text{ ksi}$$

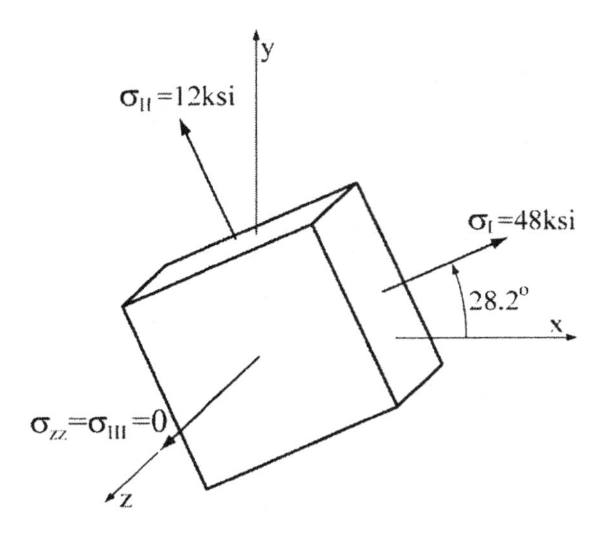

Figure 2.22 Principal stresses for the three-dimensional case and their respective orientations.

A sign cannot be given to τ_{max} without first imposing a coordinate system. Since there are two possible coordinate choices (refer back to Figures 2.5b and 2.5c), both signs are possible, and no further mention will be made of them. Although this may cause undo stress (all puns intended as always) to the analysts, the theories that use the maximum shear stress to determine the onset of yielding are independent of sign.

The orientation of the maximum shear stress element is now to be determined; because the principal axes have already been located, it will suffice to relate the maximum shear orientation relative to the principal axes. Two considerations must be observed, namely that: (1) the principal axes must conform to the right hand rule, and (2), rotations in real space must be carried out "looking down" the axis of rotation as was the case for the z-axis in x, y-plane rotations previously.

Note that I-II-III is a right-hand system, and the rotation of the II-axis is out of the plane of the page. If one rotates from the plane of σ_I to that of σ_{III}, *ccw*, the maximum shear stress state on top of the circle is encountered at 90 degrees (45 degrees in real space); that same sense rotation, at 90 degrees/2, is shown in the principal stress space below. The twisting sense of τ_{max} on the plane encountered is *cw*, fixing all shear senses.

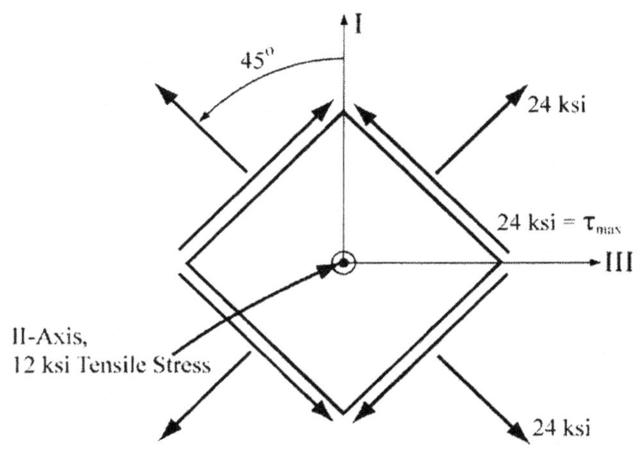

Figure 2.23 Maximum shear values and orientations.

2.8 COMPLETE THREE-DIMENSIONAL ANALYSIS

For truly three-dimensional stress states where all six independent components (three normal and three shear) are present, three-dimensional stress element sectioning and force equilibrium demands will lead to more complicated transformation equations. Unfortunately, these equations cannot represent a circle in a stress-space as was possible for the simpler cases just described. Only by considering the axis rotations leading to principal values, the three principal stress values σ_I, σ_{II}, and σ_{III} can be found as the roots to the equation:

$$\begin{aligned}&\sigma^3-\left(\sigma_{xx}+\sigma_{yy}+\sigma_{zz}\right)\sigma^2+\left(\sigma_{xx}\sigma_{yy}+\sigma_{yy}\sigma_{zz}+\sigma_{zz}\sigma_{xx}-\tau_{xy}{}^2-\tau_{yz}{}^2-\tau_{zx}{}^2\right)\sigma+\ldots\\&\ldots-\left(\sigma_{xx}\sigma_{yy}\sigma_{zz}+2\tau_{xy}\tau_{yz}\tau_{zx}-\sigma_{xx}\tau_{yz}{}^2-\sigma_{yy}\tau_{zx}{}^2-\sigma_{zz}\tau_{xy}{}^2\right)=0\end{aligned} \qquad (2.14a)$$

-or-

$$\sigma^3-I_I\sigma^2+I_{II}\sigma\ -I_{III}=0 \qquad (2.14b)$$

Since the principal values, or roots to Equation (2.14) are the same at any given point, regardless of which axis set (x, y, and z) is used to describe the nonprincipal stress components, the terms enclosed by parentheses in Equation (2.14) are always the same at that point. Not surprisingly, these values are referred to as the *stress invariants*. Using this nomenclature, the three *stress invariants* I_1, I_2, and I_3 become:

$$I_1=\sigma_x+\sigma_y+\sigma_z \qquad (2,15a)$$

$$I_2=\sigma_x\sigma_y+\ \sigma_y\sigma_z+\sigma_z\sigma_x-\tau_{xy}^2-\tau_{yz}^2-\tau_{xz}^2 \qquad (2.15b)$$

$$I_3=\sigma_x\sigma_y\sigma_z+2\tau_{xy}\tau_{yz}\tau_{xz}-\sigma_x\tau_{yz}^2-\sigma_y\tau_{xz}^2-\sigma_z\tau_{yx}^2 \qquad (2.15c)$$

Once the invariants are known, the principal stresses for a true three-dimensional stress state can then be determined using the following relationships.

$$\sigma_a=2S\left[\mathrm{Cos}\left(\alpha/3\right)\right]\ +\ \frac{1}{3}I_1 \qquad (2.16a)$$

$$\sigma_b=2S\left[\mathrm{Cos}\left(\alpha/3+120^0\right)\right]\ +\ \frac{1}{3}I_1 \qquad (2.16b)$$

$$\sigma_c=2S\left[\mathrm{Cos}\left(\alpha/3+240^0\right)\right]\ +\ \frac{1}{3}I_1 \qquad (2.16c)$$

$$R=\frac{1}{3}I_1^2-I_2 \qquad (2.16d)$$

$$S=\sqrt{\frac{R}{3}} \qquad (2.16e)$$

$$T = \sqrt{\frac{R^3}{27}} \quad (2.16f)$$

and

$$Q = \frac{1}{3} I_1 I_2 - I_3 - \frac{2}{27} I_1^3 \quad (2.16g)$$

with

$$\alpha = \mathrm{Cos}^{-1}\left(-\frac{Q}{2T}\right) \quad (2.16h)$$

It is important to note that the subscripts a-c represent the three principal values before sorting; *the maximum value will become σ_I, the middle value σ_{II}, and the smallest value σ_{III}*. Moreover, size and sign do indeed matter in this case as 0.0001MPa is always greater than a compressive -100,000,000MPa and so forth.

Finding the principal directions or Direction Cosines is also an algebraic endeavor of great fun for the fully three-dimensional state of stress (appreciating Mohr's Circle yet?). If, for example, σ_I is the principal stress whose direction we seek, then the direction cosines of the angles between the principal direction, I and the original x-, y-, and z-axes will be assigned values n, m, and n, respectively. These direction cosines are found by the simultaneous solution of the following three equations:

$$(\sigma_I - \sigma_{xx})l - \tau_{xy}m - \tau_{xz}n = 0 \quad (2.17a)$$

$$-\tau_{xy}l + (\sigma_I - \sigma_{yy})m - \tau_{yz}n = 0 \quad (2.17b)$$

$$-\tau_{xz}l - \tau_{yz}m + (\sigma_I - \sigma_{zz})n = 0 \quad (2.17c)$$

The solution of the direction cosines for all three principal stresses is possible via the following fun relationships:

$$l_i = a_i k_i \qquad m = b_i k_i \qquad n_i = c_i k_i \quad (2.18a)$$

$$a_i = \begin{vmatrix} \sigma_y - \sigma_i & \tau_{yz} \\ \tau_{yz} & \sigma_z - \sigma_i \end{vmatrix} \quad (2.18b)$$

$$b_i = -\begin{vmatrix} \tau_{xy} & \tau_{yz} \\ \tau_{xz} & \sigma_z - \sigma_i \end{vmatrix} \quad (2.18c)$$

$$c_i = \begin{vmatrix} \tau_{xy} & \sigma_y - \sigma_i \\ \tau_{xz} & \tau_{yz} \end{vmatrix} \quad (2.18d)$$

With

$$k_i = \frac{1}{\left[a_i^2 + b_i^2 + c_i^2\right]^{\frac{1}{2}}} \quad (2.18e)$$

Directions cosines for the angular location of the principal directions I, II, and III are calculated from Equations (2.18) by substituting σ_i with σ_I, σ_{II}, or σ_{III}, respectively.

Although not really covered in any detail up to this point, many of the topics in this book require the determination of stress distributions in structural elements subjected to specified loadings. In some respects, the subject of stress analysis is far too onerous to be delineated herein, and at any rate, perhaps not truly a topic to be pursued when studying the response of materials. On the other hand, the responses we seek to measure and apply in design must be determined and defined in terms of analyzed structural members, not in some purely microstructure/atomistic context devoid of practical application. Hence, we see the emergence of "battle lines" for the age old and hopefully friendly fight between the "Stress-Guesser" and "Potallurgist." With this coupling of interests in mind, we will briefly review some specific examples of utility that are used throughout this text.

2.9 STRUCTURAL STRESS ANALYSIS

Throughout the text, the concepts of stress have been used while the basic relationships common to design and not discussed in any great detail since they are usually covered by introductory texts on strength of materials or elasticity that also provide large numbers of examples, as well as the general methodologies used for their determination. This noticeable omission is intentional as the examples of interest could be considered endless. Because selected examples are repeatedly used throughout this text, a brief review of some basic relationships is certainly warranted.

Pressure Vessels: Closed-end cylindrical pressure vessels are ubiquitous in our present technological culture and offer a very useful example of biaxial-stresses. If the vessel wall is *thin* in comparison to the geometry of the vessel or $r_m/t \geq 10$, then the in-plane hoop, σ_h and axial stresses, σ_{ax} away from the ends are assumed uniform through the thickness and given by:

$$\sigma_h = \frac{Pr_m}{t} \tag{2.19a}$$

$$\sigma_{ax} = \frac{Pr_m}{2t} \tag{2.19b}$$

where *P* is the pressure, r_m the mean radius, and *t* the thickness. In the absence of applied shear stresses, the hoop and axial components are inherently principal in nature. When the vessel is internally pressurized, both stresses are tensile; if externally pressurized, both components are compressive.

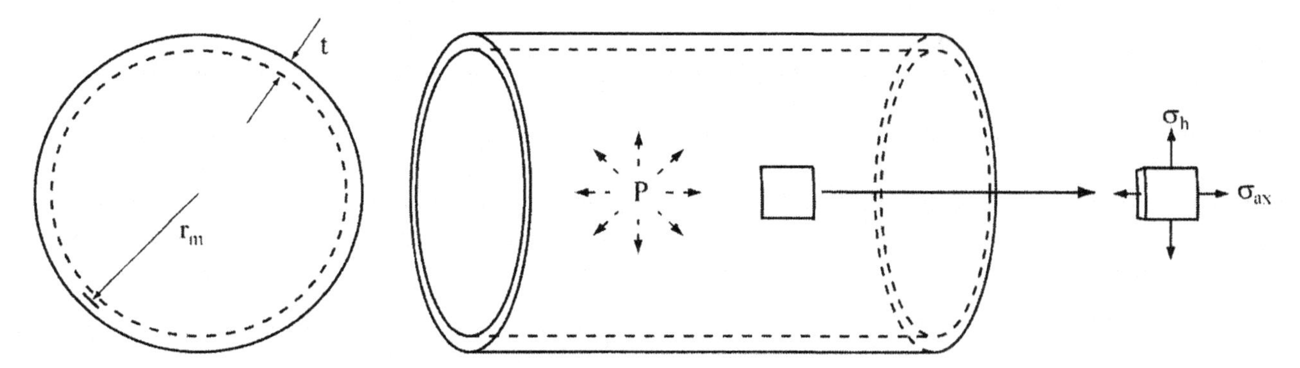

Figure 2.24 Thin-walled pressure vessel showing geometry and stress components.

Finally, if the vessel is spherical, both principals are equal in magnitude and sign and become:

$$\sigma_I = \sigma_{II} = \frac{Pr_m}{2t} \tag{2.20}$$

Beams: Beams are a common and truly crucial form of a structural member that has few limits in its applications. Moreover, many other structures and machine components are subjected to bending moments, as well as other forms of functional loading. For elastically bent beams subjected to bending moments (moments whose vectors are normal to the beam axis), the stress distribution is given by:

$$\sigma_{xx} = \pm \frac{M_b y}{I_{zz}} \tag{2.21}$$

where M_b is the moment, y is the distance from the neutral axis, and I_{zz} is the moment of inertia of the cross-section. The sign conventions implicit in Equation (2.21), as well as the resulting stress distribution can be seen below in Figure 2.25. The coordinate frame origin is located at the centroid of the cross-section area (neutral axis) and the outward-pointing x-axis is coincident with the beam axis. As shown by the figure, the maximum stress is always located at the greatest distance from the centroid or the "outer fibers" of the beam where y_{max} is conventionally renamed "*c*." The examples herein are all symmetric in their cross sections so that the centroid is the geometrical center of the area and the distance "*c*" is one-half the beam depth.

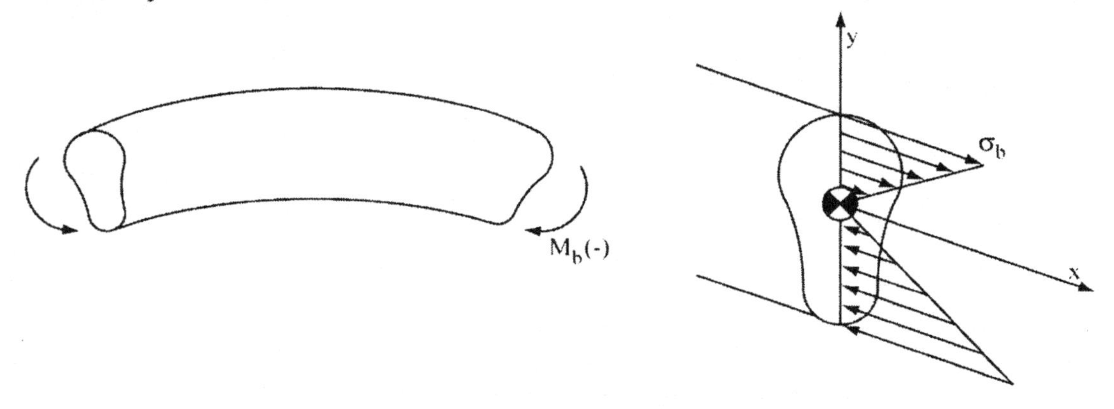

Figure 2.25 Bending stresses and sign conventions for beams.

Rods and Shafts under Torsion: Rods or shafts subjected to torsional loading are the final group of structural members and machine components to be stress analyzed (guessed). Needless to say, with all rotary power being transmitted through shafting, these are often-used structural members as well. The applied load on such a member is a moment whose vector is coincident with the member axis. The twisting moment, or torque, T when applied to a rod of circular cross-section (radius, r) rotates cross-sectional planes relative to each other without warping them. This deformation leads to an easily visualized shear strain, γ as seen in Figure 2.26. At any intermediate radial position r, the shear strain is given by:

$$\gamma = r\frac{\phi}{L} \tag{2.22}$$

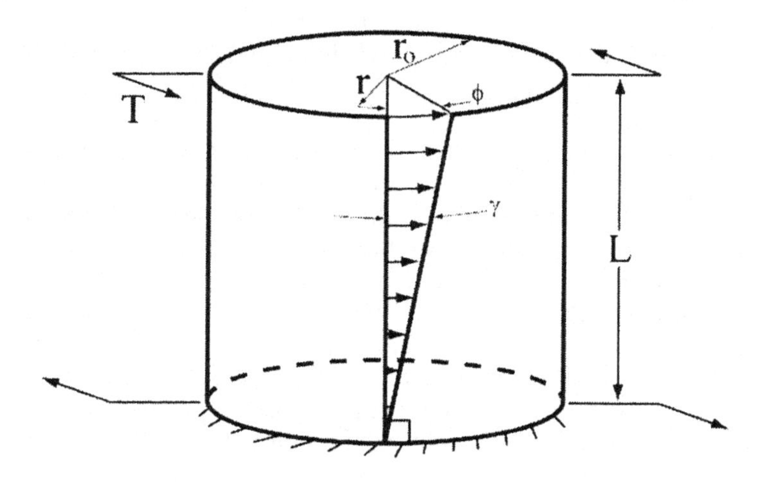

Figure 2.26 Torsional loading and the resulting angular deformation.

The shear stress τ in terms of the applied torque, T and radius, r:

$$\tau = \frac{Tr}{J} \tag{2.23}$$

where J is the polar moment of inertia. The planes of actions for the shear stress are shown below in Figure 2.27.

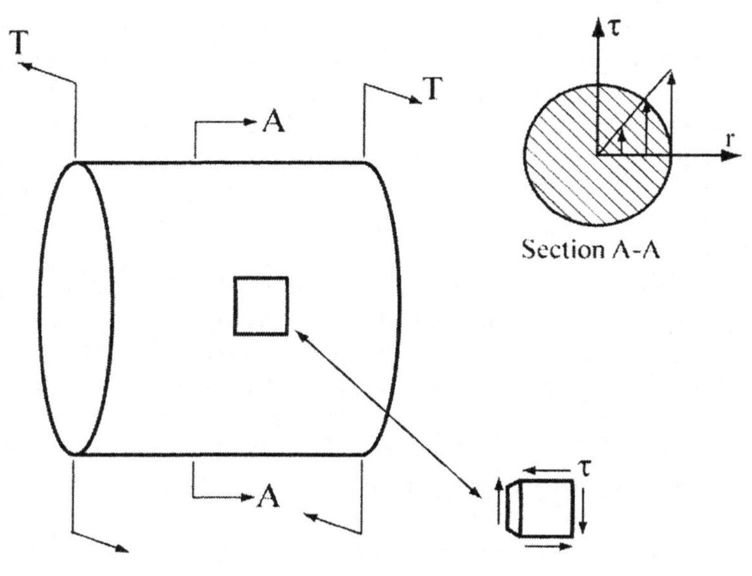

Figure 2.27 Torsional loading and stresses for a circular member.

Interestingly (and in case you were having many sleepless nights pondering this), the word Torque is a Latin-rooted word that is actually one of the few spoken Celtic words to have survived from ancient times; it was the name for the bracelets the Celts made by twisting together a bundle of wires.

2.10 PRINCIPLE OF SAINT VENANT

An understanding of the stress states in the immediate vicinity of structural discontinuities can be facilitated by considering the hypothesis attributed to St. Venant from his treatise on the torsion of shafts (circa. 1855). One expression of this principle might be the following statement: "*Statically equivalent* loadings acting over the same portion of a structural member will induce states of stress that are identical when monitored at a characteristic distance from the region of load application." Although not very clear (sounds almost like typical lawyer gobbledygook), St. Venant's principle simply tells us that in the immediate vicinity of the distributed load, it matters very much how that load is distributed, with the "immediate vicinity" being defined as a locale whose dimensions are roughly those of the region of load application. However, at some characteristic distance from the load, its distribution leads to the same stresses as would a point load: recall that the differently distributed loads with the same area must have the same resultant force and moment.

For example, consider the idealization of the "point load" used in simple beam theory; of course, there is no such load, but, rather, a distributed load applied over a very small region. The point load approximation is useful and accurate for determining stress states away from the load application region. Another analogous way of viewing this principle is to consider a rock in the middle of a flowing stream. In the immediate vicinity of the rock, the flow is visibly perturbed. However, the further you are from the rock, the less noticeable the disturbance until the flow is indistinguishable from the rest of the stream. Another example is the stress distribution in a tensile sample that we might use to test a material's ability to carry axial forces (see Chapter Four). It matters not how the sample is gripped (mated pin-and-yoke, vise-gripped, etc.), as long as the stresses and strain are monitored at some *characteristic distance* from the gripping particulars.

What is the above-mentioned characteristic distance? This distance must be related to whatever local geometries are relevant. In the foregoing example of a point load, the size of the region over which the distributed loading is applied, for example, the diameter of a circular area is the local dimension; experience indicates that the characteristic distance is four to five times that local dimension. For the tensile sample just mentioned, the local dimension is the member thickness, width, or diameter (if circular), and distances four or five times these dimensions will ensure equivalent stress states.

2.11 STRESS CONCENTRATIONS

Up to this point, the underlying assumption has been that the stresses were uniform, or changing linearly from the neutral axis. However, this is not always the case as sudden changes in geometry may lead to locally increased values of stress and strain; these geometry changes are appropriately referred to as *stress concentrations*. We can again use the analogy of a rock in the stream where the flow is perturbed and magnified (concentrated) as it goes around the obstacle. In terms of stresses, consider the tensile plate shown in Figure 2.28. A circular through-hole of radius "r" is drilled in the plate, which for the sake of mathematical simplicity, will be taken to be orders of magnitude larger than the hole such that $W >> r$.

Drilling the hole eliminates an amount of cross-sectional area equal to 2rt on the plane containing the hole; stresses on that small segment of cross-section plane, which previously contributed to vertical force equilibrium, can no longer act there. Since equilibrium must be maintained, where and how do the stresses redistribute themselves? The answer is *not* necessarily over the entire remaining cross-sectional plane. If one probes the vertically acting stresses several radii away from the hole, the presence and effect of the hole will not be seen. Rather, the redistributed stresses are found to occur in regions of close proximity to the hole, where the reduced cross-section acts to increase the compliance of the plate. This should (hopefully) sound very familiar as it is simply St. Venant's principle in action!

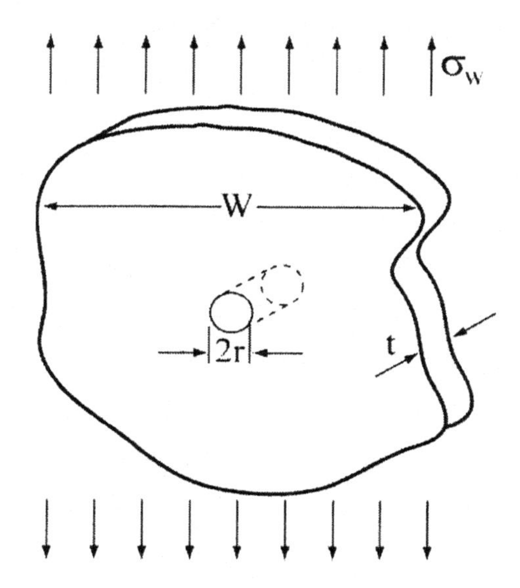

Figure 2.28 Circular hole in an infinite plate under tensile loading.

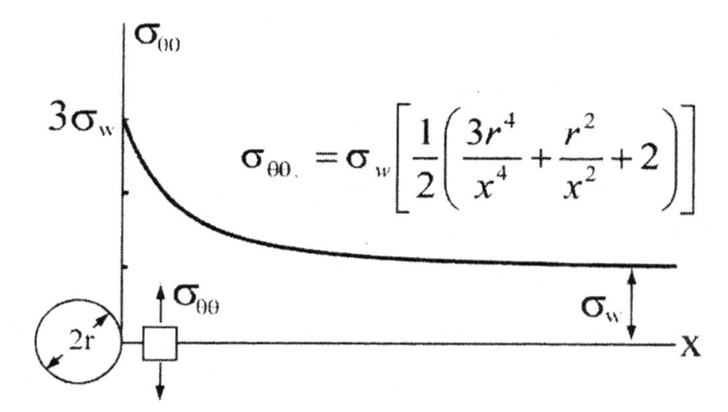

Figure 2.29 Tangential normal stress on equatorial plane for the hole-in-plate configuration.

The component $\sigma_{\theta\theta}$ is said to be concentrated at x = r and the hole is obviously a *stress concentrator* or *stress riser.* Once the observer has moved several characteristic distances away from the hole (the only pertinent "distance" in the present example is the hole radius), the stress field is the same as if the hole were not present or simply σ_w. Since the stresses away from the vicinity of the hole are not affected by the presence of the hole, an adequate description of the stress concentration phenomenon may lie in only knowing the maximum local stress value. We therefore define the *Elastic Stress Concentration Factor*, K_σ, as:

$$K_\sigma \equiv \frac{\text{Concentrated Local Stress}}{\text{Far Field (unperturbed) Stress}} \qquad (2.24)$$

For the present example, $K_\sigma = 3.0$, a numerical value that is invariant with hole size as long as $W >> r$. However, if the diameter of the hole is considered large relative to the plate width, K_σ actually decreases!

A related example would be the same tensile plate as below, but with a more general penetration; namely, a through-hole of elliptical shape as indicated below in Figure 2.30.

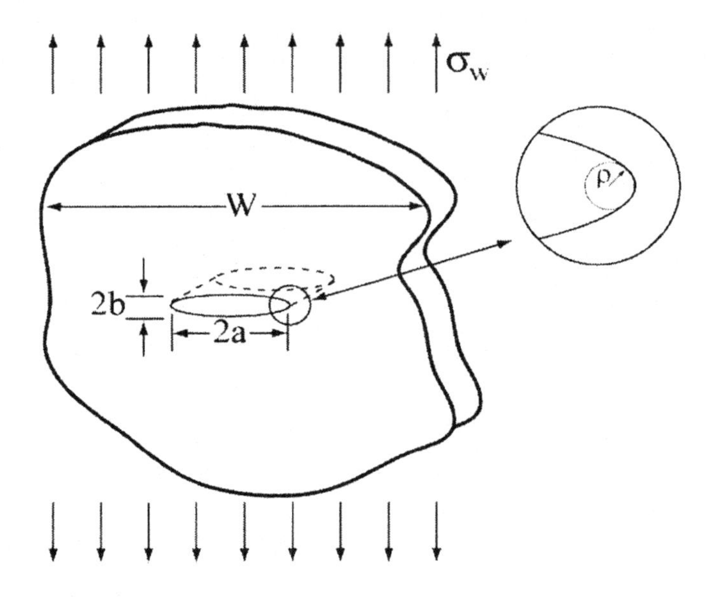

Figure 2.30 Elliptical hole in tensile plate.

The ellipse can be mathematically specified by its major and minor axis lengths, *2a* and *2b* respectively, or by one axis and a radius of curvature, ρ measured at the sharp tip. The stress concentration factor for this configuration has been found to be of the following form:

$$K_\sigma = 1 + 2\sqrt{\frac{a}{\rho}} \tag{2.25a}$$

If $\rho << a$, then Equation (2.25a) reduces to the following form:

$$K_\sigma \cong 2\sqrt{\frac{a}{\rho}} \tag{2.25b}$$

Here, one has a stress concentration factor that varies in magnitude with the *sharpness* or *suddenness* of the geometry change—that is, the value of ρ relative to that of *a*.

Example Problem 2-4: Estimate the stress concentration factor for the smallest "sharp" surface flaw that one could hope to observe nondestructively in a weld of a nuclear reactor vessel.

To begin, we must first assume that the flaw in question could be modeled adequately as a sharp ellipse. Current nondestructive evaluation technology would limit a_{min} to sizes of approximately 0.001 inches and larger. If the flaw in question was generated during welding or in service, by fatigue or stress corrosion processes, then the least value of ρ, or the sharpest possible flaw, would be around 2 or 3 interatomic spacings, or, roughly, 10 Angstrom units.

$$10\,\mathring{A} = 10x10^{-8}\,\text{cm} \cong 4x10^{-8}\,\text{inches}$$

Very approximately, K_σ for such a flaw is found to be:

$$K_\sigma \cong 2\sqrt{\frac{10^{-3}}{4x10^{-8}}} = 316$$

The fact that the stress concentration comes in at a whopping $K_\sigma = 316$, illustrates not only the necessity to consider stress concentrations in design, but also the interplay between concentrated local stress and structural material response. Does the example problem imply that safety factors in excess of 316 must be employed in design? Unfortunately, there is no clear-cut answer since any assessments of viability ultimately depends on the material to be employed in the design, the mode of failure (fracture, fatigue, creep etc.), as well as the true degree of safety required! Moreover, "safety factor" is nothing more than an ego-saving replacement for a more appropriately named term such as "ignorance factor!"

If the material is truly brittle and it fails in the elastic range of stressing, then large safety factors and/or very conservative design practices may be necessary. For brittle materials, exceeding the strength at the notch root will cause a local, but rapidly spreading failure usually thought of as *catastrophic*. If, on the other hand, the material is not *brittle,* but can in fact be permanently deformed before fracturing, such behavior is referred to as *plastic* and the solid is considered *ductile*. In this situation, the large stresses at the notch root will cause some permanent deformation that is local; the resulting stresses will then redistribute themselves (relax) and catastrophic flaw extension is unlikely. From a historical point of view, this is in part, the reason that mankind has utilized metallic materials over the millennia. Although relatively rare and expensive to refine, metals are very tolerant of the presence of flaws, and are generally insensitive to the highly concentrated strains associated with flaws.

One final point regarding elastic stress concentration factors is worth noting, as it often drives design practice in a quantitative sense. St. Venant's principle would lead one to believe that, in the immediate vicinity of a stress concentrator, the radius of curvature of the notch tip must strongly influence the local stress distribution as it is the only discernible geometric feature. By contrast, the shape of the concentrator, away from its sharp tip, is of far less significance. The previous example studied indicated that:

$$K_\sigma \propto \sqrt{\frac{1}{\rho}} \qquad (2.26)$$

The application of St. Venant's principle allows us to apply Equation (2.27) generally, provided that changes in notch tip curvature are not accompanied by gross alterations in other features of the concentrator. Notch sharpness can be understood by examining the effect of doubling a notch's radius of curvature upon the stress concentration factor.

If K_o is the original stress concentration factor and K_f is the factor after doubling the root radius, then:

$$\frac{K_f}{K_o} = \frac{1/\sqrt{2\rho}}{1/\sqrt{\rho}} = \sqrt{\frac{1}{2}} = 0.707 \qquad (2.28)$$

Thus, doubling the radius of curvature at the notch tip will reduce the stress concentration factor by about 29%. When a designer requests a "generous" weld fillet, the intention is not merely to place more weld material to increase load-bearing cross-sectional area nearly so much as it is to decrease the stress concentration factor between the joined members. Hence, size does indeed matter when it pertains to notch or fillet radii!

PRACTICE EXERCISES

1-6. The stress states in Problems 1-6 were all determined for various loaded components from planes, trains, and automobiles by an accomplished Stress Guesser who successfully completed this course. Determine the principal stresses and sketch them on an element at this orientation relative to the original x, y, and z coordinate systems. In addition, determine the complete state of stress at the maximum shear stress orientation and show them on an element relative to the principal axes.

1. $$\sigma_{ij} = \begin{vmatrix} 25 & -15 & 0 \\ -15 & -5 & 0 \\ 0 & 0 & 0 \end{vmatrix} \text{ksi}$$

2. $$\sigma_{ij} = \begin{vmatrix} 0 & 4 & 0 \\ 4 & 6 & 0 \\ 0 & 0 & 0 \end{vmatrix} \text{ksi}$$

3. $$\sigma_{ij} = \begin{vmatrix} 0 & 15 & 0 \\ 15 & 0 & 0 \\ 0 & 0 & 0 \end{vmatrix} \text{ksi}$$

4. $$\sigma_{ij} = \begin{vmatrix} 10 & -4 & 0 \\ -4 & 5 & 0 \\ 0 & 0 & 0 \end{vmatrix} \text{ksi}$$

5. $$\sigma_{ij} = \begin{vmatrix} 25 & 15 & 0 \\ 15 & -5 & 0 \\ 0 & 0 & 5 \end{vmatrix} \text{MPa}$$

6. $$\sigma_{ij} = \begin{vmatrix} 25 & 15 & 0 \\ 15 & -5 & 0 \\ 0 & 0 & -15 \end{vmatrix} \text{MPa}$$

7. The students have finally had enough with Otto Mohr and his fiendish circle. To get their well-deserved revenge, the students designed a torture machine that has the following 3-D stress state; determine the principal stresses and sketch them on an element at this orientation relative to the original x, y, and z coordinate systems. In addition, determine the complete state of stress at the maximum shear stress orientation and show them on an element relative to the principal axes.

$$\sigma_{ij} = \begin{vmatrix} -15 & 0 & 0 \\ 0 & 25 & 15 \\ 0 & 15 & -5 \end{vmatrix} \text{ksi}$$

8. Determine the principle stresses and their associated direction cosines for the following stress state associated with a failed component from a power plant coolant system.

$$\sigma_{ij} = \begin{vmatrix} -19 & -4.7 & 6.45 \\ -4.7 & 4.6 & 11.8 \\ 6.45 & 11.8 & -8.3 \end{vmatrix} \text{MPa}$$

9. A thin-walled, closed-end, cylindrical pressure vessel for jet fighter lubricant with spherical end caps is internally pressurized to 500 psi. If the mean diameter is 60.0 inches and wall thickness 1.0 inch, determine (a) the principal stresses on an element from the cylindrical portion of the tank surface, and (b) the principal stresses on an element from a portion of the spherical end caps.

10. A round shaft 2.0 cm in diameter deformed beyond design limits when it was subjected to a twisting moment of T=220 N-m and a simultaneous bending moment of the same magnitude, M=220 N-m. Define the state of stress on the outermost fiber carrying the maximum tensile bending (flexural) stress. What is the principal state there and what is the orientation of the principal values relative to the shaft geometry?

11. A brake pedal for a "real" car (Porsche of course) dear to a certain professor unfortunately failed while braking during a very sharp turn at 75+ MPH (duh!). Assuming a foot force of 300 lbs was applied to the exact center of the pedal pad, what are the principal stresses on the tops of the arm at the two cross-sections indicated? Assume both cross-sections are circular.

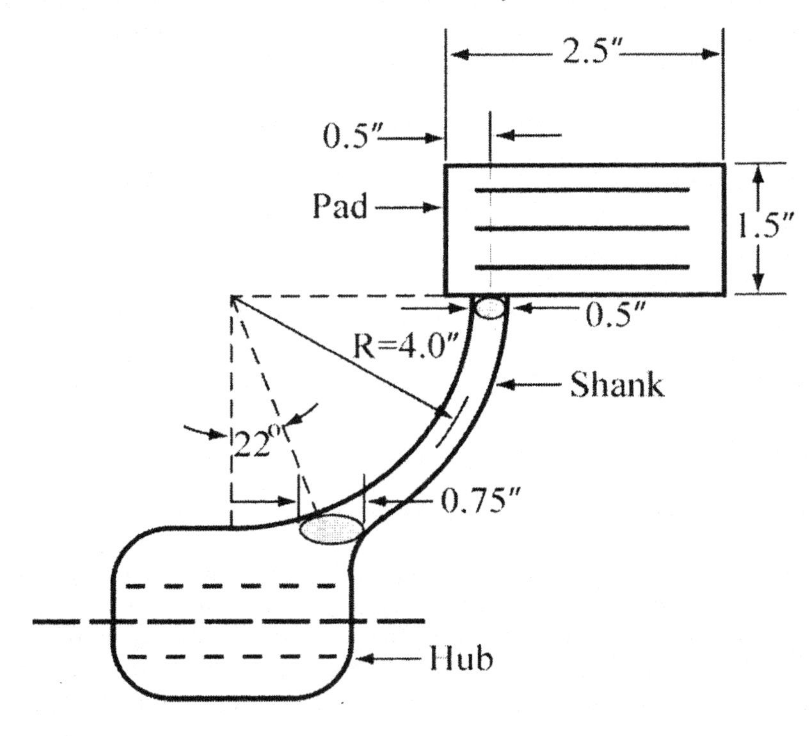

12. A stepped shaft for an aircraft compressor with the geometry shown below has a stress concentration factor of 2.3 due to the step. **Note:** the stress concentration factor is specified as magnifying the stress of the *thicker* section.

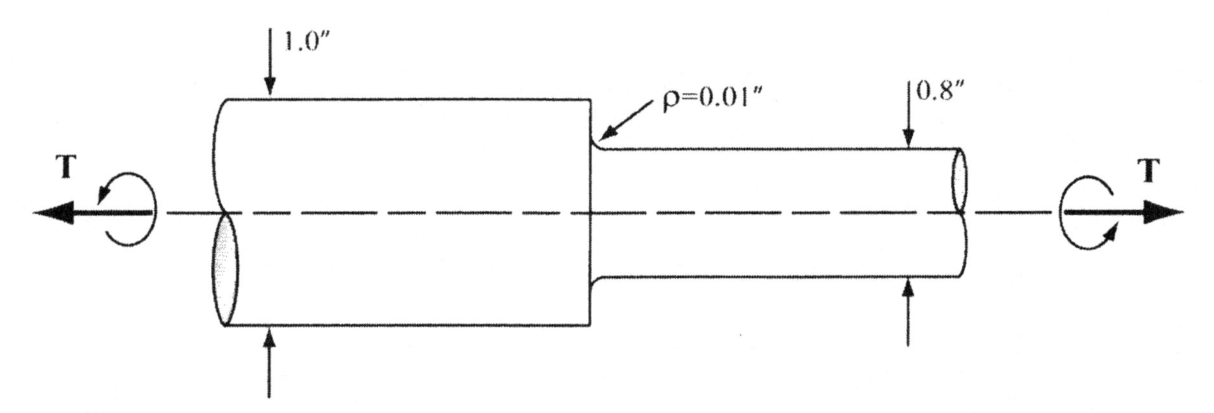

A. For a given torque **T**, determine where is the stress greatest (thick or thin section)?

B. To what value must ρ be increased if the concentrated stress is to be the same magnitude as the stress in the thinner section, for any value of **T**?

13. A large plate used in an oil rig blow-out preventer that is guaranteed to be 100% reliable (yea, right!) is subjected to a tensile load, **P** possesses a width of 20.0 inches, and a thickness of 0.25 inches. A circular through-hole of 1.0" is drilled in the center of the plate. The working stress for the plate material is 35,000 psi: this stress is not to be exceeded anywhere in the plate. What is the maximum permissible value of **P**?

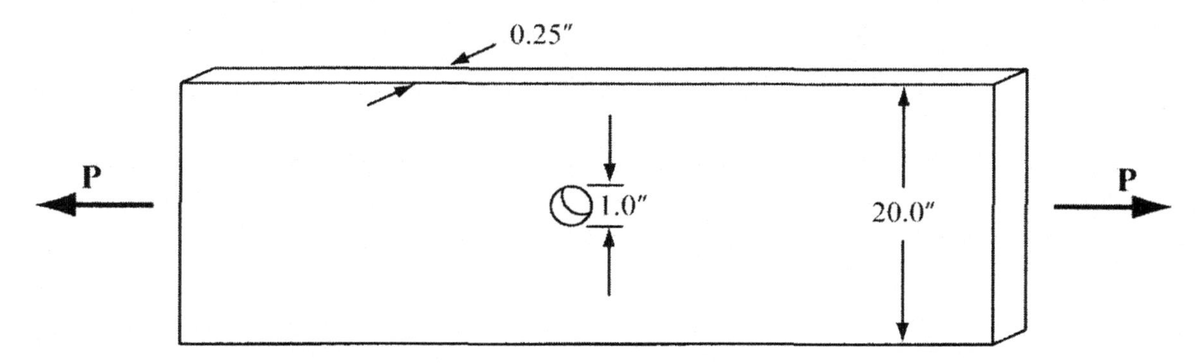

14. For shallow edge notches in the oil-pan protection "skid" plate shown below, K_σ is computed from the following formulas when subjected to tension or bending:

$$K_\sigma = 3\sqrt{\frac{t}{2\rho}} - 1 + \left(\frac{4}{2+\sqrt{\frac{t}{2\rho}}}\right); \quad \sigma_{local} = K_\sigma \sigma_{nom}$$

where:

$$\sigma_{nom} = \frac{P}{hb} \quad \text{(for tensile loading)}$$

$$\sigma_{nom} = \frac{M_b\left(\frac{h}{2}\right)}{\frac{1}{12}bh^3} \quad \text{(for bending)}$$

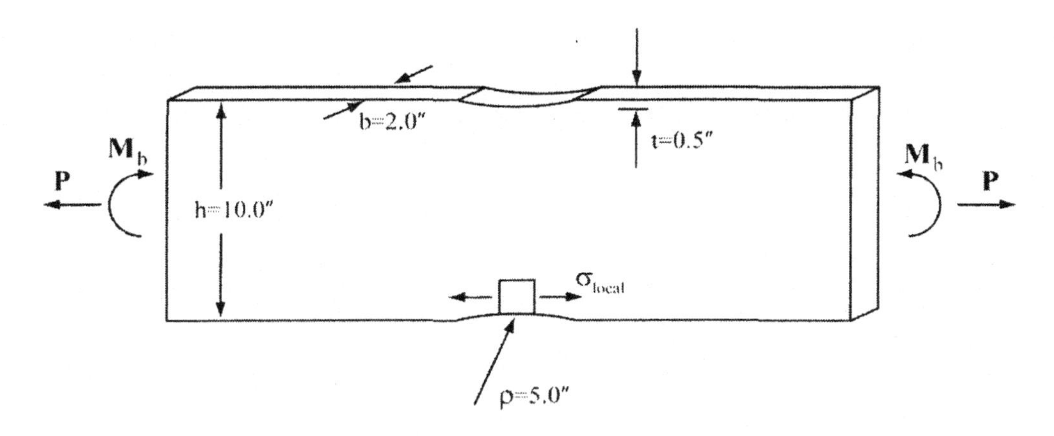

a. What is the maximum stress at the notch roots for **P** = 10,000 lbs?

b. What is the maximum stress at the notch roots for M_b = 2,000 ft. -- lbs?

c. What is the maximum stress at the notch roots (both) for the loadings of parts (a) and (b) combined?

15. For a large plate used for an elevator safety system with a relatively small circular hole ($W >> r$), it has been established that $K_\sigma = 3.0$. However, if the diameter of the hole is considered large relative to the plate width (r is close to W), $K_\sigma < 3$. Can you explain this behavior?

CHAPTER THREE

ELASTIC BEHAVIOR OF MATERIALS

3.1 ELASTIC RESPONSE

In engineering practice, the word *elastic* can actually have several interpretations. Undoubtedly, everyone has a feel for the elastic behavior of an elastomeric ring usually known as a "rubber band," especially when stung by it during an unprovoked attack; pain aside, after being stretched and used in the aforementioned way, it returns to its original size when released, and thus is considered elastic. Although most associate the word *elastic* with its ability to stretch, it is in fact, the ability to return to its original length that is of importance. Now let's consider something seemingly very different than a rubber band, in this case a solid bar. If the bar is subjected to say, tension and is therefore being stretched to a strain of ε (and carrying a tensile stress σ), there are several loading/unloading histories that are in agreement with the above notion of elastic response. Graphically, these notions are as shown below in Figure 3.1. Our main interest at this point and in this chapter will be on linearly elastic materials. The nonlinear case will be taken up in Chapter Four and the time-dependent situation in Chapter nine.

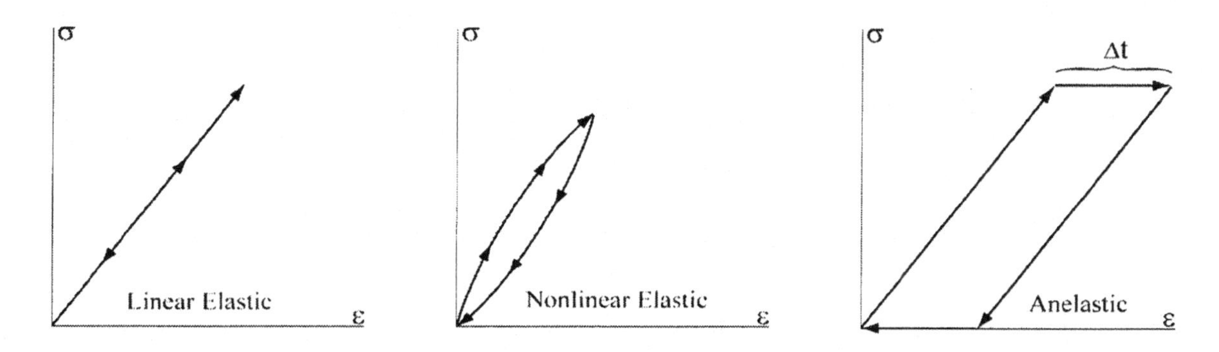

Figure 3.1 Different load histories satisfying the general idea of "elastic."

Linear elastic behavior is one of the first phenomena the student encounters in an introductory physics course that treats classical Newtonian mechanics. To begin this journey down memory lane, recall the well-known force and deformation relationship for a spring, namely $F = kx$. Here, x is the deformation and k is the stiffness that is ultimately determined by the spring geometry and material. From the just stated definition, it is clear that both the geometry and material are important considerations. Usually, the influence of geometry is easily envisioned; a thicker rubber band will be harder to stretch, but the resulting snap on an unwitting victim will no-doubt be worth the extra work! On the other hand, the contribution of the material (and its inherent properties) to the rubber band and its ability to stretch and then snap back may not be as obvious or rewarding (at least in a sadistic sense).

Perhaps the best way of separating out the influence of the geometry is by looking at the materials response as characterized by deformation or strain to a given loading/stress. In fact, if one looks at the resulting stress versus strain curve, the expected linear response is seen in Figure 3.2.

The resulting slope of the tensile (or compressive) stress-strain curve is referred to as the *Modulus of Elasticity* or *Young's Modulus* and designated as the variable, E. For those of you who are history buffs, the latter is named after Thomas Young who essentially identified this quantity in 1807. The linear response approximation as shown by Figure 3.2 is relatively accurate for materials whose moduli exceed 7 GPa (10^6 psi) in the sense that ordinary measuring devices such as micrometers or calipers cannot detect any nonlinearities. At the sub-micro level, chemical bonding and related details ultimately determine the values of E such that all steels, despite their differences in alloying, heat treatment, or state of deformation, exhibit moduli of approximately 210 GPa (30 x 10^6 psi), give or take 10% at most.

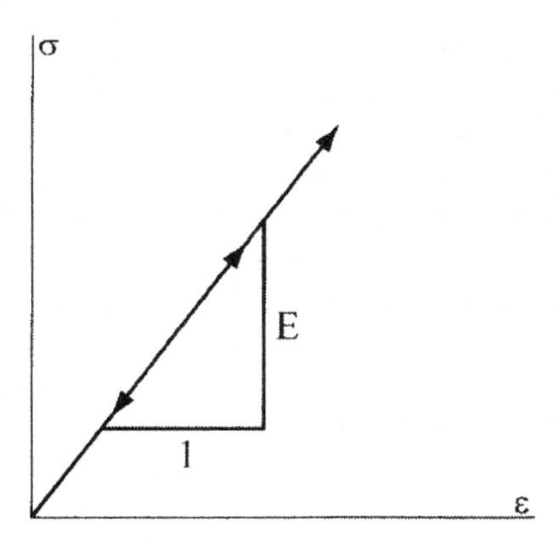

Figure 3.2 Material stiffness as defined by the Elastic Modulus, E.

Up to this point, we have only considered the response in the direction of the applied load, be it tensile or compressive. However, if one looks at the rubber band as it is stretched (or the bar if you have super vision), the width will also change by contracting (tensile load) or expanding (compressive load). Hence, when pulling (or pushing) a prismatic bar and generating axial strains $\varepsilon = \Delta l/l_o$, one can easily observe strains of a similar magnitude occurring in the transverse sense. The phenomenon of transverse strains resulting from imposed axial strains can be expressed algebraically as:

$$\varepsilon_{trans} = -\nu\varepsilon_{axial} \tag{3.1}$$

Here, the term, ν, commonly known as *Poisson's Ratio,* reflects the resulting transverse strains; as shown by Equation (3.1), Poisson's Ratio is a positive number that is usually within the range of 0.2 to 0.5, with each particular value obviously depending on the material. Values of 0.5 are usually associated with rubbery materials (Elastomers) whose stiffness is nonlinear and orders of magnitude less than typical values of E.

The shear stiffness, usually denoted as G is referred to as the *Shear Modulus* or *Modulus of Rigidity*, the latter so named since the quantity disappears for a liquid. Typically, shear moduli are less than Young's moduli. Moreover, there is no transverse strain effect in shear. For an *Isotropic Solid,* one whose elastic parameters have the same value in all directions and orientations, the elastic constants can be related in the following way:

$$G = \frac{E}{2(1+\nu)} \tag{3.2}$$

Thus, for an isotropic solid, only two of the elastic constants are independent. Solids that are precisely isotropic are limited in number; components made from pressed and sintered powders and glassy materials would be among those. Despite this limitation, the vast majority of structural materials are assumed to fit this definition, as their *anisotropy* (lack of isotropy) is relatively slight. The elasticity of truly anisotropic materials such as fiber reinforced composites, will be taken up later in this chapter.

Finally, most materials will also respond elastically to imposed shear stresses or strains; in this instance, the linear response would appear as:

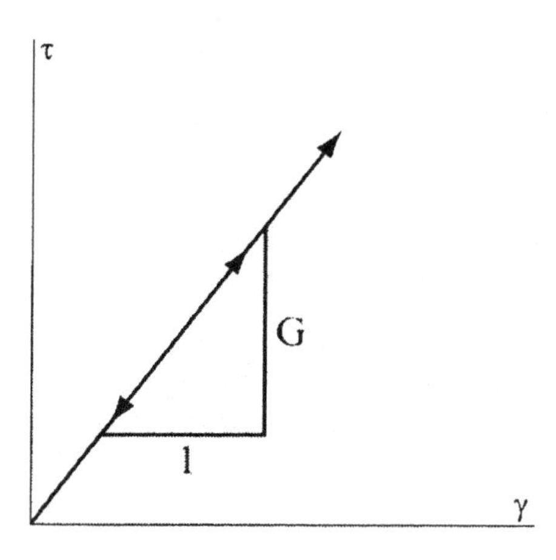

Figure 3.3 Linear elastic shear stress/strain response.

3.2 HOOKE'S LAW FOR ISOTROPIC SOLIDS

"*Ut tension, sic vis.*" No, this is not what Caesar said after consuming too much lead and wine! Robert Hooke in fact spoke this in 1678, and loosely translated, this is equivalent to "all the stretches are proportional to all the pulls, and vice-versa." Remembering that strains are caused by stresses in the same directional sense, and by transversely acting stresses as well, the above statement can be algebraically expressed as:

$$\varepsilon_{xx} = \frac{1}{E}\left[\sigma_{xx} - \nu\left(\sigma_{yy} + \sigma_{zz}\right)\right] \tag{3.3a}$$

$$\varepsilon_{yy} = \frac{1}{E}\left[\sigma_{yy} - \nu\left(\sigma_{zz} + \sigma_{xx}\right)\right] \tag{3.3b}$$

$$\varepsilon_{zz} = \frac{1}{E}\left[\sigma_{zz} - \nu\left(\sigma_{xx} + \sigma_{yy}\right)\right] \tag{3.3c}$$

$$\gamma_{xy} = \frac{1}{G}\tau_{xy} \quad \gamma_{yz} = \frac{1}{G}\tau_{yz} \quad \gamma_{zx} = \frac{1}{G}\tau_{zx} \tag{3.3d}$$

Equation (3.4) is known collectively as *Hooke's Law* for stresses and strains in an isotropic solid, and is the first example of *Constitutive Equations* that relates thermodynamic work variable pairs. Since stresses are very difficult to measure, but strains are not, it is often required that an inverted form of Equation (3.4) be available.

Mathematically speaking, the inverted form of the constitutive equations take the following form:

$$\sigma_{xx} = \frac{E}{(1+\nu)(1-2\nu)}\left[(1-\nu)\varepsilon_{xx} + \nu\left(\varepsilon_{yy} + \varepsilon_{zz}\right)\right] \qquad (3.4a)$$

$$\sigma_{yy} = \frac{E}{(1+\nu)(1-2\nu)}\left[(1-\nu)\varepsilon_{yy} + \nu\left(\varepsilon_{xx} + \varepsilon_{zz}\right)\right] \qquad (3.4b)$$

$$\sigma_{zz} = \frac{E}{(1+\nu)(1-2\nu)}\left[(1-\nu)\varepsilon_{zz} + \nu\left(\varepsilon_{xx} + \varepsilon_{yy}\right)\right] \qquad (3.4c)$$

Although relatively simple in form, the constitutive equations are quite powerful in that they easily relate both measurable (strain) and desirable (stress) quantities that are ultimately crucial to design.

Example 3-1: What elastic volume changes to a rail car hub are induced by the following applied stress state? Assume the hub is constructed from an isotropic steel with $E = 210$ GPa and $\nu = 0.28$.

$$\sigma_{ij} = \begin{vmatrix} 600 & -320 & 0 \\ -320 & 150 & 0 \\ 0 & 0 & -280 \end{vmatrix} \text{MPa}$$

Since shear stresses do not lead to changes in volume as they do not change line lengths, their existence in this problem can be ignored. Recalling the definition of normal strain, $\varepsilon = \Delta L/L$, then original and final line lengths L_o can be related:

$$L_f = L_o + \Delta L = \frac{L_o}{L_o}(L_o + \Delta L) = L_o(1+\varepsilon)$$

Moreover, the final volume can therefore be written as:

$$V_f = L_x(1+\varepsilon_{xx})L_y(1+\varepsilon_{yy})L_z(1+\varepsilon_{zz})$$

or, when expanded

$$V_f = L_x L_y L_z\left(1+\varepsilon_{xx}+\varepsilon_{yy}+\varepsilon_{zz}+\varepsilon_{xx}\varepsilon_{yy}+\varepsilon_{yy}\varepsilon_{zz}+\varepsilon_{zz}\varepsilon_{xx}+\varepsilon_{xx}\varepsilon_{yy}\varepsilon_{yy}\right)$$

If the original volume is the product of the starting lengths or L_x L_y L_z (unit volume) and the strains are of typical elastic magnitudes (<0.001), then the higher order products are negligible. Given these assumptions and the welcome simplifications that follow, the change in volume per unit volume becomes:

$$\Delta V \cong \varepsilon_{xx} + \varepsilon_{yy} + \varepsilon_{zz}$$

Calculating the strains via equations (3.4) as: $\varepsilon_{xx} = 3.03\text{x}10^{-3}$, $\varepsilon_{yy} = 0.29\text{x}10^{-3}$, and $\varepsilon_{zz} = -2.33\text{x}10^{-3}$ results in $\Delta V = 0.99\text{x}10^{-3}$ m^3. Hence, the volume changes by approximately 0.1%. This number and the corresponding magnitudes of the strains should give the reader a sense of how small these quantities really are, and perhaps more importantly, how super vision would be required to sense them for most structural materials.

3.3 THERMAL STRAINS IN ISOTROPIC SOLIDS

Up to this point, we have examined the changes to a structure when mechanical forces are applied. Interestingly, changes in the size (and possibly) the shape of a body are also conceivable where no forces act on the body. Common examples are the swelling of wood (and tissue) when water is absorbed (recall our potentially doomed boater in Chapter One), the expansion of most objects when heated, as well as their shrinking when chilled. These thermally induced changes are so profound that one can literally hear it when a radiator or baseboard heaters begin "clicking and clanking," or when a house with vinyl siding is hit by the summer sun. However, the resulting forces can also be very powerful as evidenced by the ancient Egyptians using the expansion of water soaked, wooden pegs and subsequent heating by fire to split large blocks of granite for their giant obelisks.

As discussed earlier, size change can be described in terms of normal strains, so this approach will be used to describe the effects of temperature on materials and structures. Since all materials of common engineering importance appear to expand when heated and contract when cooled (water being one notable and fortunate exception in certain instances), it is reasonable to assume that an inherent materials property is at work. In fact, the *Coefficient of Linear Thermal Expansion,* usually denoted as α, can be directly related to the temperature state and the resulting strains. However, because these dimensional changes are occurring at any temperature above absolute zero, it must be handled as a relative term with regards to some practical starting point, usually stated at "room" temperature or approximately 20°C (~70°F).

With this in mind, and for small changes in temperature denoted as dT, a similarly small strain increment, $d\varepsilon$ can be expected to occur such that:

$$d\varepsilon = \alpha dT \tag{3.5}$$

However, α will in all likelihood be functionally related to the temperature (most properties including E and ν are), so that total strains accumulated as temperature ranges between T_o and T_f must be determined by summation that when taken to the limit:

$$\varepsilon = \int_{T_o}^{T_f} \alpha(T)dT \tag{3.6}$$

From Equation (3.6), it is evident that the total thermal strain is the area under the α-T curve between appropriate temperature limits. Hence, if one were concerned with the strains induced in a ceramic coating for a gas turbine blade heated from ambient to 900°C, then the coefficient, α would definitely be dependent on temperature, and Equation (3.6) would be relevant. However, for a relatively limited range of temperature encountered in most (but definitely not all) engineering applications, α might be approximated as constant, and Equation (3.6) simply becomes:

$$\varepsilon = \alpha \Delta T \tag{3.7}$$

where $\Delta T = T_f - T_o$ and represents the change in temperature from the reference value mentioned earlier. Typically, α is a number on the order of 10^{-6} per degree ($^\circ$F or $^\circ$C). Small magnitude aside, thermal strains can often be used in an advantageous way as shown by the next example.

Example Problem 3-2: Interference fits are a common and economical way to assemble disks onto shafts such as a torque converter in an automatic trasmission. One method of assembly involves a diametral interference that is created by slightly under-sizing the hole in the disk and/or over-sizing the shaft as shown below in Figure 3.4. Because this would inherently make it hard to assemble and would create many unhappy workers, the trick is to cool and shrink the shaft while heating and expanding the wheel. For example, if the shaft shown below originally has a 1.0-inch diameter, what temperature changes would be necessary for each 0.001" in diametral interference to slip-fit the two together? Assume the wheel and shaft are composed of similar steels so that one could expect $\alpha = 7 \times 10^{-6}/^\circ$F.

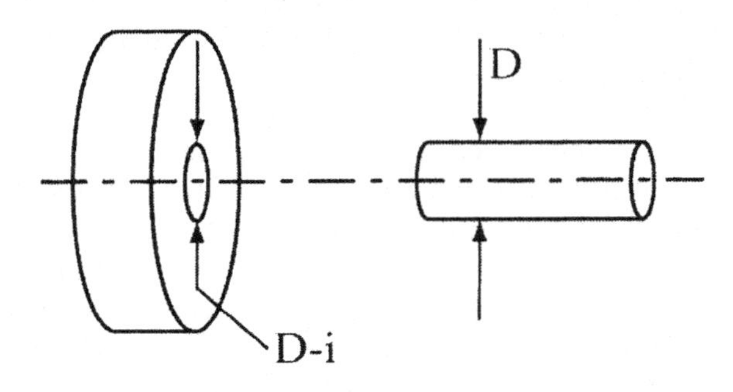

Figure 3.4 Shrink-fit assembly of a hub and shaft.

During assembly, the inner bore circumference expands from πD to $\pi(D+\Delta D)$ for a given strain $\Delta D/D$ from any heat supplied to the shaft. For the two parts to mate for the original ID=1.00", $\Delta D = 0.001$", or $0.001/1.0 = 7 \times 10^{-6}\,\Delta T$. Solving for ΔT, we find that for every 140°F temperature change, an interference fit of 0.001" can be accommodated. In reality, this temperature range can be split between the two members by heating the disk and cooling the shaft. Although not explicitly mentioned, it was assumed that α was constant over the temperature range.

3.4 CONSTRAINED THERMAL STRESSES AND STRAINS

Additional strains, and more importantly, stresses can arises when a member is subjected to some form of temperature excursion (from the reference temperature) that causes dimensional changes if no constraints exist; these constraints may be imposed by other bodies with differing temperatures and/or physical restraints and connections. To examine these situations and the ensuing ramifications for design, it is first necessary to reexamine Hooke's Law. If normal strains are caused by a combination of normal stresses and now temperature changes as just discussed, then Hooke's law (3.3) for normal stress and strain can be rewritten as:

$$\varepsilon_{xx} = \frac{1}{E}\left[\sigma_{xx} - \nu\left(\sigma_{yy} + \sigma_{zz}\right)\right] + \int \alpha dT \tag{3.8a}$$

$$\varepsilon_{yy} = \frac{1}{E}\left[\sigma_{yy} - \nu\left(\sigma_{xx} + \sigma_{zz}\right)\right] + \int \alpha dT \tag{3.8b}$$

$$\varepsilon_{zz} = \frac{1}{E}\left[\sigma_{zz} - \nu\left(\sigma_{xx} + \sigma_{yy}\right)\right] + \int \alpha dT \tag{3.8c}$$

On the other hand, the relationship between shear stresses and strains is not directly affected by changes in temperature, so the latter three expressions in Equation (3.3) are unchanged. Given this scenario, it is easy to envision how temperatures and restraints can interplay. In fact, a true restraint can stop an expanding or contracting member cold in its tracks (all puns intended as usual). Since the structure/material wants to change its length due to the combination of ΔT and α, internal forces (and therefore stresses) must also arise.

While a relatively straightforward concept (easy to say), this scenario can actually come about in two interesting and meaningful ways in engineering practice. The first and most obvious is related to actual physical restraints that somehow hold the component in place. Common examples include mechanical attachments such as bolts, lugs, adhesives, etc. to name a few. Although any form of attachment (just mentioned or still to be dreamed) will do the trick, one has to be careful because the constraints can lead to failures. For example, consider a large suspension bridge that is rigidly fixed on both sides. As the very large metal structure (could be miles in length) heats up or cools down, relatively large deformations on the order of the product of the length and $\varepsilon_{thermal}$ ($L\alpha\Delta T$) will ensue. Given the magnitude of the deformation, the resulting forces will be tremendous and the supports will ultimately fail and/or the bridge will buckle. Usually, bridges are partially "floated" in that one side is pinned and the other allowed to move. To avoid local buckling, there are also expansion seams along the span that cause that all too familiar "tha-dump, tha-dump,…" while driving.

Although not as obvious, physical restraints can also come about when there are no external attachments of any kind. Such internal restraints are indeed possible when there is either a temperature gradient across a body and/or connected materials with differing coefficients of thermal expansion. Under this very common scenario, the cooler portion (or the one with the lower value of α) restrains the expansion of the warmer section (or the one with the greater value of α). Because temperature gradients exist in almost all applications, especially automotive and nuclear where heat is generated by combustion or fission, this is a very common situation. In any of the just mentioned scenarios, severe gradients of temperature will be generated that essentially turn any material into a functionally graded one because of the differing thermal expansion. Bi-material combinations are also quite common, and can be found in toaster ovens and many mechanical thermostats; the differing expansion and the resulting bowing of bi-metallic strips is used to sense the temperature and control the device. One apparently hapless U.S. automotive manufacturer inadvertently used such a design for their brake disks that rapidly warped, much to the chagrin of many consumers including yours truly.

In terms of engineering practice, the solution procedure for these types of problems involves two main concepts. The first concept of equilibrium should hopefully be firmly ingrained in your brains by this point; with regard to general equilibrium, nothing has changed except that the thermal component may generate internal mechanical forces, even if there are no external forces applied. As such, any thermal stresses must be inherently *self-equilibrating*! If mechanical forces are also applied to the system, then they must be added to the mix, but all must still be balanced. Unfortunately, the equations of equilibrium are usually not sufficient to solve the problems as it is statically indeterminant, so knowledge of the resulting deformation must also be used. While there are many possibilities, most situations involve either zero net deformation as in fully restrained, or deformation compatibility where they distort the same or some proportionate amount even if the magnitude is unknown. The following example illustrates the solution methods just discussed and the important role constraint plays in the analysis.

Example Problem 3-3: A solid bar of circular cross-section and area A_o is held (just touching) between two rigid walls that act to constrain its horizontal motion as shown in Figure 3.5. If the building collapse sensor is fabricated from pure copper and experiences a temperature increase of 200°C, what is the state of stress in the bar? Relevant thermoelastic properties (assumed constant over the stated temperature range) for commercially pure copper include E = 120 GPa and $\alpha = 17 \times 10^{-6}/\ ^{\circ}C$. The reference temperature $T_{ref}=0^{\circ}C$

As stated earlier, the solution will require us to examine two aspects of the bars behavior, namely equilibrium and compatibility of deformation.

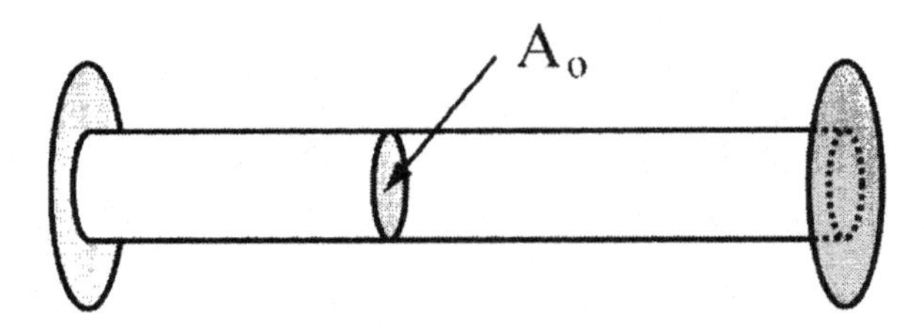

Figure 3.5 Rigidly restrained bar of circular cross-section

The first step is to determine the free-body diagram (FBD) of the bar/sensor. In order to do this, we must first consider how the bar will react to the increase in temperature. Since it will want to expand, the rigid walls will restrain this expansion and generate compressive forces and stresses as shown in Figure 3.6

Figure 3.6 Free-body diagram (FBD) of the restrained bar.

Although this can also be done by inspection, a summation of forces (stress x area) for equilibrium yields:

$$\Sigma \text{ Forces} = 0 = (-\sigma_2 + \sigma_1)\, A_o \quad \rightarrow \quad \text{Stress is uniform: } \sigma_1 = \sigma_2 = \sigma$$

Hence, in the absence of any externally applied forces, the internal stresses are the same and are simply equal to σ. Unfortunately, this does not tell us anything about the magnitude, so we must turn to compatibility.

In this case, compatibility is relatively simple as the net deformation must be zero or, $\delta_{net} = 0$ due to the rigid walls. A zero net deformation implies that any positive thermal expansion must be offset by mechanical compression by the reaction forces or:

$$\delta_{net} = \delta_{reactions} + \delta_{thermal} = \left(\frac{-P}{A}\right)\frac{L}{E} + L\alpha\Delta T = -(\sigma)\frac{L}{E} + L\alpha\Delta T = 0$$

Solving for σ, we find that the lengths all cancel and the resulting stress magnitude is $\sigma = 408$ MPa (compressive) or simply, $\sigma = -408$ MPa. It is worth noting that stresses of this magnitude are probably in excess of the elastic limit for the copper. Of course, there are no such things as truly "rigid" walls. Nonetheless, the potentially large magnitude of the resulting thermal-stresses should be noted.

In general, two basic types of constraint problems will be encountered where the material elements are either in series or parallel arrangement; most problems usually exhibit some aspects of both archetypes. To some degree, the illustrative problems given below are the analogs of real hardware, with the true complexities of actual geometry left to the practice assignments for your pleasure.

Example Problem 3-4: Consider the two-member toaster thermostat shown below that has differing material properties, lengths, and cross sections for generality. Initially, there is a gap between the two connected members and the rigid walls. As such, and in the absence of any applied loads, forces and stresses cannot develop in the material until the gap is closed. Moreover, contact can only occur when the temperature is raised. If this is the case, the loads will be the same in each component, but not the stresses because of the differing cross-sectional areas.

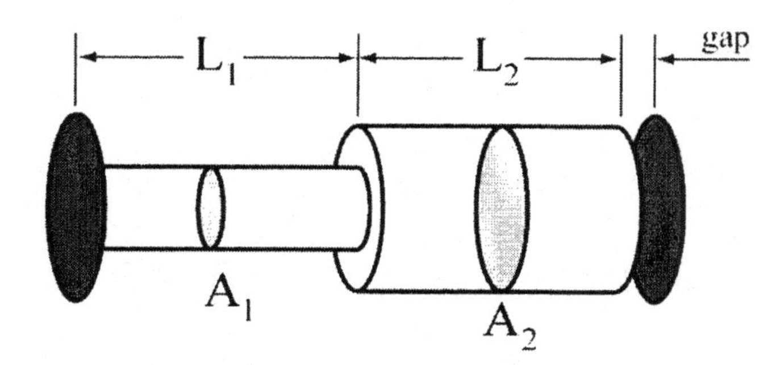

Figure 3.7 Two-member system between rigid walls with a gap

Assuming a positive change in temperature and after contact, equilibrium still rules the day and the axial forces must sum to zero per the equation and FBD below.

$$\sum F = 0 = \sigma_1 A_1 - \sigma_2 A_2$$

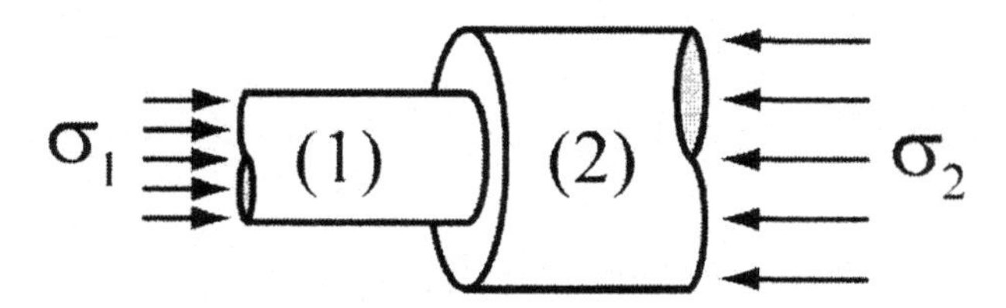

Figure 3.8 Free-body diagram (FBD) of the two material system with gap.

As stated earlier, it can no longer be assumed that the stresses are equivalent given the different cross-sectional area of each material. Once contact is made with the wall, the maximum expansion allowed would be simply equal to the gap. As such, the compatibility condition would have to be

$$\delta_{tot} = \delta_1 + \delta_2 = \varepsilon_1 L_1 + \varepsilon_2 L_2 \leq gap$$

The deformation for each component would consist of two parts, the positive thermal expansion (positive because the temperature must increase in this instance) and the compressive forces of restraint. Hence,

$$\delta_1 = -\sigma_1 \frac{L_1}{E_1} + L_1 \alpha_1 \Delta T$$

$$\delta_2 = -\sigma_2 \frac{L_2}{E_2} + L_2 \alpha_2 \Delta T$$

To complete the solution, substitute the constitutive equations into the compatibility equation, and simultaneously solve for the two stresses with the condition that δ_{net} = gap.

Example Problem 3-5: In reality, "rigid" constraints are either a fiction designed to trick the masses or a useful engineering approximation depending on your point of view. At any rate, parallel configurations offer an analytical escape from this assumption, as well as a means to analyze apparently constraint-free problems. Consider the Warp-Drive activation system shown below:

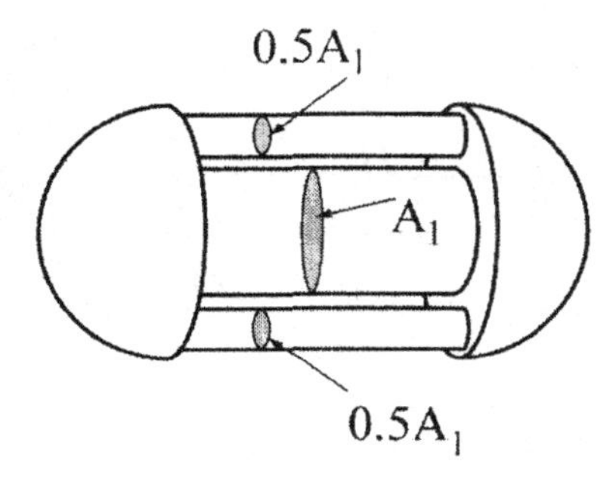

Figure 3.9 Parallel and multi-material system.

Basically, the stress/strain analysis of this activator follows along the same lines as previous examples. Equilibrium can still be used to at least set up relationships between the internal forces and stresses. However, one can not automatically set the deformation at zero (or to a gap value) since the problem is actually self-restrained.

As with the earlier examples, we will start with equilibrium, the FBD shown in Figure 3.10, and the observation of symmetry which dictates that the forces and stresses in the outer members (σ_2) are equal.

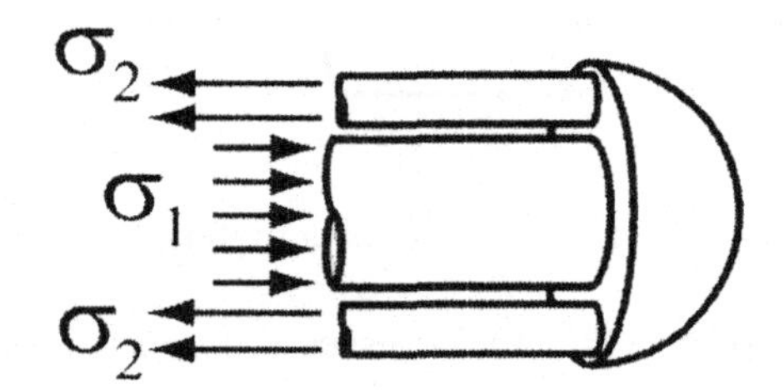

Figure 3.10 Free-body diagram (FBD) of the parallel and multi-material system.

Under these assumptions,

$$\sum F = 0 = \sigma_1 A_1 + 2\sigma_2 A_2$$

Note: at this point, the stresses have been arbitrarily drawn and will in fact, have to be of opposite sign (determined by the direction of ΔT) to self-equilibrate.

As deftly alluded to earlier, the compatibility relationships are not the same as the rigid restraints, so it cannot be assumed that the total deformation is simply zero. However, inspection of the structure and its inherent symmetry reveals that the net deformation of each member (1 and 2) must be identical; while the actual magnitude is unknown at this point, both must be the same or else it will fail and all bets (and previously defined methods) are off. Hence:

$$\delta_1 = \delta_2 = \varepsilon_1 L = \varepsilon_2 L \quad \rightarrow \quad \varepsilon_1 = \varepsilon_2$$

-or-

$$\pm\sigma_1 \frac{L}{E_1} \pm L\alpha_1 \Delta T = \pm\sigma_2 \frac{L}{E_2} \pm L\alpha_2 \Delta T$$

where the $\pm$ is left open because the signs will vary for each grouping of thermal and mechanical deformation depending on whether the temperature is increasing or decreasing. Again, the combination of the compatibility and equilibrium equations are sufficient to solve the problem.

So you do not think that we have conjured up these problems to simply enhance your suffering (ok, perhaps a little), it may be worthwhile to explore a few instances where such issues are relevant and useful. In fact, a large number of contemporary structural and machine design situations exist that feature the possibility of self-equilibrated stresses being generated between dissimilar materials due to thermal conditions. Examples include: electronic circuit boards, ceramic and semi-metallic semiconductors, and most engine parts to name just a few. Additional examples include pavements resting on different types of beds and sub-pavements, structural members and decking in bridgework, as well as ceramic protective coatings on gas turbine blading (see the next example problem), and/or just about any composite.

Example Problem 3-6: Consider theturbine blade shown below in Figure 3.11 that is protected by a ceramic thermal barrier coating (TBC); in this case, a thin coating of zirconia is used to shield the relatively tough Inconel 718 alloy blade from the extreme thermal environment. What levels of stress can be induced in the coatings as the operational temperature increases from 20°C to a maximum of 800°C? Assume that the coating thickness is constant at t = 0.254 mm and that all relevant thermoelastic properties can be assumed to be independent of temperature:

$$E_{zr} = 150 \text{ GPa} \qquad \alpha_{zr} = 10 \times 10^{-6}/^{o}C$$

$$E_{In} = 195 \text{ GPa} \qquad \alpha_{In} = 14 \times 10^{-6}/^{o}C$$

Although this is truly a three-dimensional problem, it is advisable to simplify as one-dimensional in the length sense as shown in Figure 3.11. Using this approach, the first, step is to isolate a "representative" section of the blade for the analysis as shown. Secondly, since the blade thickness, h, should be at least one order of magnitude larger than any internal ducts for cooling gases and so on, let h = 2.54 mm.

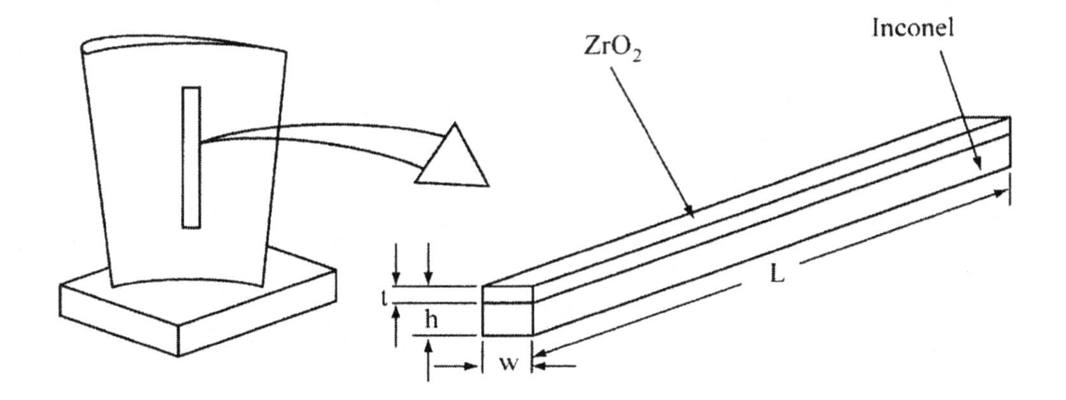

Figure 3.11 Coated turbine blade and one-dimensional simplification.

Before the analysis is started, it should be emphasized that $\alpha_{In} > \alpha_{zr}$. Because the temperature is increasing ($\Delta T\uparrow$) and both materials are therefore expanding, the higher coefficient of thermal expansion of the Inconel implies that it will be restrained by the zirconia and the zirconia, will in turn, be pulled by the expanding Inconel. Given this scenario, the zirconia will be in a state of tension and the Inconel in compression. Hence, by a brilliant observation combined with a truly keen engineering sense, we already know the signs of the two stresses and how the FBD should look as shown in Figure 3.12. If conditions were reversed and a cool down $\Delta T\downarrow$ experienced, the opposite would be true and the zirconia will be in a state of compression and the Inconel in tension.

Figure 3.12 Free-body diagram (FBD) of the coated turbine blade.

The calculations can then commence with the equilibrium equations and the already determined sense of the stresses (tension versus compression). Since the resultant forces can be determined by the product of the stresses and cross-sectional areas, equilibrium dictates:

$$\sigma_{zr}(t_{zr}w) - \sigma_{In}(t_{In}w) = 0 \quad \rightarrow \quad \sigma_{In} = 0.10\sigma_{zr}$$

Note that in this instance, the width (w) cancels out and is therefore not a factor in the solution. Furthermore, it is now inherently implied that the stresses in the zirconia and Inconel are tensile and compressive, respectively; the correct sign of the stresses must be incorporated in all subsequent calculations. Finally, one can see the influence of the greater Inconel thickness (and thus, cross-sectional area) since the corresponding stress level is only 10% of the zirconia coating.

The next step is to use the compatibility of the two layers to generate a second equation; in this instance, the layers must expand or contract the same amount, or they delaminate and system failure will ensue. Hence, mathematically the deformation must look like:

$$\varepsilon_{zr} = \varepsilon_{In}$$

-or-

$$+\sigma_{zr}\frac{L}{E_{zr}} + L\alpha_{zr}\Delta T = -\sigma_{In}\frac{L}{E_{In}} + L\alpha_{In}\Delta T$$

It is extremely important to note that the signs of the stresses and thermal expansion have been written into this relationship; because the temperature is rising (i.e., $\Delta T\uparrow$) and both materials are expanding, the thermal strains will always be positive in this instance. On the other hand, the stresses in the zirconia will be tensile (positive) and the Inconel compressive (negative) as already discussed. Finally, since the lengths (L) are the same for both materials, they will cancel for a solution independent of length; this assumption is valid as long as you are not interested in the stresses at or near the edges where the solution would not really be valid anyway.

Using the equilibrium and compatibility equations, rearranging terms, and noting the temperature change from reference or $\Delta T = 800^{\circ}C$ to $20^{\circ}C = 780^{\circ}C$, we finally arrive at the following solution:

$$\frac{\sigma_{zr}}{E_{zr}} + \frac{\sigma_{In}}{E_{In}} = \Delta T\left(\alpha_{In} - \alpha_{zr}\right) \quad \rightarrow \quad \sigma_{zr} = 434\text{MPa} \text{ and } \sigma_{In} = 43.4\text{MPa}$$

3.5 HOOKE'S LAW FOR ANISOTROPIC MATERIALS

Although the assumption of isotropy is reasonably accurate for predicting the elastic response of most metallics, ceramics, and stiff polymers, there are still a significant number of materials whose anisotropic character cannot be ignored. Historically, wood and wood laminates (plywood to be exact) were the materials most often cited for their anisotropy because the stiffness in the longitudinal growth direction (erroneously referred to as the *grain* direction), is several times higher than in the transverse sense. Due to the proliferation of engineered composite materials in the past several decades, design activity in anisotropic materials has increased markedly. Structural composites are, by definition, made of components whose individual responses are very dissimilar. Depending on how these components are arranged, the overall system can exhibit very different elastic properties in various directions. Before we go into the somewhat tedious relationships that are the hallmark of anisotropic materials, it may be a good idea to go through a relatively simple example to get things started.

Example Problem 3-7: Consider a laminated composite for aircraft wings is fabricated from component sheets (denoted as 1 and 2), which themselves are isotropic as shown in Figure 3.13. Since directionality is now an issue, the first step is to consider the response of the laminate in the x-direction; if an overall tensile stress σ_{tot} is applied in the x-sense, then:

$$\sigma_{xx,tot} \equiv E_{eff,x}\varepsilon_{xx,tot}$$

Note that an overall, or average, or *effective* modulus now relates the variables. To see how the effective modulus is fixed by the component moduli, consider the following analysis.

Assuming that only normal stresses are present, force equilibrium in the x-direction demands:

$$\sum F_x = \sigma_{xx,tot} A_{tot} = \sigma_{xx,tot}(A_1 + A_2) = \sigma_{xx,1} A_1 + \sigma_{xx,2} A_2$$

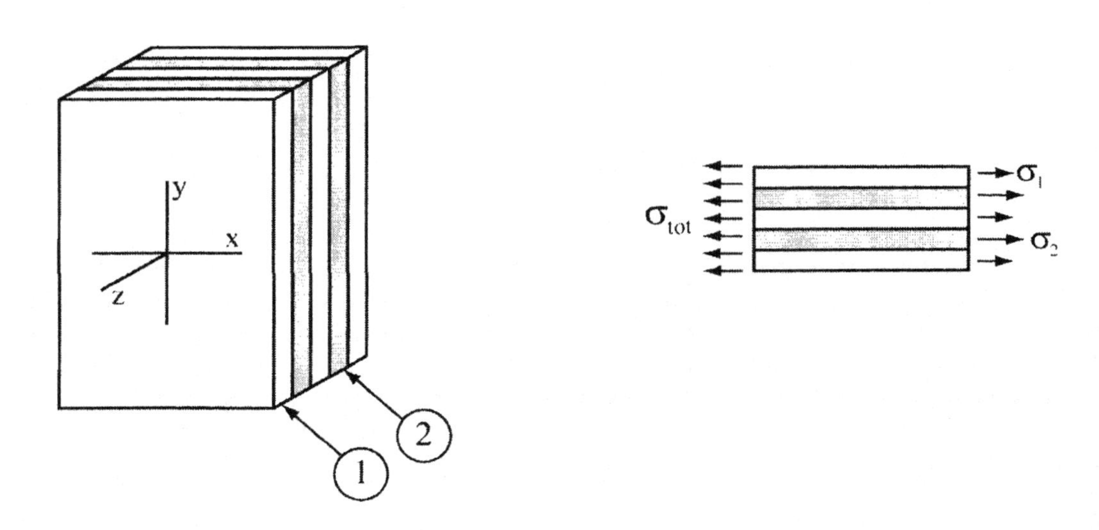

Figure 3.13 Typical laminated composite material.

As with the earlier examples, the plies must deform the same amount, lest they suffer a delamination failure. Hence,

$$\varepsilon_{xx,1} = \varepsilon_{xx,2} = \varepsilon_{xx,tot}$$

With only two equations and four unknowns, two component constitutive equations are required to solve the problem:

$$\varepsilon_{xx,1} = \frac{\sigma_{xx,1}}{E_1} \quad \text{and} \quad \varepsilon_{xx,2} = \frac{\sigma_{xx,2}}{E_2}$$

With four unknowns and an equal number of equations, panic can be avoided and the effective modulus found to be:

$$E_{eff,x} = \frac{E_1 A_1 + E_2 A_2}{A_1 + A_2}$$

Due to the symmetry of the laminate, the effective modulus would be the same in the y-direction. The effective modulus so derived is referred to as the *Rule of Mixtures* result, because it only depends on the relative amount of the two components. Furthermore, the laminate model, with equal strains assigned to both components, is the *Isostrain*, or constant strain model for a composite.

However, the composite stiffness in the z-direction will definitely be very different; every ply of the laminate carries the same stress, each ply deforms independently of the others, and the total strain is the sum of the ply strains. If each ply is the same thickness in the z-direction, the effective modulus for this example becomes:

$$E_{eff,z} = \frac{5E_1E_2}{2E_1 + 3E_2}$$

Clearly, the elastic responses of composite materials can be very direction dependent. Given this reality, the vexing question that has no doubt kept you up at night becomes: how does one express this elastic anisotropy?

As mentioned previously in Chapters One and Two, stresses and strains are mathematical entities called tensors of the second rank. The linear relationship between these parameters can only be written one way, constrained as it is by the laws of the linear algebra:

$$\varepsilon_{ij} = \sum_k \sum_l S_{ijkl}\sigma_{kl} \quad (3.9)$$

where the indices k and l take on values of x, y and z. Thus, each strain component depends on nine stress components, six of which were seen before to be independent. The coefficients S_{ijkl} are known as the *Compliances;* the larger they are, the softer the material is elastically. In the inverse form, Hooke's law can be written as:

$$\sigma_{ij} = \sum_k \sum_l C_{ijkl}\varepsilon_{kl} \quad (3.10)$$

The coefficients C_{ijkl} are known as *Stiffnesses*, and act much as moduli do. Recalling that there are nine components of stress or strain, either of the two equations given above implies 81 elastic constants. However, due to symmetries and thermodynamic considerations, most anisotropic materials have only 21 independent constants. If the discussion is restricted even further to common anisotropic materials used in load-bearing design, the number of constants is mercifully reduced to a manageable handful as seen below.

The elastic constants used to describe the response of isotropic solids such as E, G, and ν are scalar quantities, as their measured values are independent of direction in the material. In the case of anisotropy, the elastic constants have their own mathematical character; they are inherently tensors of the fourth rank, and must therefore be transformed mathematically as axes are rotated. If β_{ij} are the direction cosines between any one of the original x, y, or z and one of the new, rotated axes, x', y', or z', then the components of compliance S_{ijkl} (or stiffness C_{ijkl}) become S'_{mnop} in the rotated system or:

$$S'_{mnop} = \sum_i \sum_j \sum_k \sum_l \beta_{mi}\beta_{nj}\beta_{ok}\beta_{pl}S_{ijkl} \quad (3.11)$$

Since i, j, k, and l each take on the three values of x, y and z, Equation (3.11) should be evaluated by a computer to retain ones sanity. Fortunately, the algebra can be greatly simplified for most of the anisotropic materials that are of engineering concern as described below.

3.6 ORTHOTROPIC COMPOSITES

Generally, the anisotropic material of greatest interest for load-bearing design is the *Orthotropic* solid; this type of material is defined in terms of rotational symmetry so that the elastic coefficients remain the same in any two coordinate directions when rotated 180° about the third as shown in Figure 3.14. A relevant example might be a filament wound composite fabricated as a pressure vessel, with twice as many reinforcing filaments in the hoop versus axial direction. Other orthotropic materials include wood laminates (plywood is, but particle board is not), and perhaps surprisingly to many, rolled-metal sheet and bar stock.

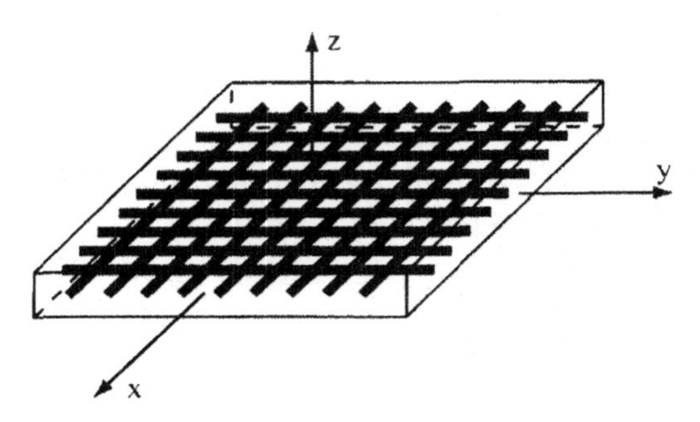

Figure 3.14 Orthotropic filament, reinforced-composite panel.

Hooke's Law in the form of Equations (3.9 and 3.10) is often expressed in a contracted engineering notation as:

$$\varepsilon_q = \sum_r S_{qr}\sigma_r \quad \text{and} \quad \sigma_q = \sum_r C_{qr}\varepsilon_r \tag{3.12}$$

In order for Equations (3.12) to be equivalent to (3.9) and (3.10), the following index *map* must be employed:

$$\begin{array}{rl} q,r = & 1;\ \ 2;\ \ 3;\ \ 4;\ \ 5;\ \ 6 \\ & \downarrow\ \ \ \downarrow\ \ \ \downarrow\ \ \ \downarrow\ \ \ \downarrow\ \ \ \downarrow \\ ij = & xx;\ yy;\ zz;\ yz;\ xz;\ xy \end{array} \tag{3.13}$$

For example, $\varepsilon_2 \equiv \varepsilon_{yy}$; $S_{12} \equiv S_{xxyy}$ and so on. The implications of Equations (3.12) are most easily seen by writing them out for the isotropic case, thus Equation (3.4) becomes:

$$\varepsilon_1 = \varepsilon_{xx} = \frac{1}{E}\sigma_1 - \frac{\nu}{E}\sigma_2 - \frac{\nu}{E}\sigma_3 = \frac{1}{E}\sigma_{xx} - \frac{\nu}{E}\sigma_{yy} - \frac{\nu}{E}\sigma_{zz} \tag{3.14a}$$

$$\varepsilon_2 = -\frac{\nu}{E}\sigma_1 + \frac{1}{E}\sigma_2 - \frac{\nu}{E}\sigma_3 \tag{3.14b}$$

$$\varepsilon_3 = -\frac{\nu}{E}\sigma_1 - \frac{\nu}{E}\sigma_2 + \frac{1}{E}\sigma_3 \tag{3.14c}$$

$$\varepsilon_4 = \varepsilon_{yz} = -\frac{\gamma_{yz}}{2} = \frac{1}{2G}\tau_{yz} = \frac{1}{2G}\tau_4 \tag{3.14d}$$

$$\varepsilon_5 = \frac{1}{2G}\tau_{xz} = \frac{1}{2G}\tau_5 \tag{3.14e}$$

$$\varepsilon_6 = \frac{1}{2G}\tau_{xy} = \frac{1}{2G}\tau_6 \tag{3.14f}$$

For an isotropic solid:

$$S_{11} = S_{22} = S_{33} = \frac{1}{E} \qquad S_{12} = S_{13} = S_{23} = -\frac{\nu}{E} \qquad S_{44} = S_{55} = S_{66} = \frac{1}{2}G \tag{3.15}$$

with all other values of S set to zero for these materials. The designation of S_{qr} as engineering parameters is certainly more reasonable, as they can be simply related to the moduli already in use.

A general response of an orthotropic solid can be written as:

$$\varepsilon_1 = S_{11}\sigma_1 + S_{12}\sigma_2 + S_{13}\sigma_3 \tag{3.16a}$$

$$\varepsilon_2 = S_{12}\sigma_1 + S_{22}\sigma_2 + S_{23}\sigma_3 \tag{3.16b}$$

$$\varepsilon_3 = S_{13}\sigma_1 + S_{23}\sigma_2 + S_{33}\sigma_3 \tag{3.16c}$$

$$\varepsilon_4 = S_{44}\sigma_4 \tag{3.16d}$$

$$\varepsilon_5 = S_{55}\sigma_5 \tag{3.16e}$$

$$\varepsilon_6 = S_{66}\sigma_6 \tag{3.16f}$$

All other S's are zero for these materials, just as for isotropic solids. However, and in direct contrast to isotropic materials, there are three different orthotropic moduli, Poisson's ratios, and shear moduli for a total of nine independent elastic constants. Thus:

$$S_{11} = \frac{1}{E_x} \qquad S_{22} = \frac{1}{E_y} \qquad S_{33} = \frac{1}{E_z} \tag{3.17a}$$

$$S_{12} = -\frac{\nu_{xy}}{E_y} \qquad S_{13} = -\frac{\nu_{xz}}{E_z} \qquad S_{23} = -\frac{\nu_{yz}}{E_z} \tag{3.17b}$$

$$S_{44} = \frac{1}{2G_{yz}} \qquad S_{55} = \frac{1}{2G_{xz}} \qquad S_{66} = \frac{1}{2G_{xy}} \tag{3.17c}$$

Equation (3.17) should make it obvious why the index formalism of Equation (3.12) is preferable; otherwise, the juggling of directional elastic parameter indices quickly gets out of hand and the unsuspecting engineer quietly goes mad.

Example Problem 3-8: The stress state indicated below in Figure 3.15 is thought to exist in a uni-directionally glass-reinforced polymer, matrix composite used for the hull of a single-seat speed boat. Determine the corresponding strain state assuming the following properties.

E_x (GPa)	E_y (GPa)	ν_{xy} (=ν_{yx})	G_{xy} (GPa)
181	10.3	0.28	7.17

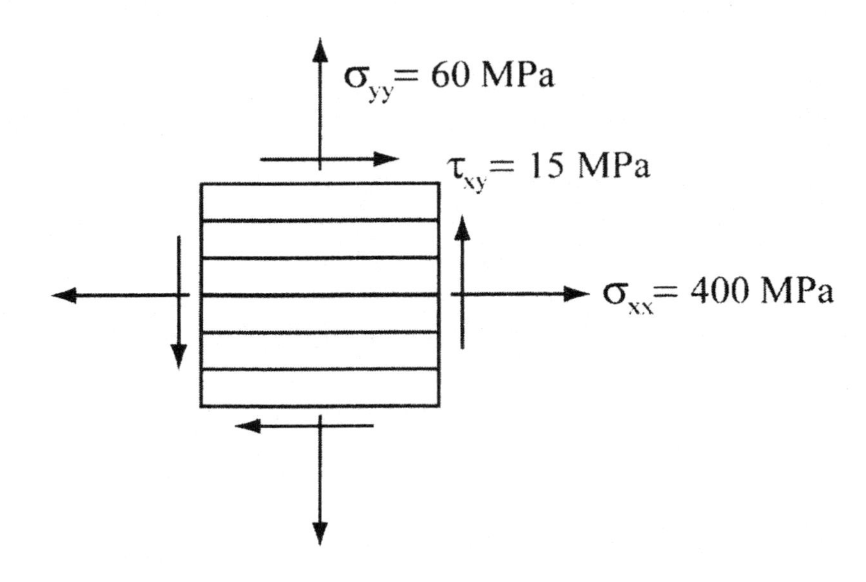

Figure 3.15 Stress state on a composite panel.

While the relationships just explored can become tedious, the calculations for this example are relatively straightforward. As such, the various constants are determined to be:

$$S_{11} = \frac{1}{E_x} = \frac{1}{181x10^9} = 5.525x10^{-12}(Pa)^{-1} = 5.525(TPa)^{-1}$$

$$S_{22} = 97.09(TPa)^{-1};\ S_{12} = \frac{-\nu_{xy}}{E_y} = -0.27(TPa)^{-1}$$

$$S_{66} = 69.7(TPa)^{-1}$$

Calculating strains from Hooke's Law via Equation (3.12):

$$\varepsilon_1 = \varepsilon_{xx} = S_{11}\sigma_1 + S_{12}\sigma_2 = 2.12x10^{-3}$$

$$\varepsilon_2 = \varepsilon_{yy} = S_{12}\sigma_1 + S_{22}\sigma_2 = 5.21x10^{-3}$$

$$\varepsilon_6 = \frac{\gamma_{xy}}{2} = S_{66}\sigma_6 = 2.09x10^{-3}$$

3.7 ELASTIC CONSTANT TRANSFORMATIONS

Equation (3.11) prescribes the manner in which anisotropic elastic constants must be transformed as axes are rotated, employing the direction cosines (β) of the angles of rotation. Although these expressions are easily evaluated via computer, they are tedious to write out in long-hand for the general case, as the summand on the right contains 21 terms. Fortunately, the equations do simplify markedly for solids whose anisotropy is reduced, such as the orthotropic solids already discussed. In fact, orthotropic sheet materials are particularly straightforward, and their analysis is as developed below.

For the *Principal Material Directions*, those that are the axes of orthotropy and contain the least number of elastic constants, Hooke's Law as described by Equation (3.12) can be expanded as:

$$\varepsilon_1 = S_{11}\sigma_1 + S_{12}\sigma_2 \qquad (3.18a)$$

$$\varepsilon_2 = S_{12}\sigma_1 + S_{22}\sigma_2 \qquad (3.18b)$$

$$\varepsilon_6 = S_{66}\sigma_6 \qquad (3.18c)$$

Note that all terms involving 3 in the subscripts have been omitted in Equation (3.18); it was assumed that the thin dimension in the 3- (or z-) direction does not allow for loading in that sense, so all stresses with 3, 4, or 5 subscripts are zero. In reality, compliances with these subscripts may have nonzero values, but will not show up in the Hooke's Law because they are coefficients for zero-valued stresses. For non-principal material directions, the strain stress law is written:

$$\varepsilon_{1'} = S_{1'1}\sigma_1 + S_{1'2}\sigma_2 + S_{1'6}\sigma_6 \qquad (3.19a)$$

$$\varepsilon_{2'} = S_{12'}\sigma_1 + S_{2'2}\sigma_2 + S_{2'6}\sigma_6 \qquad (3.19b)$$

$$\varepsilon_{6'} = S_{1'6}\sigma_1 + S_{2'6}\sigma_2 + S_{6'6}\sigma_6 \qquad (3.19c)$$

Note that in Equation (3.16), shear stresses cause normal strains, and vice versa. Additionally, the material symmetry implies that the indices can be permuted on the compliances or $S_{1'6} = S_{6'1}$, and so on. The physical relationship between the material principal directions and the nonprincipal direction orientation is shown in Figure 3.16.

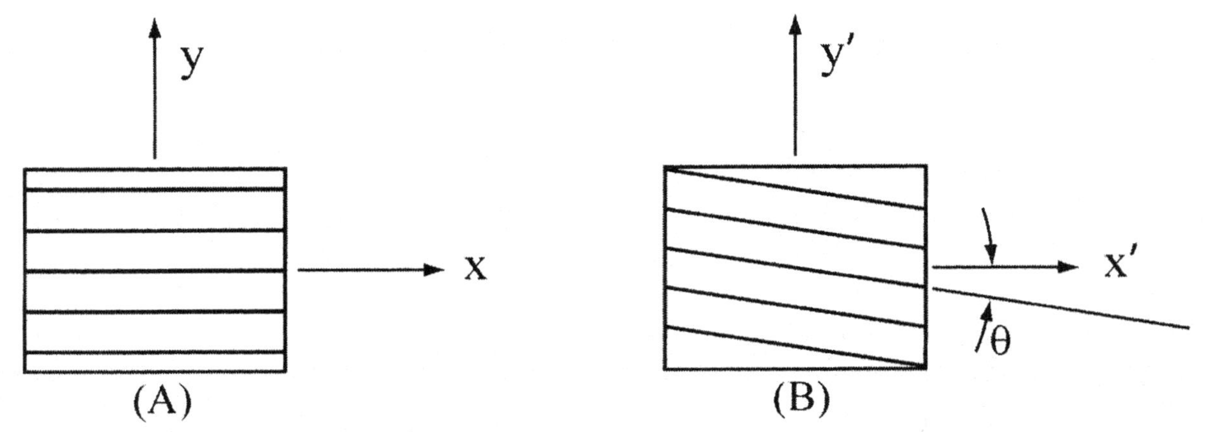

Figure 3.16 Principal orthotropic axes for (a) a unidirectionally reinforced composite, and (b) rotated through angle θ in a nonprincipal axis frame.

Applying Equation (3.11) to find each of the new, transformed compliances in Equation (3.19) and using $m = \cos\theta$ and $n = \sin\theta$, one gets:

$$S_{1'1} = m^4 S_{11} + n^4 S_{22} + 2m^2n^2 S_{12} + m^2n^2 S_{66} \tag{3.20a}$$

$$S_{2'2} = n^4 S_{11} + m^4 S_{22} + 2m^2n^2 S_{12} + m^2n^2 S_{66} \tag{3.20b}$$

$$S_{1'2} = m^2n^2 S_{11} + m^2n^2 S_{22} + \left(m^4 + n^4\right) S_{12} - m^2n^2 S_{66} \tag{3.20c}$$

$$S_{6'6} = 4m^2n^2 S_{11} + 4m^2n^2 S_{22} - 8m^2n^2 S_{12} + \left(m^2 - n^2\right) S_{66} \tag{3.20d}$$

$$S_{1'6} = 2m^3n^2 S_{11} - 2mn^3 S_{22} + 2\left(mn^3 - m^3n\right) S_{12} + \left(mn^3 - m^3n\right) S_{66} \tag{3.20e}$$

$$S_{2'6} = 2m^3n^2 S_{11} - 2mn^3 S_{22} + 2\left(m^3n - mn^3\right) S_{12} + \left(m^3n - mn^3\right) S_{66} \tag{3.20f}$$

Equation (3.20) transforms the compliances of any orthotropic sheet material in the plane of the sheet. On the other hand, the strains in Equations (3.19) transform as any other strain does according to the transformation equations of Chapter One and/or everyone's favorite, Mohr's circle for strain.

Example Problem 3-9: A tensile plate for a tilt-rotor aircraft with a circular penetration is made of a uni-directional composite material with $E_x = 38.7$ GPa, $E_y = 8.3$ GPa, $\nu_{xy} = \nu_{yx} = 0.26$, and $G_{xy} = 4.14$ GPa. Using these properties, determine the elastic compliances for an element on the periphery of the hole at 45° as shown in Figure 3.17.

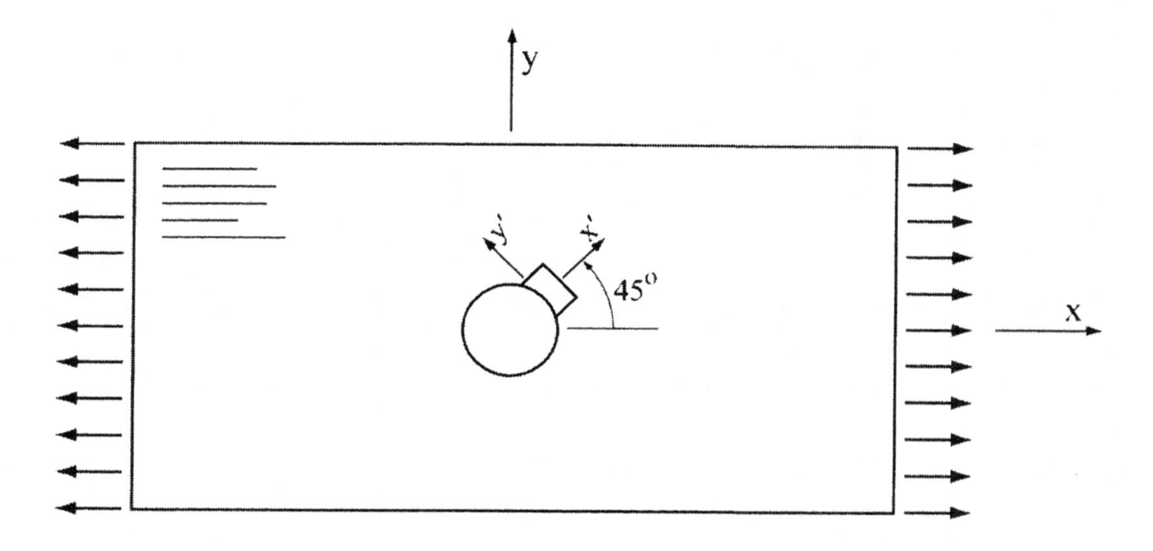

Figure 3.17 Unidirectional plate with a circular cut-out and desired orientation.

Using the various relationships just described, the compliances in the principle x, y-directions are:

$$S_{11} = \frac{1}{E_x} = 26(\text{TPa})^{-1} \quad S_{22} = 121(\text{TPa})^{-1} \quad S_{12} = -6.7(\text{TPa})^{-1} \quad S_{66} = 241(\text{TPa})^{-1}$$

The compliances in the primed coordinate senses rotated at 45° such that $m = n = 0.707$ are then from Equation (3.20):

$$S_{1'1} = 9.53(\text{TPa})^{-1} \quad S_{2'2} = 9.53(\text{TPa})^{-1} \quad S_{1'2} = -26.9(\text{TPa})^{-1}$$

$$S_{6'6} = 160(\text{TPa})^{-1} \quad S_{1'6} = -47(\text{TPa})^{-1} = S_{2'6}$$

Finally, note that $E_{x'} = E_{y'} = 10.4\text{GPa}$.

PRACTICE EXERCISES

1. The following two strain states were measured at two different points in a steel car chassis during a routine test run by the Crash Test Dummies. What are the corresponding stress states? Which case has the maximum principal stress, and what is its value? For isotropic steel, one can assume E = 30 x 10^{-6} psi and $\nu = 0.33$.

$$\text{a. } \varepsilon_{ij} = \begin{pmatrix} 900 & 200 & 0 \\ 200 & 400 & 0 \\ 0 & 0 & 0 \end{pmatrix} \text{x}10^{6} \qquad \text{b. } \varepsilon_{ij} = \begin{pmatrix} 400 & 0 & 300 \\ 0 & -700 & 0 \\ 300 & 0 & 300 \end{pmatrix} \text{x}10^{6}$$

2. Shown below are thermoelastic properties data for Inconel 713LC nickel alloy used for steam tubines. What thermal strains would be generated in an unrestrained sample of this material as it is heated from 20°C to 800°C? What stress would be induced in the same sample if it were to be rigidly constrained over the same temperature range?

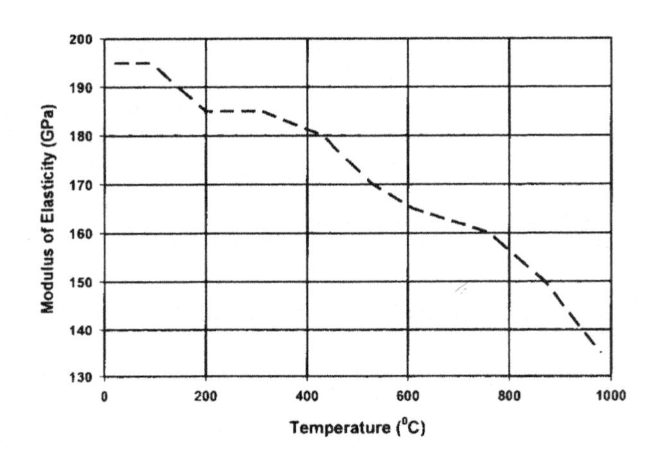

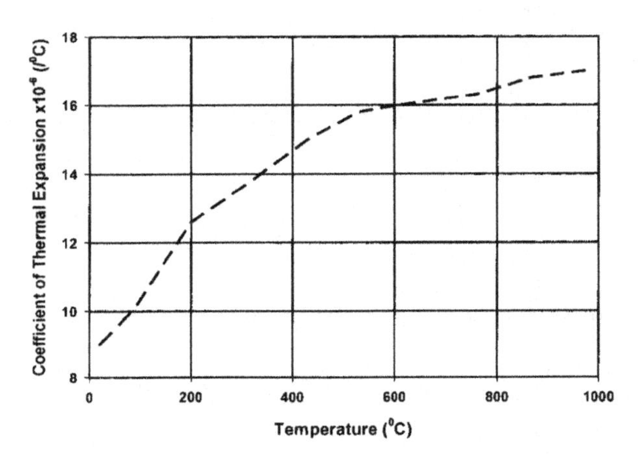

3. The aluminum and steel bars from an engine mounting system shown below are of equal length and the cross sectional area of the aluminum is twice that of the steel. At zero degrees, they just fit between two rigid walls as shown below. If the temperature of the bars is raised by 100°F, (a) what is the axial stress in the steel, (b) what is the axial stress in the aluminum, (c) what is the deflection of the interface in terms of "L," and (d) is the axial force in the steel equal to that in the aluminum? Relevant materials properties are:

$E_{st} = 30 \times 10^6$ psi $\quad \alpha_{st} = 7 \times 10^{-6}/°F \qquad E_{Al} = 10 \times 10^6$ psi $\quad \alpha_{Al} = 12 \times 10^{-6}/°F$

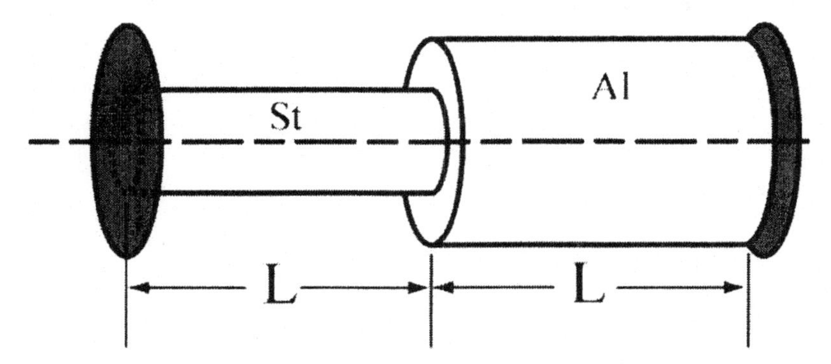

4. Jet engine cowlings use 5 μm diameter whisker filaments of SiC that are coated with 3 μm of aluminum using a chemical bonding process. The coating process is carried out at 650°C before the sample is cooled to room temperature. Elastic stresses due to thermal expansion mismatch do not begin to appreciably accumulate until the alloy has cooled to approximately 200°C due to alloy viscosity. If the stresses necessary to maintain whisker/coating compatibility are assumed to only be acting in the direction of the filament axis, what is their sign and magnitude at 20°C? Relevant materials properties are:

$E_{SiC} = 476$ GPa $\qquad \alpha_{SiC} = 4.7 \times 10^{-6}/°C$

$E_{Al} = 70$ GPa $\qquad \alpha_{Al} = 21 \times 10^{-6}/°C$

5. The infamous "Flux Capacitor" that makes time travel possible consists of an epoxy circuit board with a gold conductor post of circular cross-section. The circular button head just touches the board while the post clears the sides of the bore hole through the epoxy at 20°C. At what higher temperature would the stresses in the gold post exceed its tensile yield strength of 10 MPa and the unit malfunction (sending the hapless time traveler straight to a cretaceous period T-Rex feeding frenzy)? Assume that only the board region directly under the button heads (contact area) supports the stress transferred by the button. Relevant materials properties are:

$E_E = 3.5$ GPa $\qquad \alpha_E = 28 \times 10^{-6}/°C$

$E_G = 80$ GPa $\qquad \alpha_G = 14 \times 10^{-6}/°C$

0.2 cm diam.
Gold Post
0.5 cm
Epoxy
0.5 cm diam.

6. Revisiting an earlier example problem about a zirconia-clad Inconel turbine blade, now consider the stresses that would be induced by the same temperature change as a two-dimensional, plane stress problem. For the same increase in temperature as the example problem, what are the stresses in the zirconia if they are assumed to be the same magnitude and sign in the two directions shown ν_{In}=0.28 and ν_{Zr}=0.22?

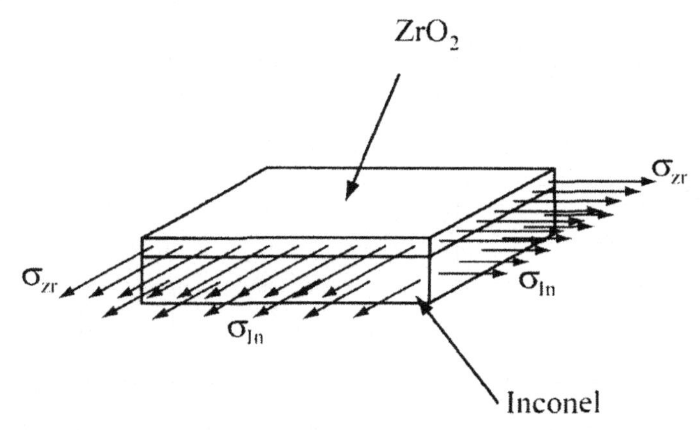

7. A structural composite designed to improve fuel economy is composed of continuous filaments aligned in the load carrying direction. The filaments are known to fracture in a brittle fashion at a stress of 2 GPa and a strain of 0.01 mm/mm. The brittle matrix containing them fails at 30 MPa and a strain of 0.25: the matrix material accounts for 75% of the bulk of the composite. What is the effective elastic modulus of the composite in the filament direction?

8. Due to exposure to an aggressive salt-water environment, a 1.0 cm diameter cable used to stop incoming aircraft on an aircraft carrier was fabricated using an epoxy-based polymer. However, carrying its design load of 400 kN proves to induce excessive deflections in the epoxy, whose modulus of elasticity is 6 GPa. Strains must be reduced by 80% by adding uni-directional glass fibers while maintaining the same overall cable diameter. If the modulus of elasticity of the glass is 69 GPa, what volume fraction of glass fiber must be added to the epoxy matrix?

9. A composite aircraft truss is fabricated from graphite fibers having an elastic modulus E = 230 GPa (note that steels are around 210 GPa) and an epoxy matrix material with a modulus of 5.5 GPa. What volume fraction of fiber must be entrained in the epoxy to achieve a fiber-to-matrix load ratio P_f/P_m = 25.0?

10. A uni-directionally reinforced graphite/epoxy composite designated T300/5208 has the following moduli and elastic parameters relative to the principal material directions: E_x = 181 GPa, E_y = 10.3 GPa, $\nu_{xy} = \nu_{yx}$ = 0.28, G_{xy} = 7.17 GPa. Find the complete state of strain present if the stress state in the x, y-plane is as given below:

$$\sigma_{ij} = \begin{pmatrix} 207 & 69 & 0 \\ 69 & 138 & 0 \\ 0 & 0 & 0 \end{pmatrix} \text{MPa}$$

11. The elastic stiffness of a material is a *response*, not a property; in particular, the stresses due to constraint have a definite effect on the apparent elastic modulus E_{app}, defined for the configuration below as the ratio of σ_{zz} to ε_{zz}.

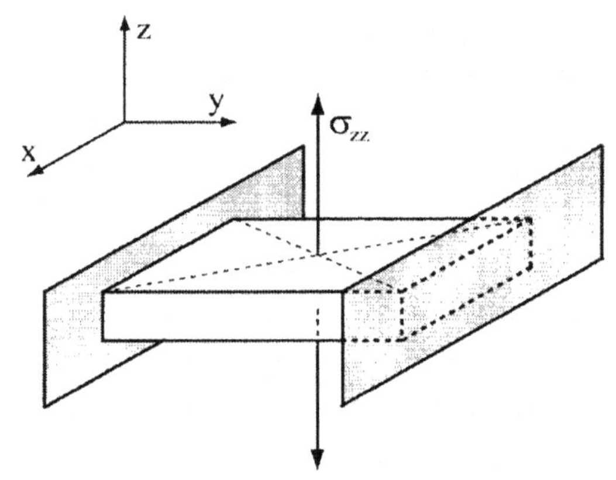

a. For the constrained material above, what is the E_{app} for complete, rigid restraint in the y-direction as indicated? (Hint: $\varepsilon_{yy} = 0$).

b. If constraint is added in the x-sense as well, or $\varepsilon_{xx} = 0$ also, what would the value of E_{app} become?

CHAPTER FOUR

TENSILE AND COMPRESSIVE RESPONSE

4.1 INTRODUCTION

Unbeknownst to many, the earliest systematic inquiries into the ability of materials to carry external loadings were carried out by the so-called natural philosophers of the Renaissance. Needless to say, these folks were not your everyday engineers as their search was for the internal structure of solids, known to us today as atoms, molecules, and so forth. By subjecting samples of common materials to tensile loads and observing their inherent resistance to deformation, these early researchers sought insights into the forces holding the constituents of matter together; long filaments of materials as disparate (and sometimes disgusting) as catgut and metals were all hung from cathedral rooftops with deadweights. The resulting deformations were then measured via graduated eyepieces affixed to crude microscopes in what was probably a very smelly environment.

Now that our knowledge of the physical nature of solids has advanced to the point where molecules, atoms, electrons, and even smaller particles are known to nearly everyone with an interest in nature, the testing of a material's tensile response fortunately rests in the capable hands of engineers. Yet, despite all of our technological and egotistical gains, tensile tests much like the odorous one of old, still remain the most common and reproducible method for the mechanical characterization of a solid. Moreover, the simple tensile (or compression) test represents a true bargain since properties useful for very complex situations can be surmised from the simplest of configurations. Given the usefulness (and perseverance) of the tensile test, this chapter describes the methodology of the tensile test, as well as an analysis of the valuable information it provides.

4.2 TENSILE TEST VARIABLES

It was just mentioned that the tensile or compressive test is a real bargain in that much can be learned from a very simple configuration. In fact, when pulling a material sample in tension (or pushing it into a state of compression), there are five important parameters that all affect the response of the material. These parameters are defined as follows:

Deformation: while usually specified as the state of strain, the deformation may be from one test or include a history of deformations encountered during fabrication processing, etc.

Loading: usually specified as the state of stress.

Time: normally specified and considered important in terms of the rate of testing; parameters such as strain or stress rate are the main factors of concern.

Testing Environment: any number of external environmental effects (all with potential design interest) may influence the tensile behavior of any given material. For example, chemically aggressive species may appear to limit or degrade strength and/or ductility. As already discussed in Chapter Three, temperature is also an environmental factor of prime concern for a number of reasons.

Material Composition and Manufacturing Methods: obviously, the chemical composition of a solid has a great deal to do with its strength and deformation characteristics. In addition, the manner in which a structural member is fabricated plays a role in determining mechanical behavior. The strength of a drawn wire will differ from that of the casting from which it came, even though the two may possess identical chemistry. Unfortunately, the scope of composition and fabrication variables is far too extensive to be fully treated in the present text.

4.3 TEST PROCEDURES

Although a tensile or compressive test appears to be so straightforward that the procedures should seem intuitively obvious, the need for both repeatability and validity of results places some constraints on specimen geometry and loading particulars. For example, the specimen ends must possess greater strength than the portion of the specimen actually being tested, lest the gripping mechanism "bite" off the ends in some manner. Such considerations usually lead to the classic "dog bone" specimen, which is shown below in Figure 4.1 for both a flat and cylindrical configuration.

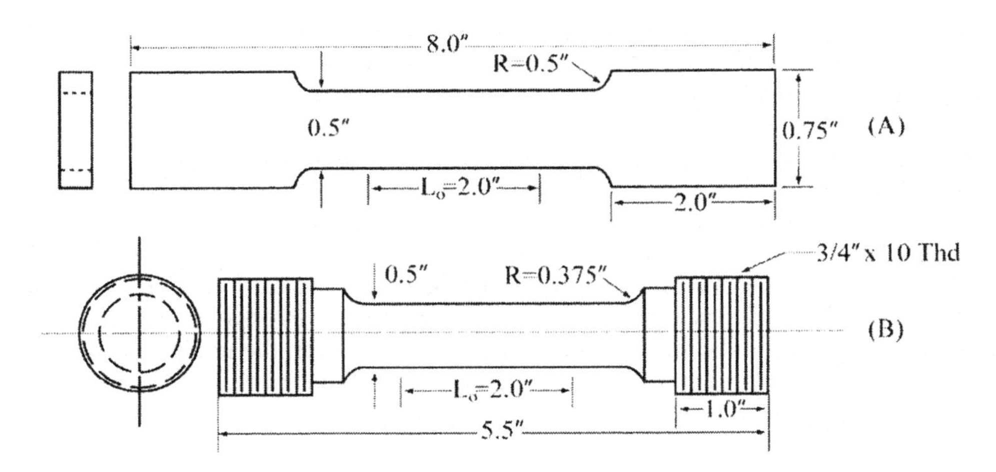

Figure 4.1 Standard ASTM (A370) tensile specimens for (A) sheet materials, and (B) bars or plates.

Despite the overall simplicity of the test, a considerable amount of effort has been devoted to defining test sample geometries that provide accurate stress/strain states while simultaneously minimizing material consumption such as the specimens of Figure 4.1. The American Society for Testing and Materials (ASTM) promotes standard testing procedures with Standard E8 covering all aspects of tensile testing of metals. As is often the case, alternative standards may apply to other materials and will be discussed in subsequent chapters.

4.4 TEST MEASUREMENTS

For the tensile test specimens of Figure 4.1 (or any other prismatic bar loaded in pure tension), it is assumed that the *tensile stress* or σ (referred to as the *Nominal* tensile stress) is given by one of the world's most complicated equations:

$$\sigma = \frac{P}{A_o} \tag{4.1}$$

where P is the load (lbs or N) and A_o is the original (nominal) cross-sectional area. In addition, the *nominal strain* or ε (induced by a nominal stress or vice versa) is also arrived via another complicated relationship as discussed in previous chapters:

$$\varepsilon = \frac{\Delta L}{L_o} \tag{4.2}$$

where ΔL is the change in length and L_o is the original length as defined in Figure 4.2. It should be noted that most modern tension test machines operate by deforming the specimen via rotating screws or hydraulic rams so the load (P) is generated by the resistance of the specimen to the applied deformation. In such a situation, the deformation or strain is the independent variable, whereas the load, or stress, is the dependent variable. Usually, the tests are conducted at a constant temperature and deformation rate. Finally, it should also be noted that any subscripts referring to direction (x-, y-, and z-axes) have been omitted in Equations (4.1 and 4.2), since only a single uni-axial load is involved and directionality is obvious. However, transverse strains are also present as will be discussed shortly.

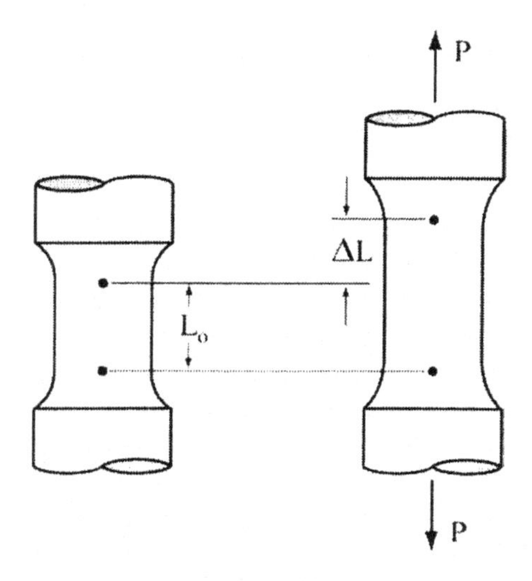

Figure 4.2 Loading and deformation of a tensile specimen.

4.5 ELASTIC RESPONSE MEASURES

A designer of load-bearing structural elements needs to have answers to important questions such as: "how large can my stress magnitudes be in this material before failure?" or, "if I have a stress of a given magnitude, what deformations are to be expected?" The numerical values of quantities we use to answer these and other important questions are often collectively referred to as *Mechanical Properties*. The term "property" implies a characteristic that is intrinsic to the solid itself. As we shall soon discover, the parameters necessary to utilize solids in load-bearing design are usually sensitive to stress- or strain-rate, temperature, microstructure, and so on. In short, the parameters of concern are more precisely *Mechanical Responses* as their numerical values are sensitive to extrinsic influences.

A wide variety of useful engineering solids respond elastically in a linear fashion for small imposed strains as shown in Figure 4.3. Because we are dealing with a material in the elastic range (no permanent deformations), the loading and unloading paths in the stress strain space are coincident. The tensile response equation for such a solid is written:

$$\sigma = E\varepsilon \tag{4.3}$$

Equation (4.3) should look very familiar as it is the tensile form of Hooke's Law; the constant of proportionality E is known as *Young's Modulus* or the *Modulus of Elasticity* and is a measure of the solid's stiffness.

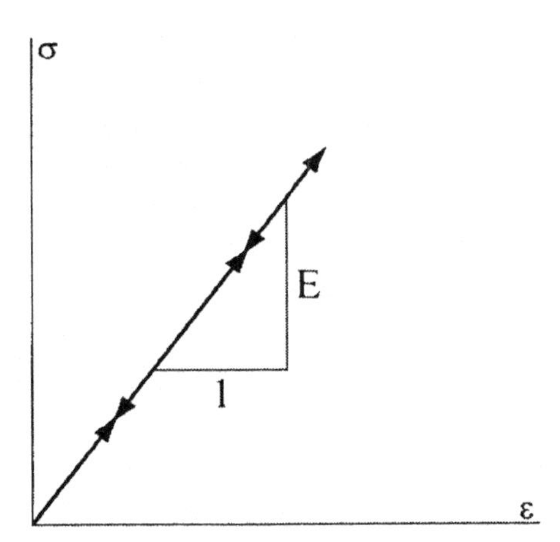

Figure 4.3 Linear response for loading/unloading paths in tension.

For relatively stiff materials, say $E > 10^6$ psi (7 GPa), the linear approximation of Equation (4.3) is usually valid. Of course, nature never truly works in a simple manner (or makes life easy) as Equation (4.3) would imply. Nevertheless, within the accuracy and resolution of measuring instruments such as micrometers, dial gages, and electrical resistance strain gages, the assumed linearity (and validity) of Equation (4.3) can be demonstrated for most materials of engineering importance. Hence, applying Equation (4.3) in design practice is perfectly permissible since the measuring instruments used in fabrication are usually no better than those used in materials testing. However, it should also be noted that the modulus of elasticity is sensitive to the chemistry of the solid, the rate of load or deformation application, and the temperature, to name just a few significant external variables.

As discussed in Chapter Three using the "deadly" rubber band example, it is a common observation that as one stretches a solid, its cross-section is reduced in size. Because a tensile strain is produced in the load direction, the thickness/width in the transverse direction(s) decrease, thus generating an additional strain. The transverse strain effect is measured by what is now known as Poisson's ratio or:

$$\nu \equiv -\frac{\varepsilon_{trans}}{\varepsilon_{axial}} \tag{4.4}$$

Here, ν is *Poisson's Ratio* that is a positive number that is typically within the following range of $0.2 \le \nu \le 0.5$. It should be noted that Equation (4.4) is only defined for uni-axial tension or compression with typical values (along with elastic modulus) given in Table 4.1. The variations and ranges in the data of Table 4.1 clearly demonstrate the influence of composition variations, as well as the ever present (and annoying) experimental errors and inconsistencies.

Up to this point, nothing has been said about how materials behaviors can change. For many materials, most notably alloys under load, two significant situations will be encountered as one moves up the tensile stress strain response plot as shown in Figure 4.3; either the sample will break while elastically stressed or some form of nonlinear behavior will ensue. Figure 4.4A illustrates the type of behavior known as *Brittle*, at least in the engineering sense. Brittle behavior is noted in most ceramics, concrete, glass, naturally occurring materials such as stone, some polymers, and in a few metals such as cast iron. Whether or not a material exhibits brittle behavior may depend on the temperature at which it is tested; lower temperatures tend to promote brittleness and this may have contributed to the sinking of the Titanic in the icy North Atlantic. The maximum tensile stress is referred to most commonly as the *Tensile Strength* (abbreviated hereafter as TS), although it is sometimes called the Ultimate Strength or the Ultimate Tensile Strength (UTS), the last being somewhat redundant.

Table 4.1 Tensile elastic constants of selected solids at room temperature.

Material	**E (psi)**	**E (GPa)**	ν
Al Alloys	$9.9\text{-}11.5 \times 10^6$	68-79	0.32-0.34
Brasses	$14\text{-}16 \times 10^6$	97-110	0.33-0.36
Ti Alloys	$15\text{-}17 \times 10^6$	104-117	0.33-0.34
Steels	$28\text{-}32 \times 10^6$	193-221	0.26-0.30
Na/Ca Glasses	$7\text{-}12 \times 10^6$	48-83	0.21-0.27
SiC	68×10^6	469	0.23
Al_20_3	47×10^6	325	0.16
Concrete	$1.4 \times 5.8 \times 10^6$	10-40	0.11-0.21
Nylon	4.1×10^5	2.8	0.4
Bakelite	$7\text{-}10 \times 10^5$	5-7	0.43
Polyvinylchloride	5.1×10^5	3.5	0.4

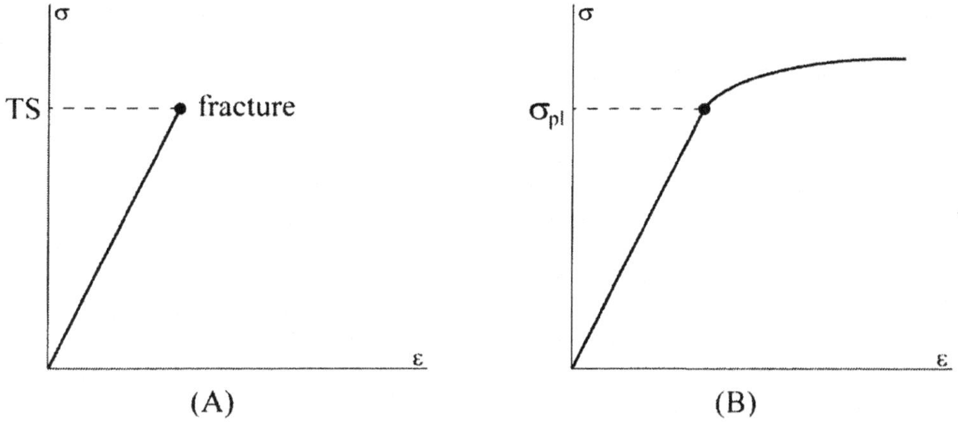

Figure 4.4 Possible consequences of excessive stress/strain on a linear elastic material exhibiting (A) brittle fracture and (B) nonlinear deformation (defined from the Proportional Limit, σ_{pl}).

If on the other hand, the response progresses as shown in Figure 4.4b, the deviation from stress-strain linearity becomes a point of some concern, particularity if the stress analysis formulas applied in design are based on Hookean linearity. As shown in the Figure, the stress value corresponding to the uppermost linear value is labeled, σ_{pl} and is referred to herein as the *Proportional Limit.* It should be noted that σ_{pl} is not a strength value, but merely an indication that stress and strain are no longer simply related via the elastic modulus. Perhaps for that reason, and since it is not a material strength measure, the value of σ_{pl} is rarely catalogued for designer consumption.

From a practical design standpoint, the nonlinear response (Figure 4.5) and the determination of an effective modulus can be relatively difficult to characterize quantitatively. One possible method would be to determine the local tangent modulus via the slope of the σ-ε curve at any point. The Initial *Tangent Modulus*, E_{tani}, would represent the slope as both σ and ε → 0; it is the highest value stiffness, but does not necessarily relate any relevant stress or strain points. Other Tangent Moduli, E_{tan}, must be specified at a particular stress level, possibly one of design interest such as any arbitrary σ_{des} in Figure 4.5. Although fully analytical theories of nonlinear elasticity exist, they do not necessarily lend themselves to an easy interpretation of tensile results as just shown. As such, we have spared you the mathematical complexities because they are beyond the scope of the present treatment anyway.

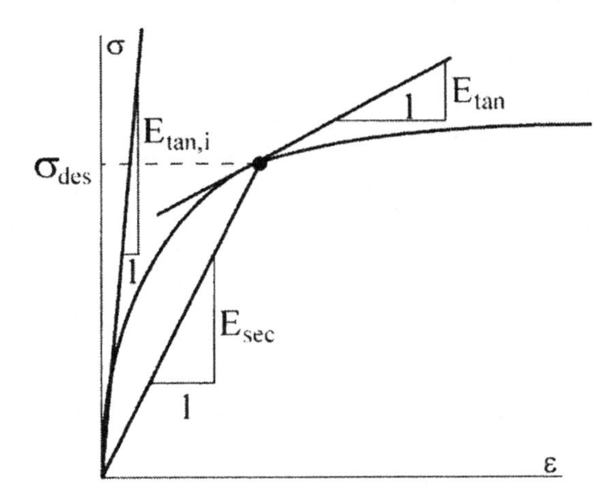

Figure 4.5 Nonlinear elastic response and moduli.

Example Problem 4-1: Figure 4.6 shows a tensile stress/strain curve for what is considered to be a linear polypropylene that was obtained at 60°F and a strain rate of 4.2 x 10^{-3} mm/mm. Determine the initial tangent modulus and the secant modulus at a design stress of 6.5 MPa.

Graphical construction lines are shown on the graph corresponding to a tangent line to the curve passing through points (0, 0 MPa) and (0.0065, 1.25 MPa) as well as a secant line from the origin (0, 0 MPa) to the point (0.012, 6.5 MPa). The slopes of these lines are indicated on the graph, with an initial tangent modulus of 0.19 GPa and the secant modulus of 0.54 GPa.

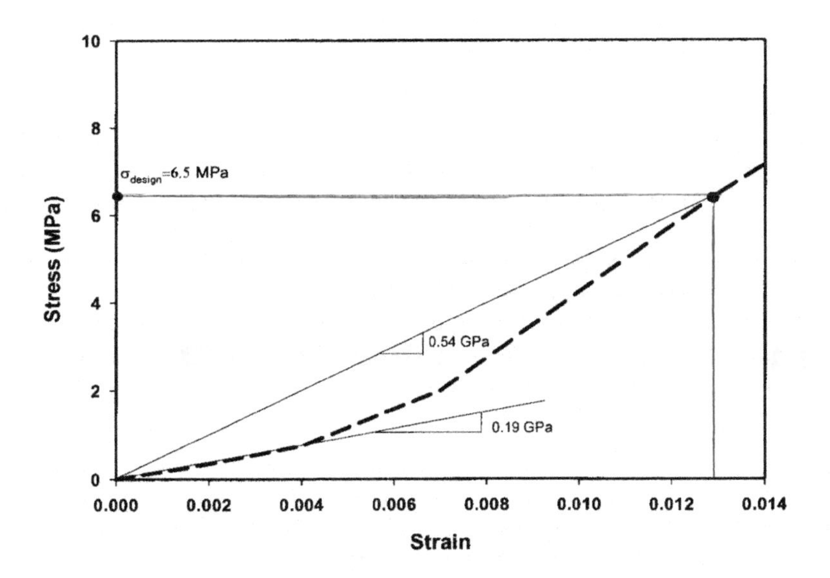

Figure 4.6 Tensile stress strain curve for linear polypropylene.

4.6 INELASTIC (PLASTIC) TENSILE RESPONSE

If the material is not brittle and considered to be approximately linear elastic, its response beyond the elastic region is of obvious interest to the design engineer. The catchall "obvious" is used because most manufacturing operations would not be possible if the material remained linear and fully elastic. After all, the very simple, but extremely useful paperclip would be nothing more than a wire if it was not taken beyond the elastic range. As such, it is important to consider the nonlinear response of many materials such as the typical curve shown below in Figure 4.7.

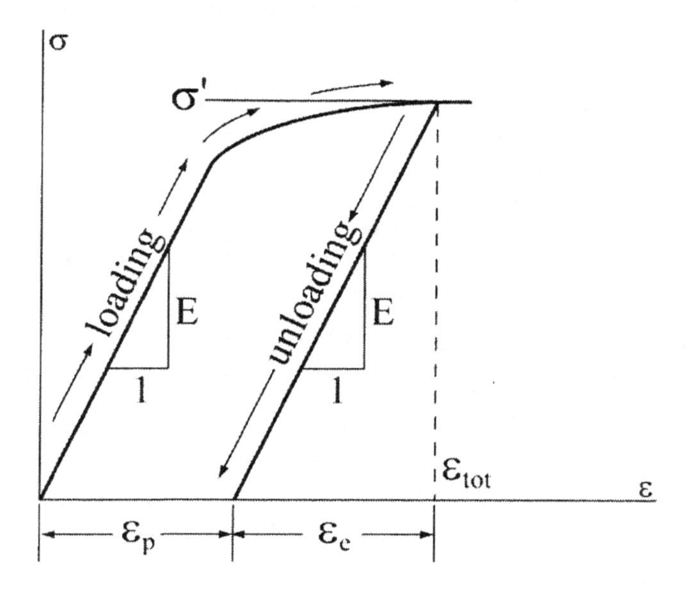

Figure 4.7 Tensile response curve with a loading/unloading path beyond the linear elastic range.

As shown in the response curve, the material behaves in a linear elastic fashion with stress and strain directly related by the elastic modulus, E up to a point. Interestingly, the slope of the unloading path is still the elastic modulus, even when loaded beyond the linear region. However, upon unloading to a zero stress from the previous value of σ', one observes a permanent strain, ε_p in the material; ε_p is referred to as a *Plastic Strain*, where plastic refers to a material's ability to be permanently deformed. The term *plastic* is often erroneously used to describe organic polymer solids since some of the early ones were indeed "plastic." Materials that can be deformed plastically are *ductile*, as opposed to *brittle* materials that clearly cannot. For the ductile material shown in the figure, the total strain state at σ' is therefore composed of both an elastic and plastic component:

$$\varepsilon_{tot} = \varepsilon_e + \varepsilon_p = \frac{\sigma'}{E} + \varepsilon_p \tag{4.5}$$

Here, ε_p is the plastic or permanent strain and ε_e is a recoverable (upon unloading) elastic strain; since ε_e is generally of the order 10^{-3} at most, it may become negligible if ε_p is relatively large as is often the case. While the ductile response curve of Figure 4.7 is typical for most metallics and some polymers, it should be noted that ductility is both a load rate (deformation) and temperature sensitive response, with even some ceramics exhibiting slight ductility at elevated temperatures. Unless otherwise specified, a tensile response curve is taken to be the room temperature response of what would be considered a quasi-static test, namely one performed in a time span of at least tens of seconds up to a few minutes. The potential effects of rates that are extremely fast or slow will be discussed further on.

For obvious reasons, the onset of plastic deformation at a specified stress magnitude is of fundamental importance to design. If a precision machine tool is to remain elastic in service so that specific tolerances are maintained, then the stress (or strain) levels in the machine must avoid the plastic region of the curve. Conversely, if paper clips are to be fabricated from a roll of wire, the engineer designing the fabrication process would soon be selling pencils on the street (or worse) if the deformations were wholly elastic! Clearly, the onset of plastic deformation is very important and is generally referred to as *Yielding*. However, yielding is often used synonymously with failure, even though it may not be the case as was seen with the paper clip. Moreover, it is easier to name the phenomena than it is to quantitatively specify its onset for most engineering materials.

Because yielding essentially heralds the onset of permanent strains in a material, one approach would be to load/unload the sample while progressively raising the stress level until a measurable plastic strain was noted. Aside from the tedium and expense involved in such a testing scheme (still better than selling pencils on the street), the repetitive loading itself might actually influence the yield response. Because of these difficulties, an operational definition of yielding has been almost universally adopted as a solution to the above dilemma. This approach is known as the *0.2% Offset Yield Method.*

The method is actually (and fortunately) quite simple to implement. In a "nutshell," an "offset" strain value of 0.002 (0.2%) is assumed to be the minimum amount of permanent or plastic strain at yielding. At this value of strain, a line is drawn up to the stress strain curve using the elastic slope, E as shown in Figure 4.8; the stress value determined by the intersection of the line and curve is the *Offset Yield Strength*, or the *0.2% Offset Yield Strength*, hereafter designated σ_y. Although this is somewhat like stating who is buried in Grant's Tomb, it should be noted that when the material is stressed to σ_y and unloaded, a permanent strain of approximately 0.002 has therefore been induced.

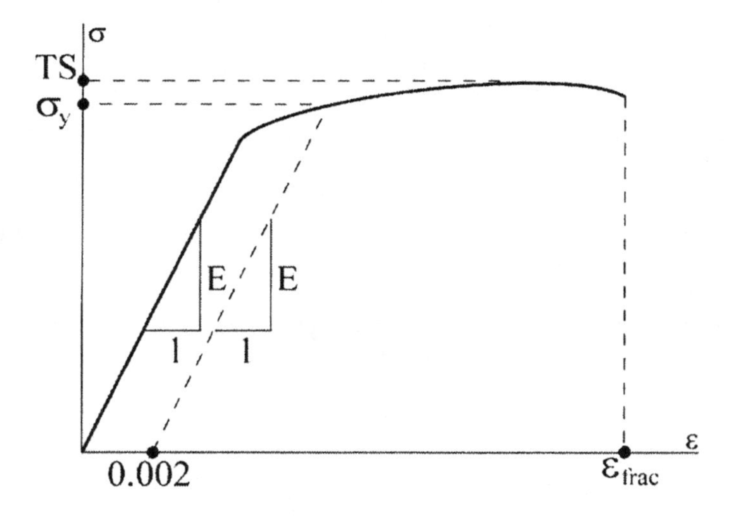

Figure 4.8 Typical stress strain curve and the 0.2% offset yield strength.

Figure 4.8 also shows other measures of design significance in the plastic range. The maximum supportable stress before complete failure occurs is know as the *Tensile Strength* or *TS*:

$$TS = \frac{P_{max}}{A_o} \tag{4.6}$$

In addition, the ductility of the material or the extent to which it can be permanently deformed, can be specified by the *Strain At Fracture*, or ε_f. Conventionally, ductility is expressed as the *Percent Elongation*:

$$\% \text{ Elongation} = \frac{L_f - L_o}{L_o} \text{ x } 100\% = \varepsilon_f \text{ x } 100\% \tag{4.7}$$

where L_f and L_o are the final and original gage lengths, respectively. A related quantity for expressing ductility is the *Percent Reduction in Area* where:

$$\% \text{ Reduction in Area} = \frac{A_o - A_f}{A_o} \text{x} 100\% \tag{4.8}$$

Both ductility measures can be computed from post-fracture measurements as long as the plastic strains (at fracture) are large compared to the elastic component so that the final specimen dimensions are reflected by the fragments. While both quantities are certainly valid, it is more difficult to measure the changing area or radius so the use of the *Percent Elongation* is more common.

Example Problem 4-2: Shown below is the tensile stress-strain curve for aluminum alloy 2024-T3 used for lawn mower housings. Given the stress strain shown, determine: (a) Modulus of Elasticity, (b) Proportional Limit, (c) Yield Strength (0.2% Offset), (d) Tensile Strength, and (e) Percent Elongation.

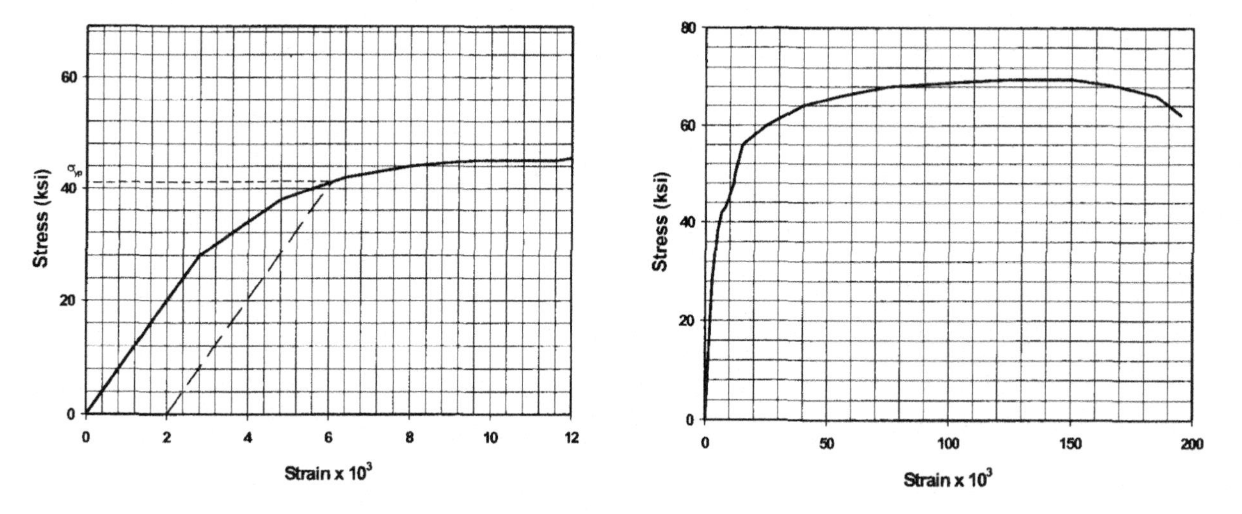

Figure 4.9 Stress strain curve for 2024-T3 showing linear region (left) and overall response until failure on the right.

All of the values can be easily obtained graphically or numerically from the stress strain curve shown in Figure 4.9. Using the enlarged linear section, the elastic modulus can be obtained from the slope of the line; at an elastic strain of 0.001, the stress is measured as 10,000 psi. Therefore, the elastic or Young's modulus E = 10,000/0.001 = 10 x 10^6 psi. By inspection, the proportional limit or σ_{pl} = 28,000 psi. Similarly, the yield strength or σ_y = 42,000 psi via the parallel line starting at the offset strain of 0.2%. The highest portion of the curve reveals an ultimate tensile strength of TS = 69,000 psi. Finally, the strain at failure or % Elongation = 0.19 x 100 = 19%.

4.7 STRAIN HARDENING

Consider a typical stress strain curve for a ductile material as shown in Figure 4.10. If this material was loaded for the first time to a total strain, ε_a within the plastic range, the observed 0.2% offset yield-strength would still be σ_{y1} as expected. Moreover, if the sample was now unloaded from the total strain ε_a, the path would terminate at a new strain, ε_b with the elastic slope (of *E*) retained. If the sample was then reloaded back up to the same stress level, the new stress strain curve would have its origin at ε_b. Unloading would take it back down to ε_b as no additional plastic strains would be induced, even though σ_{y1} was clearly exceeded. If additional permanent or plastic strains are to be induced, they would require a higher stress level because a new 0.2% offset yield strength of σ_{y2} now exists. Furthermore, one can see that $\sigma_{y2} > \sigma_{y1}$; plastic straining can raise the yield strength or actually *Harden* a ductile material.

Strain Hardening, or the somewhat older term *Work Hardening*, refers to the degree to which a ductile response has a positive slope to its σ-ε curve in the plastic region. The existence of a positive slope means that additional increments of strain require ever-increasing stress. To observe this first hand, try bending a wire back and forth a few times. For most materials, the second bend will be more difficult than the first and the pivot point will begin to heat up as the material work hardens. On the other hand, a material that displays a complete lack of strain hardening, or has a relatively flat tensile response curve in the plastic region, is referred to as *Perfectly Plastic*.

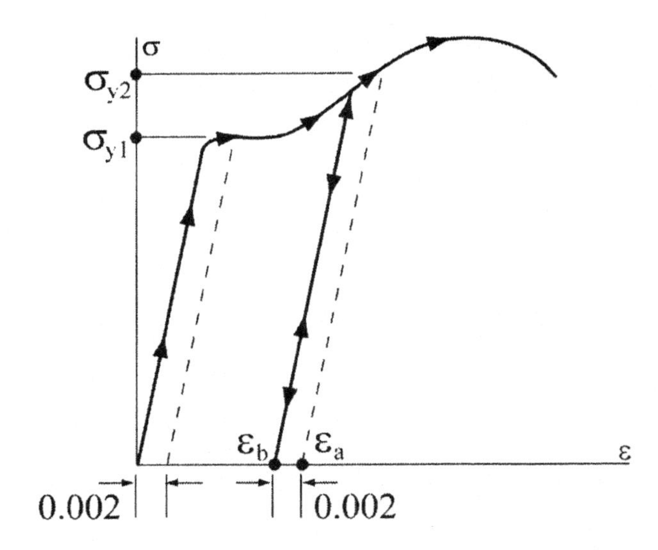

Figure 4.10 Increased yield strength obtained through plastic deformation.

Strain or work hardening is a response that may either help or hinder the engineer in a specific application. One very positive aspect of strain hardening is that the alloy may be strengthened by plastic deformation. For example, cartridge brass, the name given to an alloy of 70 wt. % of Cu, 30 wt. % Zn, can be commercially obtained with various degrees of strain hardening as noted in Table 3.2.

Table 3.2. Typical strain hardened 70/30 brass (INS C26000).

Temper Designation	Material Condition	Yield Strength (ksi)	% Elongation
HO	¼ hard	40	43
HO2	½ hard	52	23
HO4	Hard	63	8
HO	Extra hard	65	5

While the alloy is hardened and/or the yield strength increased, the ductility tends to decrease for many metallics. Hence, one of the negative aspects of strain-hardening is that ever-increasing forces (and machinery power) must be applied as the processing proceeds. Finally, it should be noted that strength increases obtained via plastic deformation are thermodynamically unstable. Raising the temperature of a strain-hardened metallic to about one-half of its melting point and slowly cooling to room temperature will often lower the strength and increase the ductility. Such thermal treatments that lower the yield strength or "soften" the material (enhanced ductility) are known as *Annealing* heat treatments. The ability of most metals to be repeatedly deformed, annealed, deformed, and so on, almost indefinitely is one of the paramount reasons we utilize these relatively rare and costly materials.

4.8 YIELD POINTS

For some alloys, yielding is not a gradual and homogeneous process, instead, consisting of sharp discontinuities in material stiffness as the elastic limit is exceeded in one region, then another, and so forth. A typical example of discontinuous yielding is shown in Figure 4.11 that depicts the initial segment of the stress strain response for 0.20 wt. % carbon steel in the annealed state. Note that the *Lower Yield Point,* σ_{yp} is equivalent to the offset yield strength of $\sigma_y \approx 48{,}000$ psi in this case. As seen in the figure, the yield point phenomena represent a sudden loss of load-bearing resistance at the stress and strain of point A, commonly referred to as the *Upper Yield Point.* In order to practice "safe engineering," the lower yield point stress should be considered a reasonable gage of yield strength and may be used in the same context as the offset yield strength. On the other hand, the upper yield point is not a satisfactory design parameter as it is sensitive to the testing machine, the recent thermal and mechanical history of the material, and could lead to design instabilities if slightly exceeded in service.

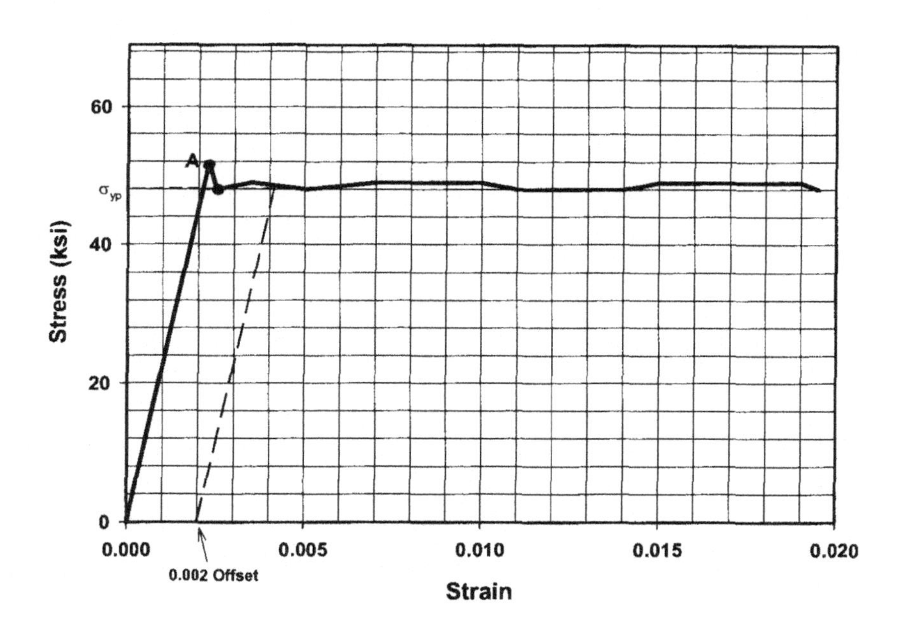

Figure 4.11 Discontinuous yielding behavior of annealed 1020 steel.

4.9 INHOMOGENEOUS STRAINS AND YIELDING

For conventional specimen geometries, the normal stresses and strains generated in a tension or compression test are usually thought of as being uniform over the constant gage section, A_o. Because stresses are generated by a material's inherent resistance to deformation, uniform states of stress and strain will only be present when this resistance is uniform. In the elastic range, stress and strain uniformity is usually apparent, at least when averaged over material volumes of 10^{-3} in^3 or greater. However, the onset of yielding gives rise to deformation modes that are extremely sensitive to slight and local variations in plastic deformation resistance as seen in most metallics, particularly the commonly used multiphase alloys.

An example of nonuniform plastic deformation at yield is shown in Figure 4.12; the initial appearance of the bands in Figure 4.12 coincided with the upper yield point of Figure 4.11. The lines on the specimen gage section, which became visible after uniformly polishing (before straining) to a bright finish, are located at roughly 45 degrees to the load axis. Visible to the naked eye, these lines are the results of hundreds of thousands of dislocation lines cascading onto the specimen free surface in regions of relative softness. These regions of non-uniform deformation are referred to as Lüders bands and can be conceptualized as the barely visible manifestation of the slip process by which ductile materials plastically deform. Figure 4.13 is a schematic depiction of one such slip-band. The constant strength portion of the stress-strain curve in Figure 4.11 (following the lower yield point B) continues until the plastic strain in the entire gage section is uniform.

Figure 4.12 Lüders bands in a yielded 1020 annealed steel.

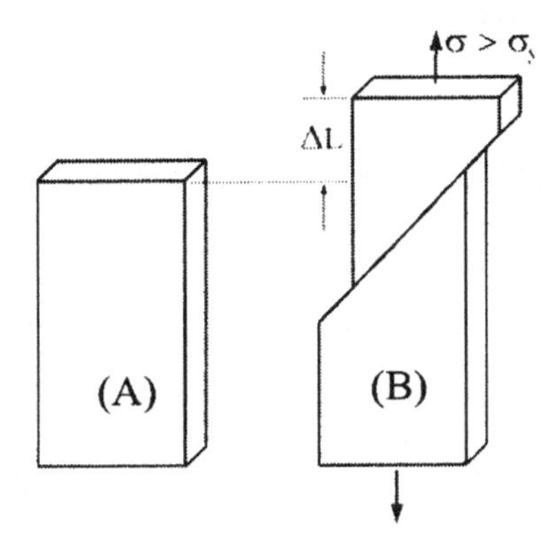

Figure 4.13 Exaggerated slip deformation of a tensile specimen (a) before yielding, and (b) after yielding.

4.10 FRACTURE IN TENSION

As shown by the slip bands just discussed, plastic flow in solids is induced by the presence of shear stresses; the more ductile the solid, the greater the likelihood that the final fracture will exhibit shear-induced failure. Figure 4.14 contrasts the fracture appearance of low-ductility high-strength steel with that of a similar alloy that is of relatively low strength, but high ductility. Note that the fracture plane for the low ductility, high strength steel (achieved via heat treatment) is normal to the tensile axis and flat in appearance. In contrast, the more ductile and medium strength steel has failed in the traditional "cup and cone" mode. As plastic strains approach their maximum value, they tend to become localized at one point along the gauge section as seen in Figure 4.15.

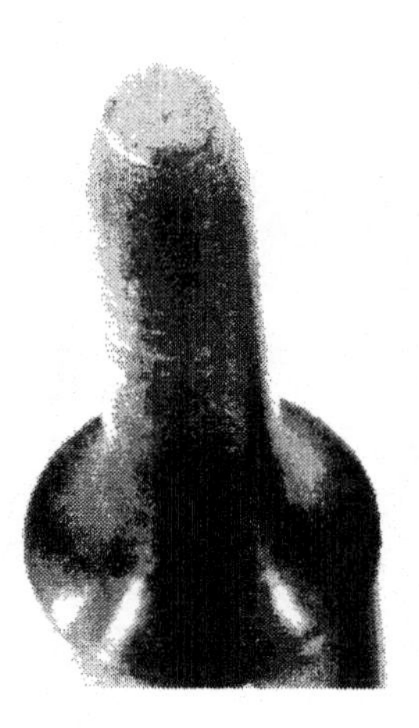

Figure 4.14 Tensile fractures for (left) medium carbon annealed steel with tensile strength of 92 ksi and (right) die steel with tensile strength of 242 ksi.

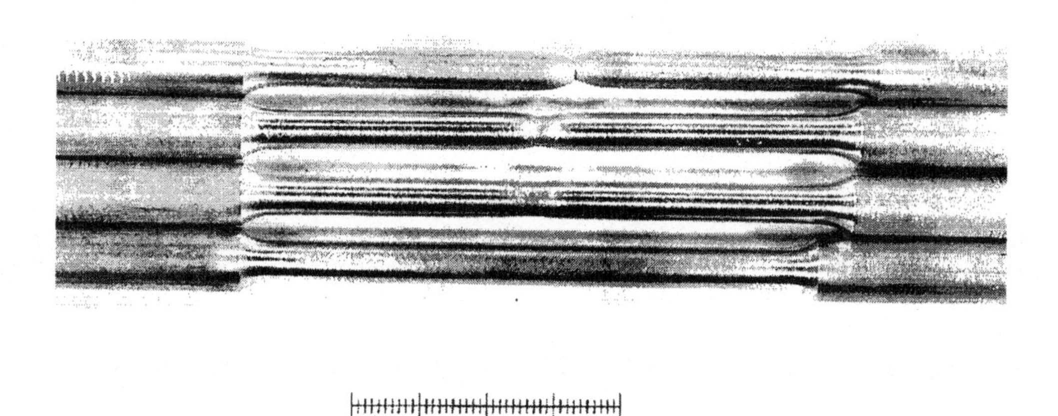

0 1 2 3 4

Figure 4.15 Development of localized plastic strain or "necking" in a relatively ductile 6061-T6 aluminum alloy.

Internally, the high degree of plastic strain has created voids that further reduce the true cross-section sustaining the load that in turn, tends to localize the severest strain even further. A sectioned aluminum tensile sample complete with a localized strain-induced "neck" and internal porosity is shown in Figure 4.16. The necked down portion of the gage section is no longer subjected to a simple uniaxial tensile stress state; the pinched-in neck induces a state of tri-axial tension, which in turn, opens up a flattened crack centered and normal to the specimen axis. As the tensile test proceeds, the crack eventually becomes the bottom of the "cup" or top of the truncated "cone" depending on which side of the specimen you are attached to. After the centerline crack has developed, the remaining cross-sectional ligament is essentially a tube with diminishing wall thickness. Since the relatively thin wall (approaching a membrane) cannot support triaxial stresses, the stresses dissipate and the wall then fails in shear giving rise to the sides of the "cup" and "cone." The cup wall is a *Shear Lip* and is often found on fracture surfaces of failed machine and structural elements; fortunately for anyone doing a failure analysis, its presence serves to identify regions on a fracture surface associated with ductile overloads.

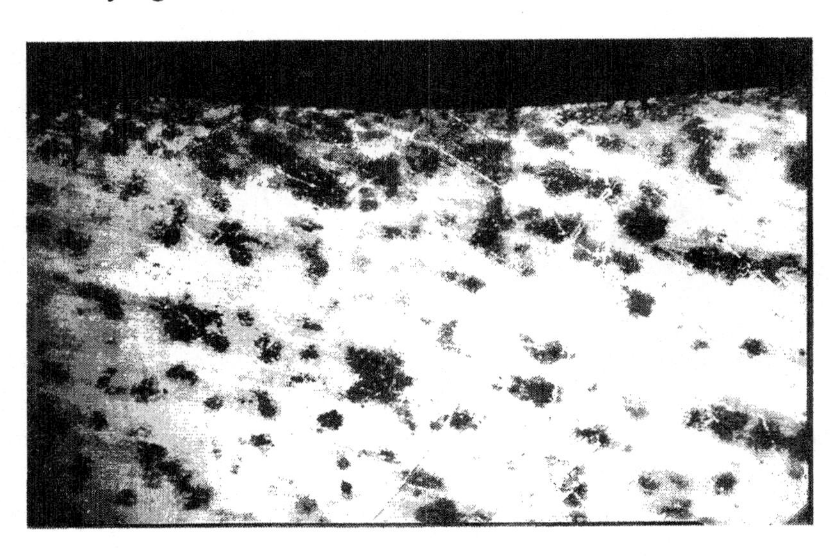

Figure 4.16 Sectioned neck region of a 6061-T6 aluminum tensile sample showing internal porosity.

4.11 BRITTLE STRENGTH

It is important to note that the usual methods of tensile testing for ductile materials (metals and polymers) cannot be easily or more importantly, accurately adapted to brittle materials such as ceramics. In particular, the stress concentrations connected with tensile tests must be removed from the proverbial and literal equation as brittle materials are extremely prone to failure at such localized sites. Thus, the usual gripping arrangements (mating threads, vise grips, or pin holes, etc.) are not permissible lest the specimen fail prematurely and at a stress level that is potentially unknown and certainly not uniform. As such, the applied loads must be smoothly applied over a relatively large amount of specimen surface area; this usually implies that the loads must be transmitted by a relatively compliant intermediate material such as a stiff elastomer. In addition, the test section must also be gradually tapered such as in an hourglass fashion. Because only the limit of the elastic range of strength or *Tensile Strength* is sought for brittle materials, an extensive gage section as opposed to an hourglass neck is unnecessary. Finally, alignment becomes very important as any deviations may cause additional bending stresses that can significantly lower strength estimates.

Before we move on, it should be noted that one of the fundamental differences between ductile and brittle materials is the propensity to yield. While brittle materials can in fact yield, it is rarely seen since brittle fracture will usually occur from tensile stresses long before the shear stress states are sufficient for yielding to commence. Some clever people have actually machined ceramics as they would with alloys by "tricking" the material into high states of shear. However, this is rare and difficult to do so alternative methods of shaping and cutting such as lasers, grinding, and so on, are often used.

At first glance, compressive strengths appear to be more easily determined for brittle materials since the samples need only be compressed. However, there are two experimental difficulties that must be overcome to obtain valid compressive test results. The first problem is to avoid sample buckling; parallel specimen ends and truly axial load application are definitely necessary in this instance. In addition, the specimens should be no longer than about four diameters as a compressed cylinder will tend to increase in diameter as it is loaded due to the Poisson Ratio effect. This radial swelling is restrained by friction at the specimen ends where they contact the loading plates. Should this frictional resistance to radial expansion be excessive, a barrel-shaped configuration results, and the outer periphery may carry tensile stresses that will cause fracture since the compressive strength of brittle materials usually exceeds the tensile strength by an appreciable amount. To avoid such a deleterious state, end friction must be minimized by using devices such as air bearings, discs of lower friction materials, easily deformed elastomers, and/or other anti-friction schemes. Given these concerns, the seeming ease of performing compressive tests begins to evaporate. Thus, the compressive strength cannot simply be substituted for the tensile strength. Some discussions about the ratio between the two measures and the reason for the divergence will be discussed in greater detail in Chapter Seven.

4.12 BENDING STRENGTH MEASURES

Because simple tensile tests do not look very promising for truly brittle materials (unless you are a true glutton for punishment), a new test methodology is ultimately required. In this case, it is perhaps good to recall that the bending stresses, σ_b, in a beam subjected to a pure bending moment, M_b, vary linearly from zero (at the neutral axis of the cross-section) to maximum values at the outer edges as shown below in Figure 4.17.

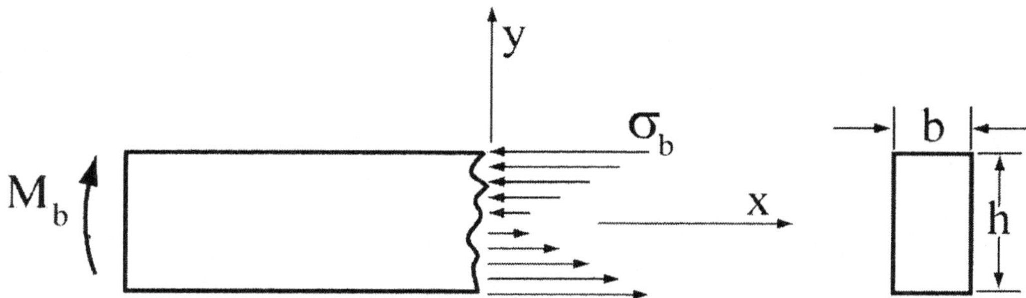

Figure 4.17 Bending stress distribution for a pure bending moment, M_b.

As reviewed earlier, the bending stress and moment of inertia for a rectangular section are:

$$\sigma_b = \frac{M_b y}{I_{zz}} \qquad I_{zz} = \frac{1}{12} bh^3 \tag{4.9}$$

Since brittle solids usually exhibit greater strength in compression than tension, failure will be caused by the tensile stresses in the distribution; the maximum tensile bending stress for the rectangular cross-section is then found for y =-h/2 or:

$$\sigma_{max} = \frac{6M_{max}}{bh^2} \tag{4.10}$$

Although it is often erroneously referred to as the *Tensile Strength* of a brittle solid, the maximum tensile bending stress computed in Equation (4.10) is more correctly called the *Modulus of Rupture* (MOR) in bending. Interestingly, the modulus of rupture and the Tensile Strength are not exactly the same; the differences are due to the very nature of the fracture process to be discussed further in Chapter Seven.

In contrast to the tensile test discussed in earlier sections, MOR measurements are relatively straightforward to carry out experimentally. Moreover, simple beams of rectangular cross-section are the easiest to fabricate, and no special gripping is necessary to load them assuming care is taken at the contact/loading points. The most desirable loading scheme, commonly referred to as four-point for obvious reasons, is shown below. For this often used configuration, the mid-span of the test beam is one-half of the total length, L. Because couples are created on each side by the opposing reactions, the moment M_b and ensuing stresses are constant over the midsection. As will be discussed in greater detail later, a constant and relatively large, tensile stress region is essential if the strength and all of the inherent defects that influence it, are to be adequately characterized.

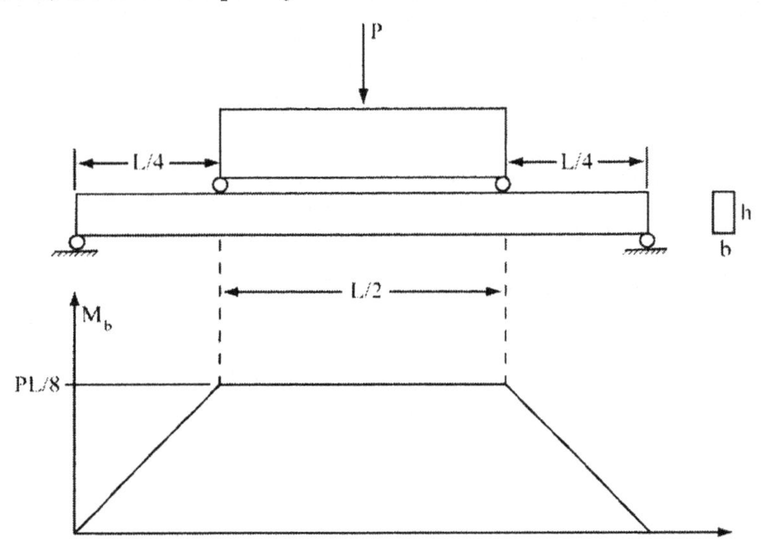

Figure 4.18 Four-point bend configuration for MOR testing.

4.13 HARDNESS MEASURES AND METHODS

Hardness is a qualitative term that may have different meanings to different engineers and scientists. In the present context, hardness will be defined as a mechanical response that describes the resistance of a surface to superficial plastic penetration by a relatively hard indenter. During the indentation process using either a ball or cone pressed on the surface, much of the subsurface bearing stress will consist of a hydrostatic pressure which does not contribute to plastic flow as there is no shear. The remainder of the bearing stresses will be in a state of uniaxial compression, thus causing the material to yield and flow plastically underneath the indenter. Since only a fraction of the apparent bearing stress will cause the surface to deform plastically, the total bearing stress may appear to be several times the ultimate strength.

The reader may be pleased to discover that indentation hardness tests have been thoroughly standardized over the years so the practices and procedures are well established. In fact, the *Brinell* and certainly the *Rockwell* systems may sound familiar and are commonly used in the United States. Originally developed at the turn of this century by Johan A. Brinell, the hardness test that bears his name features a hardened steel ball (diameter, D = 10.0 mm) as the indenter; in instances where stronger materials such as quench/temper strengthened steels are under scrutiny, a harder ball of tungsten carbide (WC) may be appropriate. As convention would have it, the loading on the indenter is usually set at 3000 kg and produced by levered weights or fluid pressure acting on a piston. After loading the indenter onto the specimen surface for 30 seconds, the load and indenter are removed, the diameter, *d* of the plastically formed crater measured via microscopy and the Brinell hardness number computed as:

$$\text{BHN}=\frac{\text{Indenter Load (3000 or 500 kg)}}{\text{Crater Surface Area}}=\frac{2P}{\pi D\left(D-\sqrt{D^2-d^2}\right)} \tag{4.11}$$

To ensure that a sufficient amount of material is available to constrain the indenter, the specimen edge should be no closer than 2.5d to the periphery of the crater and the specimen thickness must exceed the value of *d*. Usually, the measured BHN for many ductile materials, most notably the mild structural steels, is about three times the tensile strength. However, the true value of the Brinell test, or any other indentation hardness test for that matter, is the correlations between tensile or yield strengths and hardness numbers such as BHN. As would be expected, these numbers have been worked out experimentally with correlation tables and/or graphs usually provided with the equipment. Finally, the reader should realize that resulting ratio of d/D is practically bounded; ratios greater than 0.5 reflect craters with excessively vertical walls that absorb too much of the applied load as friction. If necessary, as is often the case for softer aluminum and magnesium alloys, the ratio can be reduce by adopting a lighter load of 500 kg. Furthermore, d/D <0.2 to 0.3 may indicate insufficient deformation to truly sample the material's response.

Named after its developer Stanley P. Rockwell, an alternative indentation hardness test was introduced two decades after the Brinell method. For this very common test, the depth of the indenter penetration or *d* (in mm) is the measurement made using various standard indenter geometries and loads. Once the surface is marked by the indenter and measured, the Rockwell hardness number can be calculated as:

$$\text{RHN} = A_1 - A_2 d \tag{4.12}$$

where the coefficients A_1 and A_2 depend upon the indenter/load pair. In actual practice, one never directly measures *d* and computes the RHN from Equation (4.12). Instead, Rockwell testing machines have their readouts directly calibrated in RHN units rather than a penetration depth; a particular choice of indenter/load pair is equivalent to a *Scale*, denoted as A scale, B scale, C scale, and so on. For example, the B scale is commonly used for medium strength metallics and features a 1/16 inch diameter hardened steel ball as indenter and a load of 100 kg. The C scale corresponds to a conical diamond indenter and a load of 150 kg and is commonly used for high-strength steels, especially those that can be strengthened by quench-temper heat treatment. As in the case of the Brinell test, the worth of the Rockwell test lies in previously determined correlations between RHN and conventional tensile strength measures. By way of example, a Rockwell hardness number of 95 on a B scale (RHB = 95) in steel would correspond roughly to a Tensile Strength of 100,000 psi, where a RHC = 40 might imply a Tensile Strength of 180,000 psi.
Each test method clearly has its strengths and weaknesses (as usual, all puns intended). The Brinell test samples with a much larger volume of material than the Rockwell, making it more appropriate for metals with inclusions and pores. Moreover, it also tends to be insensitive to surface conditions provided the surface is cleaned of loose scale, gross dirt, and so on. However, the crater left behind may be objectionable when a specific surface finish is desired. On the other hand, the Rockwell technique is more flexible with a wide variety of indenters and loads to choose from. It also leaves a much smaller indentation mark behind, making it more suitable for finished surfaces. However, greater care must be exercised in preparing clean and smooth surfaces for testing. Finally, the Rockwell method is ill-suited to inhomogeneous materials such as segregated and porous aluminum alloy castings to name a few.

4.14 ENERGY MEASURES IN TENSION

As it turns out, there are many design applications where the simple properties such as stiffness, strength, or ductility are not sufficient by themselves to determine the optimum material. A prime example would be a design whose functions include the absorption of work, or kinetic energy, in addition to other types of load-bearing capability such as a car bumper. Of course, the government never really helps (at least with simplifications) so at one time, federal regulations mandated that the bumper system be able to withstand a 5 mph impact without permanently deforming. The original and presumably good idea was to eliminate the multitudinous insurance claims from minor mishaps in parking lots that were forcing up insurance premiums at a grand rate. All other factors being equal, the selection of material to be used in the bumper would lean towards those best able to absorb energy elastically. On the other hand, if a driver found himself/herself out of control and hurtling towards a massive rock or tree at 50 mph, rebounding elastically would not be a really good idea. Instead, one might wish that the front end of the car be capable of absorbing the greatest possible amount of energy plastically. The question now becomes: how does one evaluate or develop a material that might at least exhibit an optimum combination of the two desired properties.

In order to determine this measure, an analysis must consider the deformation and work involved. If a constant force, *F* acts upon a system which undergoes slight translation, ds in the direction of *F*, the increment of work done on the system, dW, is:

$$dW = F \cdot ds \tag{4.13}$$

Let our "system" be a unit cube (L = 1.0) of material, carrying a normal stress, σ as shown in Figure 4.19.

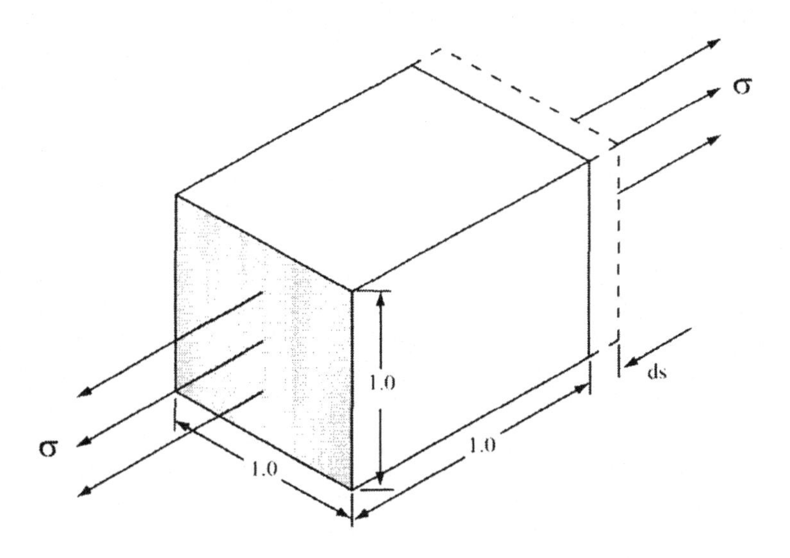

Figure 4.19 Normally stressed unit cube.

Given this arrangement, the force then equals $F = \sigma(1.0)^2 = \sigma$; the deformation in the stress direction, ds, is then $ds = d\varepsilon(1.0) = d\varepsilon$. Accordingly, the work done on the unit volume-sized system becomes:

$$dW = F \cdot ds = \sigma \, d\varepsilon \tag{4.14}$$

If there is no heat flow in or out of our unit cube system, the first law of thermodynamics tells us that:

$$dW = du = \sigma\, d\varepsilon \tag{4.15}$$

Thus, we can then define the *Elastic Strain Energy Per Unit Volume Stored* in tension as a new variable, u such that:

$$u = \int \sigma d\varepsilon \tag{4.16}$$

For a Hookean material, $\sigma = E\varepsilon$ or $d\varepsilon = d\sigma/E$ and Equation (4.16) then predicts a stored elastic energy of

$$u = \frac{1}{E}\int \sigma' d\sigma' = \frac{\sigma^2}{2E} = \frac{1}{2}\sigma\varepsilon = \frac{1}{2}E\varepsilon^2 \tag{4.17}$$

for any elastic stress, σ. Since the linear elastic range is defined by the proportional limit, σ_{pl}, the maximum elastic strain energy that can be stored or the *Modulus of Resilience, u_r*, becomes:

$$u_r = \frac{(\sigma_{pl})^2}{2E} \tag{4.18}$$

While the use of calculus might stir fear in the hearts and minds of the readers, the integral of Equation (4.16) is nothing more than the area under the stress strain curve; in fact, the last representation of energy stored in a linear elastic material given in Equation (4.17) is just the triangular region of the stress strain curve. For deformation beyond the linear elastic region in a ductile solid, the area under the stress strain curve must be graphically or numerically integrated to assess energy absorption given the complexities of actual response curves. Once calculated, the energy absorbing capability of a ductile material when strained to fracture is referred to as the *Toughness* as shown in Figure 4.20.

Because complete stress strain data is not always available or trustworthy (a long story for another day), toughness calculations can become, shall we say, "tough." In these situations, an estimate of a material's true toughness can be made by computing the *Modulus of Toughness, T*:

$$T = \frac{\sigma_y + TS}{2}\varepsilon_f \tag{4.19}$$

Equation (4.19) approximates the area under an actual stress-strain curve as a rectangle whose height lies halfway between the yield strength and the ultimate tensile strength, and whose base is the strain at fracture as shown in Figure 4.20.

Nonlinearly elastic materials also possess elastic energy storage capabilities, but they are not easily described by parameters such as the modulus of resilience via Equation (4.18). Instead, the basic definition of work-done/energy-stored must be applied graphically to the stress strain response curve. Using this approach, the energy stored upon loading is indicated by the shaded area under the curve in Figure 4.21a. Upon releasing the load, and returning the material to zero loads elastically, the area under the unloading curve represents recoverable energy or *Resilience* as depicted in Figure 4.21b. Note that the *Modulus of Resilience* and *Resilience* are the same only if the material is linearly elastic and loaded to the proportional limit stress. Clearly, a nonlinear material cannot have a proportional limit or exhibit a modulus of resilience; such a material will actually exhibit many levels of resilience since there is no single specified stress from which it is to be unloaded. Finally, if the work absorbed by the material exceeds what is recovered as resilience, the difference becomes "trapped" within the solid and is referred to as a *Hysteresis* or energy permanently absorbed per load/unload cycle as shown in Figure 4.21c. In reality, the hysteresis absorption may have been converted to heat or acoustical energy that dissipates and/or irreversible micro-damage in the form of cracks. Materials that can absorb energy and covert it to heat or some other form of benign dissipation are good candidates for shock and vibration dampers.

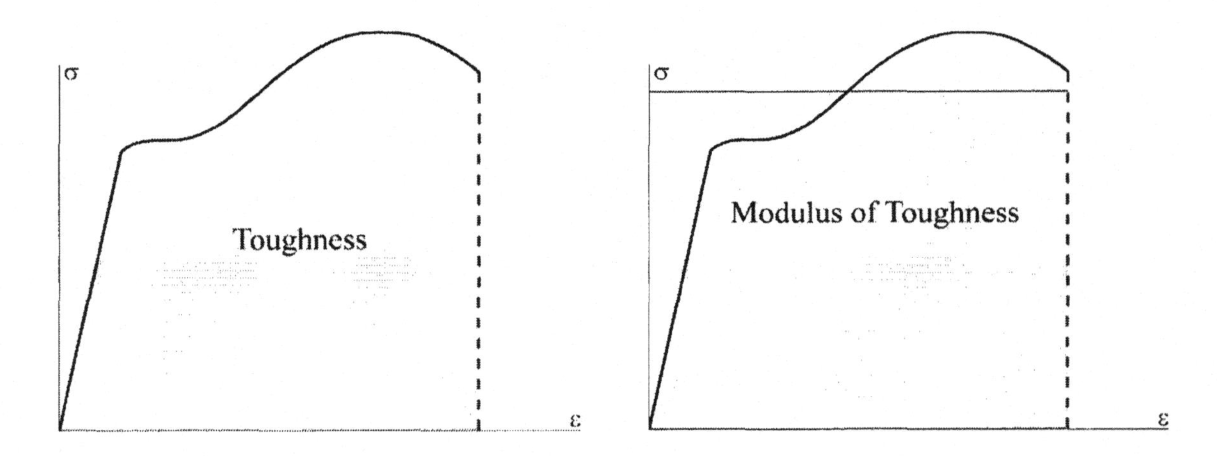

Figure 4.20 Toughness or energy absorbed per unit volume to fracture.

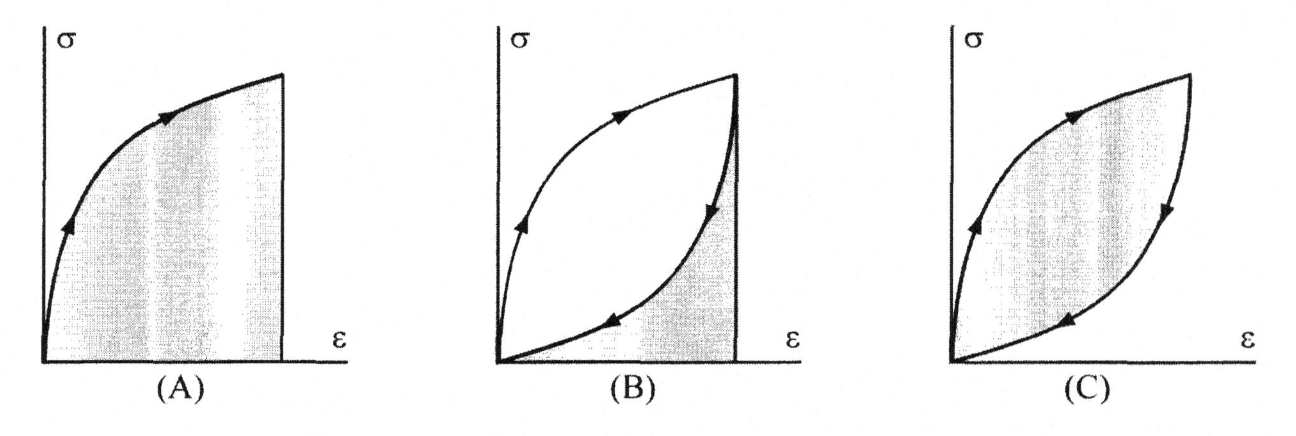

Figure 4.21 (A) Energy per unit volume absorbed in loading, (B) energy per unit volume recovered upon elastic release, or resilience, and (C) energy permanently absorbed, or hysteresis.

Example Problem 4-3: Shown below in Figure 4.22 is a loading/unloading response curve for polypropylene obtained at 60°F and at a strain rate of 4.2×10^{-3}. The top curve was obtained upon loading to 7.7 MPa, the bottom one upon uploading. Given this data, Find: (a) the work input upon loading to 7.7 MPa, (b) the elastic energy (resilience) recovered upon unloading, and (c) the hysteresis lost to the material during the load/unload cycle.

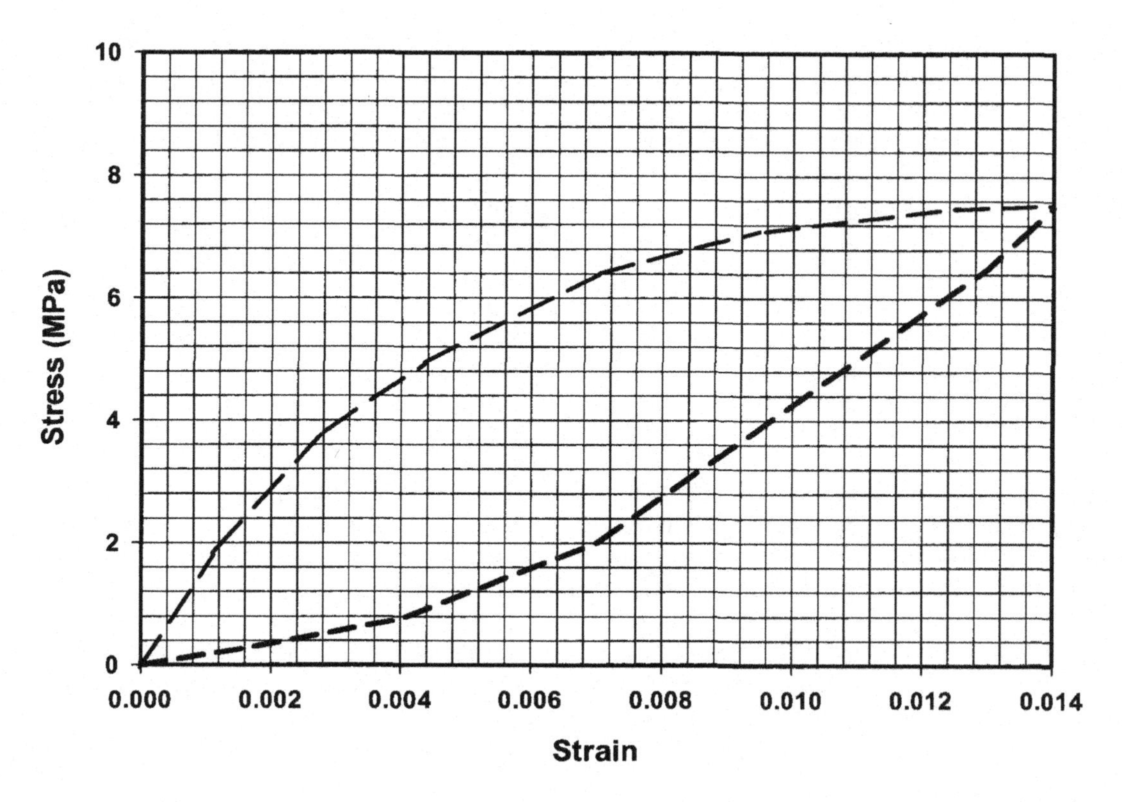

Figure 4.22 Tensile stress strain curve for polypropylene type material.

An examination of the response curve shows that each square on the graph grid measures 0.25 MPa by 0.0005 m/m, with a represented energy of 0.25 x 0.0005 = 0.125 kJ per square. Thus:

a. By a direct, tedious, and admittedly crude count, there are approximately ~447 squares under the loading curve, for a sum total of 55.9 kJ/m^3.

b. By direct count, there are also ~332 squares under the unloading curve for a total resilience of approximately 41.6 kJ/m^3.

c. The hysteresis loss is the difference between (a) and (b) or 4.65 kJ/m^3.

PRACTICE EXERCISES

1. The Fellowship of the Ring has decided to use some basic engineering know-how (learned from this very class) to find a way to destroy the One Ring to Rule them All. A standard specimen that is 0.505 inches in diameter with a gage length of 2 inches was used for a tensile test. The following data was recorded for the Ring alloy with fracture occurring at an elongation of 0.018 inches:

Load (lb)	Elongation (in)	Load (lb)	Elongation (in)
0	0	7200	0.007
1000	0.001	8000	0.008
2100	0.002	8300	0.009
3200	0.003	8600	0.010
4200	0.004	8700	0.012
5200	0.005	8750	0.014
6200	0.006	8800	0.018

Plot the stress strain diagram and determine: (a), approximate proportional limit, (b) yield strength based on the 0.2% offset method, (c) modulus of elasticity, (d) modulus of resilience, (e) tensile strength, and (f) percent elongation. Note: it may be advisable to plot two graphs: one to emphasize the elastic and near-yield region and the other showing the full response over the entire strain range.

2. At the heart of any good automobile engine is a piston return spring that is often fabricated from SAE 1020 annealed steel. During a tension test of this alloy, the following data were recorded with fracture occurring at an elongation of 0.500 inches:

Load (lb)	Elongation (in)	Load (lb)	Elongation (in)
1200	0.0002	8500	0.0068
1800	0.0004	8400	0.0080
2400	0.0006	8200	0.0098
3000	0.0008	8200	0.020
3600	0.0010	8400	0.030
4200	0.0012	9400	0.050
4800	0.0014	10,600	0.075
5400	0.0016	11,400	0.100
6000	0.0018	12,300	0.150
6500	0.0020	12,800	0.200
7000	0.0023	13,500	0.300
7400	0.0026	13,700	0.380
8000	0.0032	13,100	0.450
8300	0.0038	11,500	0.500

A standard specimen for an impact resistant alloy for tank armor with an initial diameter of 0.505 in. and a gauge length of 2 inch was used for the tensile test. In addition, the diameter at fracture was 0.391 in. Using the data given, plot two stress strain diagrams: one showing the entire stress strain curve and the other the stress-strain results up to a total elongation of 0.0098 in. Once plotted, determine the following mechanical properties: (a) proportional limit, (b) upper yield point, (c) lower yield point, (d) 0.2% offset yield stress, (e) modulus of elasticity, (f) modulus of resilience, (g) tensile strength, (h) Percent elongation, (i) gauge length at fracture, and (j) percent reduction in area.

3. Shown below is the stress-strain response curve for cartridge brass (70% Cu, 30% Zn, or UNS alloy C26000) in the HO1 temper used for sniper rifles that are specifically designed to ruin someone's day.

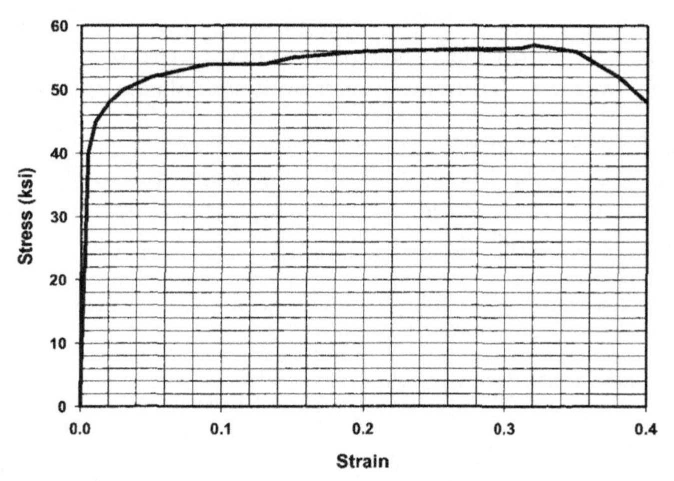

More refined data in the elastic and near-yield regime are given as:

Stress (psi)	Strain
13,072	0.00108
26,144	0.00217
39,200	0.0033
44,120	0.0055
47,390	0.013

Using this data and graph, determine the following: (a) modulus of elasticity, (b) proportional limit, (c) 0.2% offset yield strength, (d) ultimate tensile strength, (e) modulus of resilience, (f) toughness, (g) percent elongation.

4. For the aircraft worthy aluminum alloy 5052, in the -0 temper state shown below, find: (a) tensile strength, (b) percent elongation, and (c) toughness.

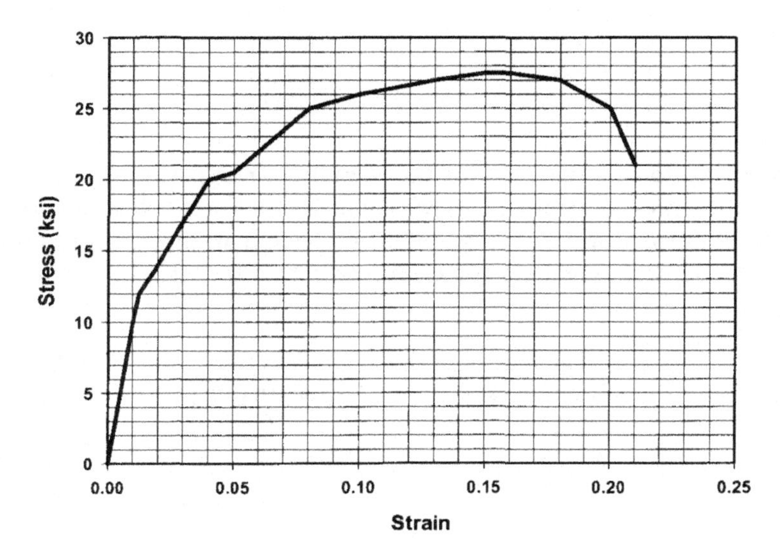

5. In a tensile test of an alloy used for ballistic-missile launch tubes on "Boomer" submarines (talk about the potential for ruining someone's day), the yield strength was determined to be σ_y= 36,000 psi based on an offset strain of 0.15%. In addition, the modulus of elasticity was found to be E = 15 x 10^6 psi. Given this data, find: (a) the total strain corresponding to the yield strength, (b) what portion of the total strain in part (a) is elastic, and (c) what portion is plastic? (d) at a tensile stress of 42,000 psi, the total tensile strain was found to be 0.009 in/in.; determine the plastic portion of this total strain.

6. A quantum stabilizer made of cold-worked columbium is 15 in. long, and has a rectangular cross-section ¼ in. by ¾ in. For a tensile load of 5000 lb., determine the total change in: (a) length, (b) cross-sectional dimensions, and (c) volume. For the analysis, assume the alloy exhibits elastic behavior with E = 22.7 x 10^6 psi and ν = 0.28.

7. In a tension test of cast iron used for engine blocks, the following stress strain values were recorded:

Stress (psi)	Strain (in/in)	Stress (psi)	Strain (in/in)
0	0	13,000	0.00075
2500	0.00010	16,000	0.0010
5000	0.00022	20,000	0.0014
8000	0.00040	23,000	0.0017
9500	0.00050	27,000	0.0025

For this data, determine: (a) the stress strain curve, (b) initial tangent modulus, (c) secant modulus for a stress of 8000 psi, (d) tangent modulus for a stress of 8000 psi, (e) tensile strength, and (f) percent elongation.

8. Using the following data for rubber used for automobile engine mounts, construct a stress strain diagram and determine: (a) secant modulus and the tangent modulus at a service stress of 200 psi, (b) resilience, and (c) hysteresis loss per cycle in units of in-lb/in.^{3}

LOADING		UNLOADING	
Stress (psi)	Strain (in./in)	Stress (psi)	Strain (in./in)
0	0.00	300	4.25
60	0.40	240	4.15
120	1.25	130	3.90
180	2.42	120	3.13
240	3.55	100	2.00
300	4.25	60	.070
		0	0.00

9. Numerical values of the tensile response of the aluminum-magnesium-silicon alloy 6061 used for light-weight competition bicycle components are tabulated below for heat treatments –0, –T4, and –T6. Assuming elastic, linear strain hardening behavior, sketch all three tensile response curves on the same graph, and estimate the toughness for each heat treatment. Hint: Idealize the graph as linear elastic, linear plastic.

Alloy Form	Elastic Modulus (psi)	Tensile Strength (ksi)	Yield Strength (ksi)	% Elongation
6061-0	10x10^6	17	7	25
6061-T4	10x10^6	33	19	22
6061-T6	10x10^6	42	37	12

10. A structural member in a nuclear reactor is to be fabricated from a cold worked zirconium alloy for which E = 14 x 10^6 psi and σ_y = 58 ksi. The design constraint of the member is that it must carry a tensile load of 60,000 lbs. Using this information, determine: (a) if the member is to carry no greater than one-third of the yield strength of the material (safety factor of 3.0), what is the minimal cross-sectional area required, (b) using an entirely different design code, the 2.0-foot-long member must elongate no more than 0.03 inches under load; what minimal cross-sectional area is necessary to meet the elongation restriction, and (c) if both restrictions (a) and (b) apply simultaneously, what minimal cross-sectional area must be specified for the design?

11. A axle steel compression specimen, 0.5 inches in diameter and with a 0.5 inch gage length was loaded in compression with the following results: The proportional limit and yield loads were found to be σ_{pl}=6,000 and σ_y=6,200 lbs. respectively. At the proportional limit, the axial elongation was −0.00050 inches, and the transverse dilation was 0.00012 inches. Using this information, determine: (a) proportional limit, (b) yield strength, (c) modulus of elasticity, (d) modulus of resilience, and (e) Poisson's ratio.

12. A concrete cylinder 8.0 inches in diameter and with a gage length of 10.0 inches was subjected to compressive loading. The following compressive loads and axial dilatation were recorded:

Load (lb)	Dilatation (in)	Load (lb)	Dilatation (in)
0	0	36,000	0.00421
4,000	0.00026	40,000	0.00609
8,000	0.00053	44,000	0.00692
12,000	0.00081	48,000	0.00922
16,000	0.00113	52,000	0.01177
20,000	0.00158	56,000	0.01575
24,000	0.00206	58,000	0.01847
28,000	0.00260	60,000	Failed at this load
32,000	0.00327		

Plot the complete compressive stress-strain curve and find: (a) ultimate strength in compression, (b) percent axial contraction (opposite of elongation) at fracture, (c) initial tangent modulus, (d) secant modulus for a stress of 650 psi, (e) tangent modulus for a stress of 650 psi, and (f) toughness.

13. Experiments show that for ductile metals, there is no volume change in the plastic range of straining. Neglecting the small elastic volume change that would occur from the total strain, compute the value of Poisson's ratio for zero volume change. Subtle hint: do a thought experiment wherein you apply an axial strain of your choice, observe a transverse strain, calculate the before and after volumes, and demand the change to be zero!

14. A sample of titanium carbide used for cutting tools is thought to have a tensile strength of 900 MPa. Referring to the beam testing configuration of Figure 4.18 with L = 9 cm, b = 0.5h = 0.6 cm, what load P would cause the beam to fracture if the bending- or flexural-strength (also called the *modulus of rupture in bending*) is equivalent to the tensile strength?

CHAPTER FIVE

YIELDING FOR MULTIAXIAL STRESS STATES

5.1 INTRODUCTION

Up to this point, we have examined states of both stress and strain with little consideration to what they imply in terms of design behaviors. Obviously, other forms of analysis must also be employed to determine the upper limits to the elastic stresses and strains, as well as the potentially bad behaviors when exceeded. For ductile materials destined to be in structural applications, the limiting values of stress and strain would help ensure that plastic deformation does not take place. Conversely, design situations might arise in which inelastic behavior is appropriate as so poignantly discussed with the paper clip example in an earlier chapter. Because design procedures will likely be concerned with both sides of the yield boundary, it is worthwhile to explore current theories dealing with the onset of yielding.

5.2 YIELDING IN PURE SHEAR

Before we delve into the actual theories for predicting yield behaviors, it is useful to first examine particular elastic states as we approach the proportional and yield limits. Consider the elastic stresses and strains in a member subjected to pure shear that is considerably different from the uniaxial stress state encountered under tensile or compressive loading. Pure shear states are most easily generated by applying a torque or twisting moment, M_t, to a bar of circular cross-section as shown in Figure 5.1.

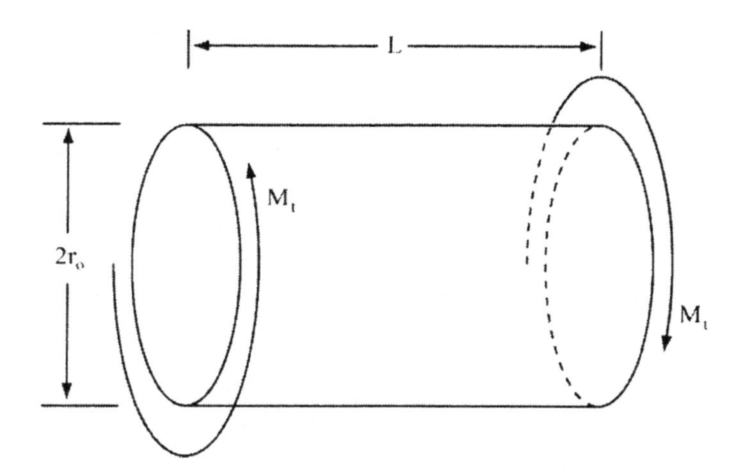

Figure 5.1 Circular bar subjected to a twisting torque, M_t.

The well-known exact solutions for the shear stresses and strains involved in the purely elastic case assume the following form:

$$\tau = \frac{M_t r}{J} \tag{5.1a}$$

$$\gamma = \frac{r\phi}{L} = \frac{1}{G}\tau \tag{5.1b}$$

where M_t is the twisting moment or torque, r is the radius, and ϕ is the *Angle of Twist* in radians through which one cross-section rotates relative to another at some arbitrary distance, L away. In addition, the polar moment of inertia, J, is defined as:

$$J = \frac{\pi}{2} r_o^4 \qquad \text{Solid Shaft} \qquad (5.2a)$$

$$J = \frac{\pi}{2}\left(r_o^4 - r_i^4\right) \qquad \text{Tube} \qquad (5.2b)$$

Figure 5.2 represents a portion of the torque versus angle-of-twist response for a solid round bar of 6061-T6 aluminum alloy. Not unexpectedly, the response has the same general appearance as a typical tensile curve complete with linear-elastic and elastic-plastic regions.

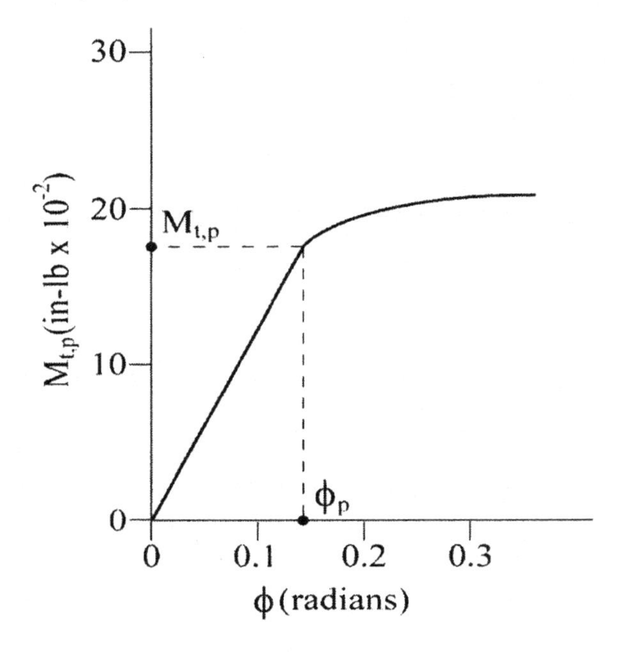

Figure 5.2 Torque versus angle of twist for a solid shaft of 6061-T6 aluminum.

As discussed in earlier chapters, Equation (5.1) applies only to the linear-elastic portion of the curve up to the proportional limit torque, M_{Pl}. In addition, the *Shear Modulus*, *G*, is given by:

$$G = \frac{\tau}{\gamma} = \frac{M_{t,p} L}{\phi_p J} \qquad (5.3)$$

Although the nonlinear response beyond the proportional limit renders the stress-torque relation in Equation (5.1) inaccurate, the strain relation remains valid for values on the order of 0.10 or less.

Given the stress gradients present through the cross-section, a better configuration for determining the response beyond the proportional limit would be a thin-walled tube of circular cross-section (Figure 5.3) because the shear stress distribution is assumed to be uniform across the thickness.

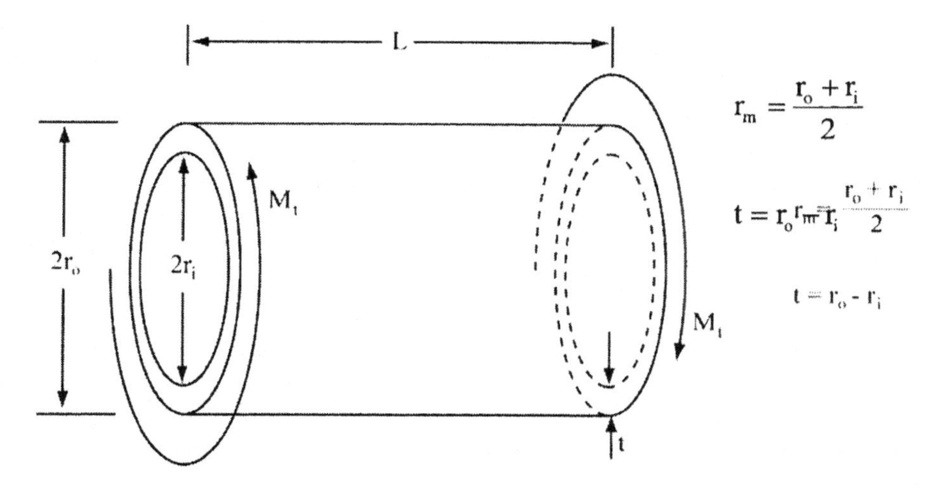

Figure 5.3 Thin-walled tubular specimen used to determine shear response.

For this configuration, the shear stress distribution is defined by the following relationship that reflects the thin-walled nature via a mean radius, r_m:

$$\tau = \frac{M_t}{2\pi r_m^2 t} \tag{5.4}$$

Interestingly, an examination of Equation (5.4) reveals that it is numerically indistinguishable from Equation (5.1) if the thin-walled assumption is valid ($r_m/t \geq 10.0$) so that the strains could be specified in Equation (5.1) using either $r = r_o$ or $r = r_m$.

Torque versus angle-of-twist data for a portion of the response of 6061-T6 aluminum alloy is shown in Figure 5.4. Included in the figure is a plot of the data converted to stress and strain using Equation (5.4). Since the onset of yield is no clearer than seen with the simple tensile test, the yield strength response, τ_y, was determined using the 0.2% offset method. Using Figure 5.4, the offset shear yield strength for the 6061-T6 tested was found to be $\tau_y = 22{,}900$ psi. As with the tensile test discussed in earlier chapters, other strength measures such as the ultimate shear strength and ductility could also be determined from the complete τ-γ curve. Unfortunately, thin-walled tubes tend to buckle at large plastic strains, thus limiting the usefulness of the test method.

Interestingly, the same 6061-T6 aluminum alloy was found to have a tensile yield strength of $\sigma_y =$ 40,000 psi. It should therefore be immediately and intuitively obvious (hopefully) that yield strength is not an inherent property of a ductile metal. Instead, it appears that the yielding envelope will be a function of the applied stress state that could include any combination of normal and/or shear stresses. As such, it would be impractical to define yield strengths experimentally for all stress states of interest unless you had a lot of time on your hands and an infinitely patient employer (good luck on that one!). Rather, some form of theory is required so that the designer can accurately predict yielding based on complex stress states while still using data collected from a relatively simple test.

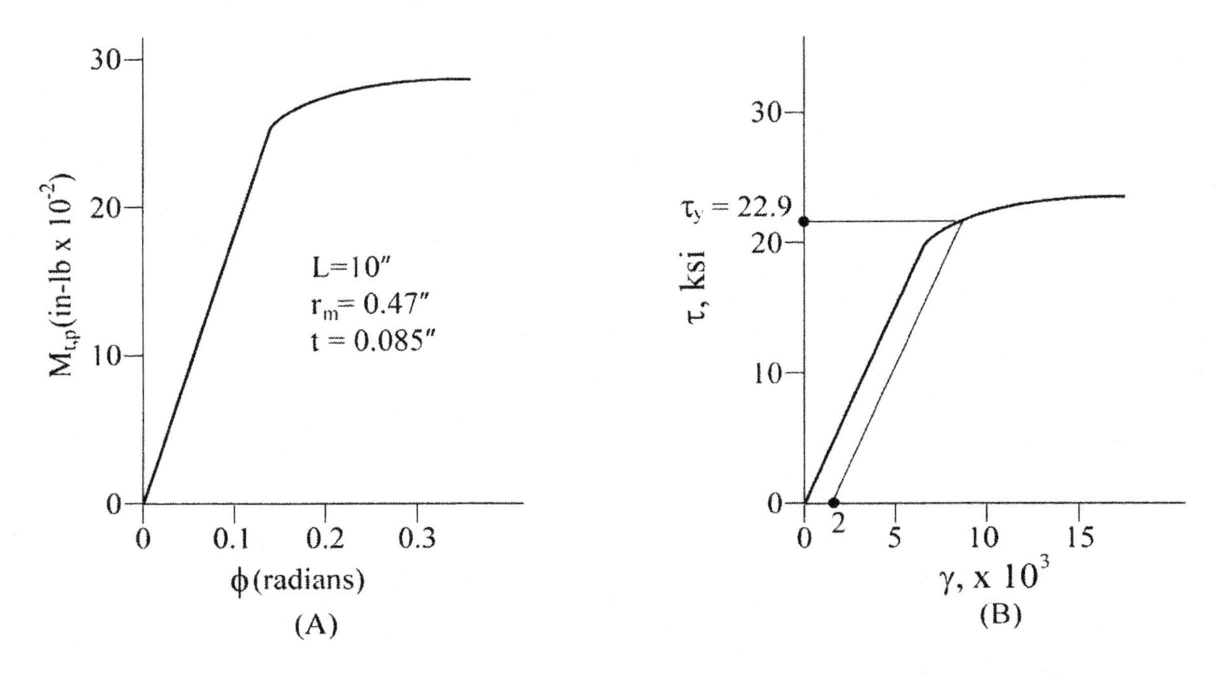

Figure 5.4 (A) Torque versus angle of twist for a thin-walled 6061-T6 tube, and (B) shear stress versus shear strain for 6061-T6 aluminum alloy.

5.3 YIELDING FOR ARBITRARY STRESS STATES

As already shown in Chapter 2, any combination of normal and shear stresses that may exist at a point within a component can be reduced to three principal values or σ_I, σ_{II}, and σ_{III}. In order to simplify the analysis, we will first only consider the plane stress situation shown in Figure 5.5 so that one of the principal stresses is zero; in this case $\sigma_{III} = 0$. Although this assumption will simplify the analysis, the conclusions reached will be equally applicable to fully three-dimensional stress states as well.

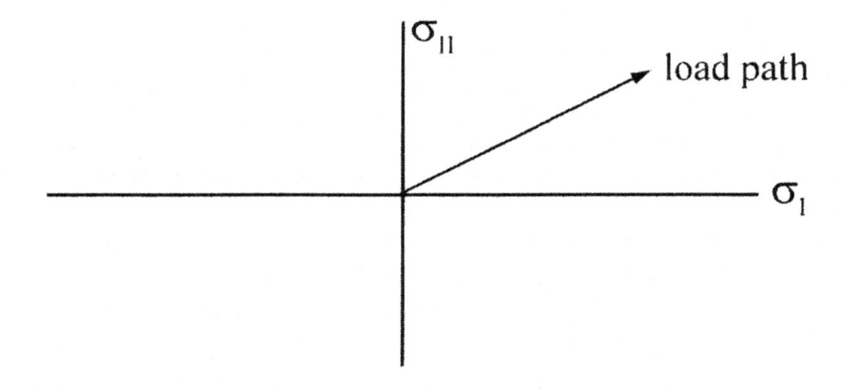

Figure 5.5 Plane stress load path in a two-dimensional principal stress space.

Using the stress space shown in the figure, it is reasonable to expect that the structure will initially behave elastically, at least near the origin. As the forces and moments applied to the structure are increased, the stresses and their principal values would grow. Eventually, this load path must correspond to yielding, at least for the plastic capable material now under consideration. A viable theory for yielding must therefore, predict this critical point for the stress-state shown, as well as for all possible load paths; the resulting locus of all yield points will be a line in the two-dimensional load space and a surface in three dimensions. While this locus is perhaps easy to envision, the proverbial "devil" is always in the details. Nonetheless, there are a number of viable approaches that yield (puns intended as usual) very similar results as described in the next sections.

5.4 MAXIMUM SHEAR STRESS THEORY

It has been observed (and mentioned countless times earlier) that yielding has been associated with the presence of shear stresses in a structure under load. Hence, it is not surprising that the yield theory proposed by Tresca and Guest and appropriately named the Maximum Shear-Stress Theory (MSST), recognizes that shear stresses induce crystalline slip or plastic deformation. In fact, the theory states that "yielding occurs when the greatest of the three possible maximum shear-stresses equals a critical value, *K*, typical of the material."

Recalling (no doubt with great pleasure) the three Mohr's circle possible and their respective maximum shear stresses, the theory may be algebraically expressed as:

$$\frac{1}{2}(\sigma_I - \sigma_{II}) = \pm K \tag{5.5a}$$

-or-

$$\frac{1}{2}(\sigma_{II} - \sigma_{III}) = \pm K \tag{5.5b}$$

-or-

$$\frac{1}{2}(\sigma_{III} - \sigma_I) = \pm K \tag{5.5c}$$

Note that the theory does not recognize sign differences, so it is assumed that the response to tensile and compressive stresses is equivalent; the plus/minus signs on the right side of Equations (5.5) are therefore necessary because the differences in principal values may turn out to be of either sign.

To evaluate the critical value, K for any given material, the tensile test is employed owing to its simplicity; σ_I = P/A while $\sigma_{II} = \sigma_{III} = 0$. Under these conditions, it can be easily shown that $\sigma_I = \sigma_y$ at yielding, where σ_y is still the yield strength determined by the 0.2% offset or an equivalent method. By using the simple tensile test and its results, Equations (5.5a) and (5.5c) predict quite simply:

$$\frac{\sigma_y}{2} = K \tag{5.6}$$

Notice that the quantity $\sigma_y/2$ is the maximum shear stress in tension at yielding. Substituting the critical value of *K* into equation (5.5), the Maximum Shear Stress Theory or MSST becomes:

$$\sigma_I - \sigma_{II} = \pm\sigma_y \tag{5.7a}$$

-or-

$$\sigma_{II} - \sigma_{III} = \pm\sigma_y \tag{5.7b}$$

-or-

$$\sigma_I - \sigma_{III} = \pm\sigma_y \tag{5.7c}$$

For plane stress states, the locus of yield points may be located by dividing the σ_I and σ_{II} principal stress space into the segments as indicated in Figure 5.6; the three regions in the principal stress space are characterized by the following orderings of the principal components:

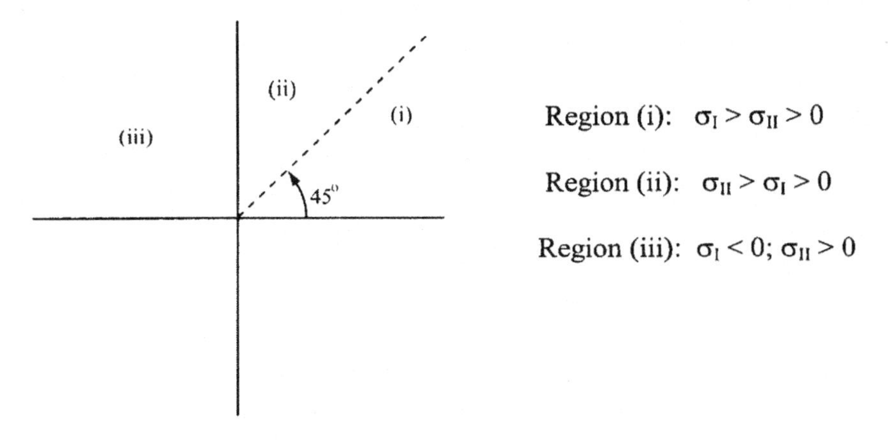

Figure 5.6 Three regions for placing the MSST locus of yield points.

The corresponding Mohr's circles, with the greatest maximum shear stresses indicated are given in Figure 5.7.

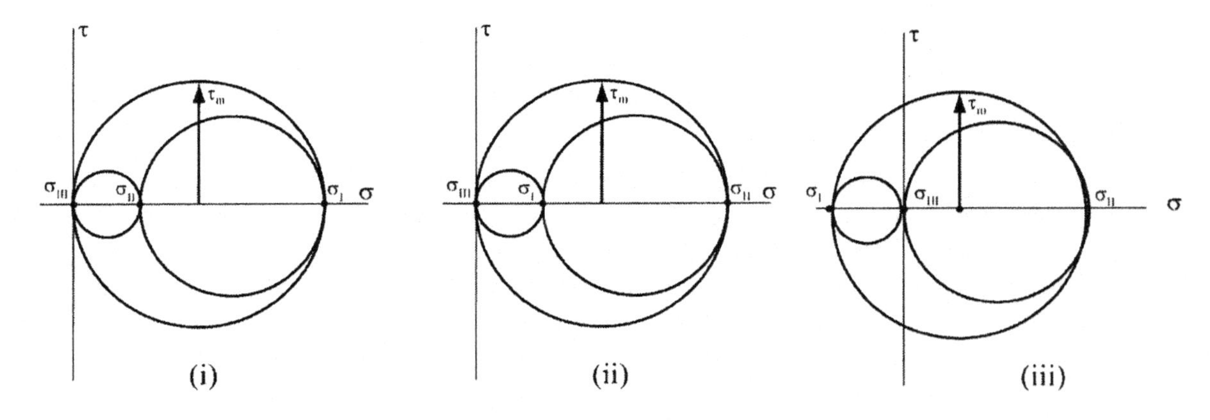

Figure 5.7 Mohr's circles for designated regions i, ii, and iii.

For region i, Equation (5.7c) defines yielding, because $\sigma_I = \sigma_y$. Similarly, Equation (5.7b) governs yielding in region ii, and finally, Equation (5.7a) for region iii. The remaining two quadrants of the principal stress space can be examined by the reader for the position of the yield locus with the final result shown in Figure 5.8.

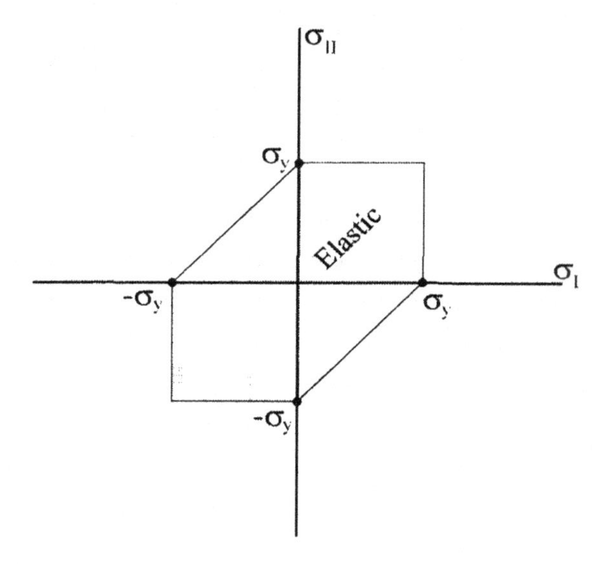

Figure 5.8 Locus of yield points as predicted by the Maximum Shear Stress Theory (MSST).

All principal stress states within the yield locus are wholly elastic, with any states defined outside being *elastic-plastic* or *Yielded.* It must be noted that the yield locus of Figure 5.8 is for the plane stress state only; the locus of yield points in a three-dimensional space is the surface of a hexagonal right cylinder whose axis is inclined and coincident with the line $\sigma_I = \sigma_{II} = \sigma_{III}$ as shown in Figure 5.9. Interestingly, the line where $\sigma_I = \sigma_{II} = \sigma_{III}$ represents a state of hydrostatic stress where yielding can never occur because shear stresses are never present ($\sigma_I - \sigma_{II} = 0$ etc.).

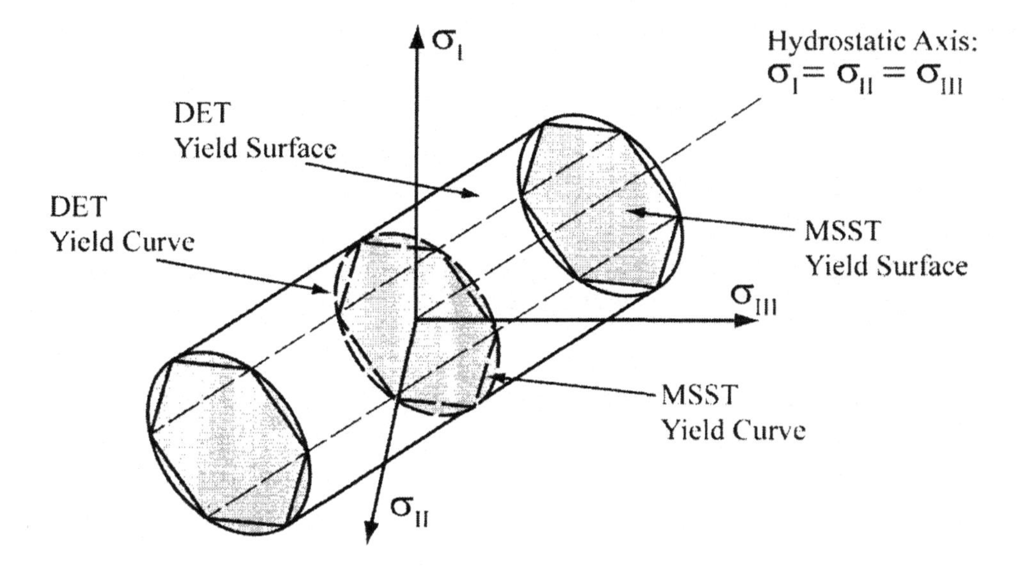

Figure 5.9 Potential yield-surfaces for a three-dimensional stress state.

Before we proceed, it is important to note that although looking somewhat complicated, the MSST theory can be relatively straightforward (and yes, easy) to implement, especially if another simple rule is adopted. If it is assumed that the material is isotropic and therefore yields at the same level in all directions, then directionality is not an issue. If we then also assume that the principal stress order is $\sigma_I > \sigma_{II} > \sigma_{III}$, then the maximum shear stress will ALWAYS be defined by the I-III plane. As such, only Equation (5.7c) is required to determine if yielding has occurred, regardless of the stress-state since the largest shear will always initiate yielding first! Otherwise, one must use all three versions of Equation (5.7) and then select the largest shear values and/or compare to the directional yield strengths. The sanity saving benefits of this approach is illustrated in Example Problem 5-1.

Example Problem 5-1: Your first assignment at the ACME tank company is to produce a cylindrical, closed-end pressure vessel as shown in Figure 5.10. The vessel wall is made from a rolled plate 0.25 inches thick, with a mean cylinder radius of 2.5 inches. If the vessel is fabricated from a typical medium strength steel whose reported tensile yield strength is estimated to be σ_y =100,000 psi, what internal pressure P will initiate yielding in the vessel wall?

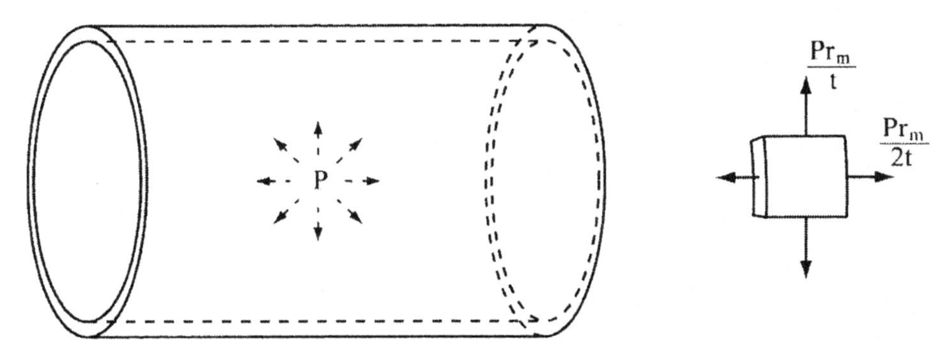

Figure 5.10 Thin-walled pressure vessel under pressure and the resulting stress state.

At any point on the cylinder away from the ends, there are two normal-stress components as already discussed in Chapter 2; one component is in the hoop or circumferential direction while the other orients in the axial sense as also shown in Figure 5.10. With no other stress components present, especially shear, the applied stresses are principal values and can be readily calculated as:

$$\sigma_I = \sigma_{hoop} = \frac{Pr_m}{t} = 10P \qquad \sigma_{II} = \sigma_{axial} = \frac{Pr_m}{2t} = 5P \qquad \sigma_{III} = 0$$

with the reasonable assumtion that $\sigma_{III} = 0$ given the thin wall and its corresponding inability to sustain any stress. Under the assumed stress state, the MSST applied to all three planes (or cases i, ii, and iii) gives us:

i (I, II Plane) $\quad \sigma_I - \sigma_{II} = 10P\text{-}5P = 5P = 100{,}000$ psi $\quad \rightarrow \quad P_y$=20,000 psi

ii (II, III Plane) $\quad \sigma_{II} - \sigma_{III} = 5P\text{-}0 = 5P = 100{,}000$ psi $\quad \rightarrow \quad P_y$=20,000 psi

iii (I, III Plane) $\quad \sigma_I - \sigma_{III} = 10P - 0 = 10P = 100{,}000$ psi $\quad \rightarrow \quad P_y$=10,000 psi

It is apparent that an internal pressure of P = 20,000 psi will yield the vessel on the I, II and II, III planes. However, a lesser pressure of P = 10,000 psi will initiate yielding on the σ_I -σ_{III} -plane; thus, the lowest value of pressure that will initiate yielding anywhere in the vessel wall will be a value of P = 10,000 psi.

In general, Equations (5.7) will generate three different answers, all of which indicate a failure on a different plane. If the material is anisotropic, then the three planes must be examined for the proper choice. However, If we assume that the material is isotropic in yielding and that the principal stress order is $\sigma_I > \sigma_{II} > \sigma_{III}$, then the maximum shear stress will ALWAYS be defined by the I-III plane, or:

$$\sigma_I - \sigma_{III} = 10P = 100{,}000 \text{ psi} \quad \rightarrow \quad P_y = 10{,}000 \text{ psi}$$

5.5 DISTORTION ENERGY THEORY

As its name implies, the Maximum Shear Stress Theory was developed on the observations that the yielding was somehow related to shear slip and the underlying shear stresses. Although certainly true, the yield locus is composed of straight lines and abrupt corners that do not bode well for modeling nature since it rarely exhibits either. With this in mind, there have been efforts over the years to develop a more realistic theory to describe yielding in ductile materials. Eventually, an alternative theory for predicting yielding was independently proposed by von Mises, Huber, and Hencky, with von Mises apparently winning the "name game" to a point since he is usually associated with the theory. Unfortunately for poor von Mises's ego, the theory is also frequently referred to as the *Octahedral Shear Stress* or the *Distortion Energy Theories*. Not coincidently, it is the last name and the concept of *Distortion Energy* we shall pursue, since the idea also has applications beyond the current topic of yielding.

To begin with, the term "distortion" implies shear or some form of shape change, but not necessarily changes in volume. Because plastic flow involves no volume change, like-minded parameters could be useful in predicting the onset of plastic deformation as long as we can separate the contributions from the applied stresses. Along these lines, the principal stresses can be visualized as consisting of two parts: one component causes distortion or shape change while the other hydrostatic stress or pressure induces all of the volume changes that occur during deformation. The name of the latter is no accident as hydrostatic pressure produces only volume changes and no shear. Algebraically, this can be written as:

$$\sigma_I \equiv \sigma_{hp} + \sigma_1 \tag{5.8a}$$

$$\sigma_{II} \equiv \sigma_{hp} + \sigma_2 \tag{5.8b}$$

$$\sigma_{III} \equiv \sigma_{hp} + \sigma_3 \tag{5.8c}$$

where σ_{hp} is the *Hydrostatic* or mean component and σ_1, σ_2, and σ_3 are the *Deviator* or distortion components that should not to be confused with the principal stresses, σ_I, σ_{II}, and σ_{III}. Assuming that the principal values are known for any given loading situation, four unknowns exist in Equation (5.8).

Fortunately, additional information is provided by examining the volume changes that can be expected under the applied stresses and the underlying strains. Assuming an initial volume of material of V, the fractional change in volume of a uniformly strained element, or ΔV/V must be the first strain invariant. If no volume change is allowed, then the normal strains in question must be the distortion strains ε_1, ε_2, and ε_3 such that:

$$\frac{\Delta V}{V} \equiv 0 = \varepsilon_1 + \varepsilon_2 + \varepsilon_3 \tag{5.9}$$

Since yielding is achieved by increasing the elastic stresses and strains to their limit, the distortion strains can be related to the associated stresses by Hooke's Law:

$$\varepsilon_1 = \frac{1}{E}\left[\sigma_1 - \nu\left(\sigma_2 + \sigma_3\right)\right] \tag{5.10a}$$

$$\varepsilon_2 = \frac{1}{E}\left[\sigma_2 - \nu\left(\sigma_1 + \sigma_3\right)\right] \tag{5.10b}$$

$$\varepsilon_3 = \frac{1}{E}\left[\sigma_3 - \nu\left(\sigma_1 + \sigma_2\right)\right] \tag{5.10c}$$

Substituting Equations (5.10) into (5.9), the following result is obtained for constant volume deformations:

$$\sigma_1 + \sigma_2 + \sigma_3 = 0 \tag{5.11}$$

Adding the three principal values in Equations (5.8) and substituting Equation (5.11) into the sum, the hydrostatic pressure component is found to be:

$$\sigma_{hp} = \frac{\sigma_I + \sigma_{II} + \sigma_{III}}{3} \tag{5.12}$$

Note that the pressure component is indeed the mean (normal) stress, as well as one-third the first stress invariant discussed in Chapter Two. To obtain the *Strain Energy* associated with the distortion stresses and strains, it is necessary to first consider the total strain energy of the system, denoted as U_t:

$$U_t = \frac{\sigma_I \varepsilon_I}{2} + \frac{\sigma_{II} \varepsilon_{II}}{2} + \frac{\sigma_{III} \varepsilon_{III}}{2} \tag{5.13}$$

Recalling that each principal strain is also proportional to all three principal stresses (via Hooke's Law), the total strain energy per unit volume becomes:

$$U_t = \frac{1}{2E}\left[\sigma_I^2 + \sigma_{II}^2 + \sigma_{III}^2 - 2\nu\left(\sigma_I\sigma_{II} + \sigma_{II}\sigma_{III} + \sigma_{III}\sigma_I\right)\right] \tag{5.14}$$

In a similar fashion, the strain energy per unit volume due to the pressure component is:

$$U_{hp} = \frac{\sigma_{hp}\varepsilon_{hp}}{2} + \frac{\sigma_{hp}\varepsilon_{hp}}{2} + \frac{\sigma_{hp}\varepsilon_{hp}}{2} = \frac{3}{2}\sigma_{hp}\varepsilon_{hp} \tag{5.15}$$

Each individual pressure strain component can also be found from Hooke's law:

$$\varepsilon_{hp} = \frac{1}{E}\left[\sigma_{hp} - \nu\left(\sigma_{hp} + \sigma_{hp}\right)\right] = \frac{\sigma_{hp}}{E}(1-2\nu) \tag{5.16}$$

Thus, for the strain energy per unit volume associated with the hydrostatic pressure component, one obtains:

$$U_{hp} = \frac{(1-2\nu)}{6E}\left[\sigma_I^2 + \sigma_{II}^2 + \sigma_{III}^2 + 2\sigma_I\sigma_{II} + 2\sigma_{II}\sigma_{III} + 2\sigma_{III}\sigma_I\right] \tag{5.17}$$

The total strain energy per unit volume minus that associated exclusively with volume changes, or U_d is:

$$U_d \equiv U_t - U_{hp} = \frac{1+\nu}{6E}\left[\left(\sigma_I - \sigma_{II}\right)^2 + \left(\sigma_{II} - \sigma_{III}\right)^2 + \left(\sigma_{III} - \sigma_I\right)^2\right] \tag{5.18}$$

Finally, and based on the underlying assumptions just espoused, the *Distortion Energy Theory* (DET) for yielding can now be considered to occur at any point in a structure when the local distortion energy reaches a critical value, K considered typical of the material. Mathematically, the DET theory becomes:

$$\left(\frac{1+\nu}{6E}\right)\left[\left(\sigma_I - \sigma_{II}\right)^2 + \left(\sigma_{II} - \sigma_{III}\right)^2 + \left(\sigma_{III} - \sigma_I\right)^2\right] = K \tag{5.19}$$

As before, the critical material parameter, K is evaluated by a simple tensile test, where $\sigma_I = \sigma_y$ at yield and $\sigma_{II} = \sigma_{III} = 0$. Equation (5.19), for the tension test then simplifies to:

$$\left(\frac{1+\nu}{6E}\right)\left(2\sigma_I^2\right) = \frac{1+\nu}{6E}\left(2\sigma_y^2\right) = K \tag{5.20}$$

In the principal stress space, the distortion energy theory of yielding can now be related to the 0.2% offset tensile yield strength as:

$$\left(\sigma_I - \sigma_{II}\right)^2 + \left(\sigma_{II} - \sigma_{III}\right)^2 + \left(\sigma_{III} - \sigma_I\right)^2 = 2\sigma_y^2 \tag{5.21a}$$

In terms of the applied normal and shear stresses in a Cartesian coordinate system, the DET can also be expresses as in a more palatable way for those who still fear Mohr's circle:

$$\left(\sigma_x - \sigma_y\right)^2 + \left(\sigma_x - \sigma_z\right)^2 + \left(\sigma_y - \sigma_z\right)^2 + 6\left(\tau_{xy}^2 + \tau_{xz}^2 + \tau_{zy}^2\right) = 2\sigma_y^2 \tag{5.21b}$$

As with the Maximum Shear Stress Theory discussed in the last section, shear stresses are still important as they appear as the squared differences between the principal values. However, and this is a "biggie," all three shear planes (smallest, medium, and maximum) are now included in the analysis. Hence, all three (Mohr's) circles and the maximum shear associated with each must now be considered in the final analysis.

For states of plane stress ($\sigma_{III} = 0$), Equation (5.21a) becomes:

$$\sigma_I^2 + \sigma_{II}^2 - \sigma_I \sigma_{II} = \sigma_y^2 \tag{5.22}$$

According to the Distortion Energy Theory in the plane stress-state, the locus of yield points is an ellipse inclined at 45 degrees as shown by Figure 5.11. When the DET is compared to the MSST, it is evident that both are very similar (in full agreement at some locations), at least in the overall size and shape of the envelope. However, and as mentioned earlier, the DET and its smooth boundary is a more realistic model of materials behaviors. In three-dimensions, the distortion energy theory predicts that all principal stress values corresponding to the onset of yielding lie on the surface of a right circular cylinder whose axis is coincident with the line $\sigma_I = \sigma_{II} = \sigma_{III}$ as previously shown in Figure 5.9.

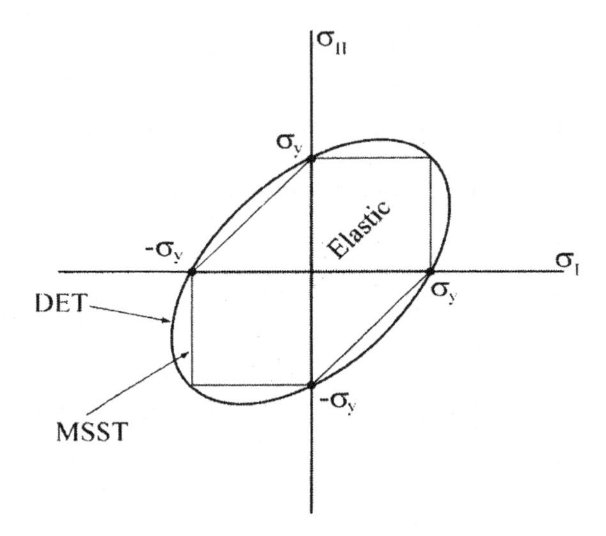

Figure 5.11 Yield point loci as predicted by both Distortion Energy and Maximum Shear Stress Theories for plane stress ($\sigma_{III} = 0$).

Example Problem 5-2: You still are working for the ACME Company and your assignment is the same as in Problem 5-1, except that the maximum allowable internal pressure is to be determined via the DET. Since the pressure vessel wall is in a state of plane stress, the DET takes on the following form:

$$\sigma_I^2 + \sigma_{II}^2 - \sigma_I \sigma_{II} = \sigma_y^2$$

Substituting the known values of stress in terms of the unknown pressure (P), Equation (5.22) becomes:

$$\left(\frac{Pr_m}{t}\right)^2 + \left(\frac{Pr_m}{2t}\right)^2 - \left(\frac{Pr_m}{2t}\right)\left(\frac{Pr_m}{t}\right) = (100,000)^2$$

or

$$(10P)^2 + (5P)^2 - (5P)(10P) = (100,000)^2$$

After rearranging terms and solving for the unknown pressure, P:

$$P_y = \frac{2000}{\sqrt{3}} = 11,547 \text{ psi}$$

Note that this result is 15.5% higher than that obtained when the Maximum Shear Stress Theory was employed to solve the exact same problem; for the pressure vessel used in Examples 5-1 and 5-2 with $\sigma_{II} = \sigma_I/2$, and both $\sigma_I > 0$ and $\sigma_{II} > 0$, the load path intersects the two boundaries as shown in Figure 5.12.

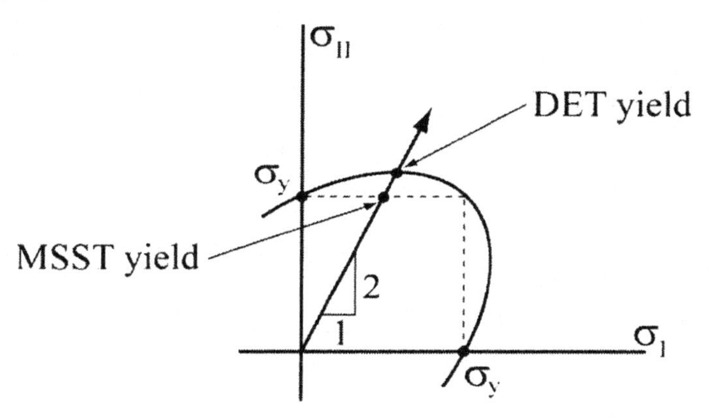

Figure 5.12 Yield point loci as predicted by DET and MSST for the thin-walled vessel of Example 5-1 and 5-2 ($\sigma_I > \sigma_{II} > 0$).

Hence, the MSST is conservative in that it will predict lower pressures for the onset of yielding as already shown in Figure 5.11. On the other hand, both DET and MSST will always agree for the uniaxial and equibiaxial cases.

5.6 INTERNAL FRICTION THEORY

The yield theories considered in the previous sections were both based on the assumption that the yield strengths in simple tension and compression were equal. Although this is certainly a reasonable assumption for most isotropic metals, there are materials which have differing yield strengths in tension and compression. One theory that considers this potentially significant difference is the internal friction theory.

As the name implies, the Internal Friction Theory (IFT) is based on the assumption that the critical shear-force causing yield is swayed by the presence of internal friction forces that are in turn influenced by the normal stress acting on the shear planes. Under this mind- and tongue-twisting assumption, not to mention the realities of force friction relationships, the critical shear stress also becomes a function of the normal stress acting on the shear planes. This assumed linear relationship can be written as:

$$\tau_{max} = A\sigma_n + B \tag{5.23}$$

where A and B are functions of the tensile and compressive yield strengths. In order to fully define the theory, one must use the dreaded Mohr's circle(s) once again to examine the possible biaxial principal stress states. As with earlier examples, we must temporarily suspend our useful and labor saving rule that $\sigma_I > \sigma_{II} > \sigma_{III}$ in order to establish the yield envelope.

Case I: $\sigma_I \geq \sigma_{II}$ and σ_I = +ve (positive) and σ_{II} = -ve (negative)

Referring to Figure 5.13 and your fond memories of earlier Mohr's circle discussions, it can be seen that the following definitions are valid:

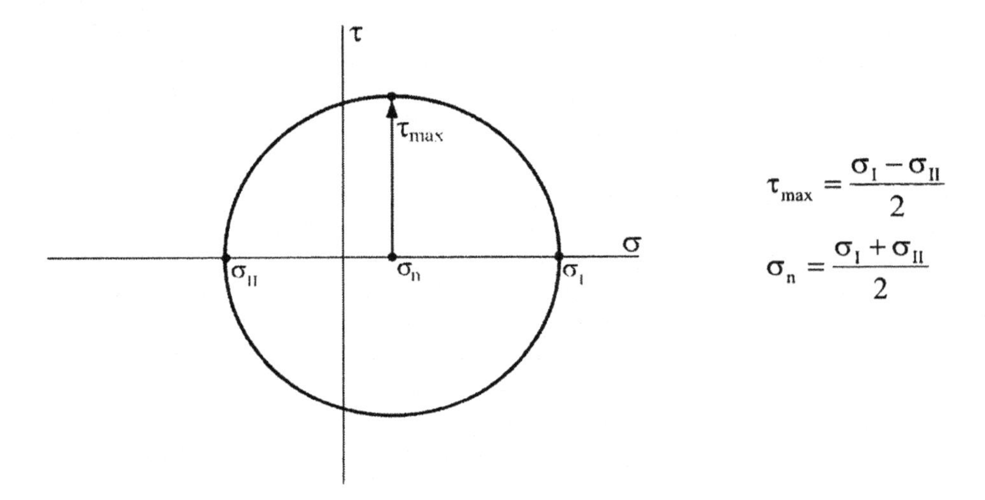

$$\tau_{max} = \frac{\sigma_I - \sigma_{II}}{2}$$

$$\sigma_n = \frac{\sigma_I + \sigma_{II}}{2}$$

Figure 5.13 Principal stress states for Case I.

Substituting these relationships into Equation (5.23), one obtains:

$$\sigma_I(1-A) - \sigma_{II}(1+A) = 2B \tag{5.24}$$

Noting that for the case of $\sigma_{II} = 0$, $\sigma_I = \sigma_{yt}$ at yield, where σ_{yt} is the tensile yield strength. In addition, for the other case of $\sigma_I = 0$, then $\sigma_{II} = \sigma_{yc}$ at yield where σ_{yc} is the compressive yield strength. Implementing these conditions in Equation (5.24), one obtains:

$$\sigma_{yt}(1-A) = 2B \tag{5.25a}$$

$$\sigma_{yc}(1+A) = 2B \tag{5.25b}$$

Solving these equations simultaneously results in:

$$A = \frac{\sigma_{yt} - \sigma_{yc}}{\sigma_{yt} + \sigma_{yc}} \tag{5.25c}$$

$$B = \frac{\sigma_{yt}\sigma_{yc}}{\sigma_{yt} + \sigma_{yc}} \tag{5.25d}$$

Finally, substituting the values of A and B from Equations (5.25c) and (5.25d) into Equation (5.24) and simplifying, the internal friction theory for this stress state is given as:

$$\sigma_I - \left(\frac{\sigma_{yt}}{\sigma_{yc}}\right)\sigma_{II} = \sigma_{yt} \tag{5.26a}$$

Case II: $\sigma_{II} > \sigma_I$ and σ_{II} +ve and σ_I –ve

For this case of simply reversing the labeling of Figure 5.13, one easily obtains:

$$\sigma_{II} - \left(\frac{\sigma_{yt}}{\sigma_{yc}}\right)\sigma_I = \sigma_{yt} \tag{5.26b}$$

Case III: $\sigma_{II} > \sigma_I$ with both σ_I and σ_{II} +ve and $\sigma_{III} = 0$

$$\sigma_I = \sigma_{yt} \tag{5.26c}$$

Case IV: $\sigma_I \geq \sigma_{II}$ with both σ_I and σ_{II} +ve, and $\sigma_{III} = 0$

$$\sigma_{II} = -\sigma_{yt} \tag{5.26d}$$

Case V: $\sigma_{II} \geq \sigma_I$ with both σ_I and σ_{II} -ve, and $\sigma_{III} = 0$

$$\sigma_{II} = \sigma_{yc} \tag{5.26e}$$

-and-

$$\sigma_I = -\sigma_{yc} \tag{5.26f}$$

Equations 5.27 expressing the *Internal Friction Theory* of failure can best be illustrated through the use of a failure envelope such as shown in Figure 5.14.

It should be noted that we are once again faced with a somewhat unrealistic theory given the straight borders and abrupt corners. In this regard, it is not really surprising that the IFT simply reduces to the Maximum Shear Stress theory for equal tensile and compressive yield strengths.

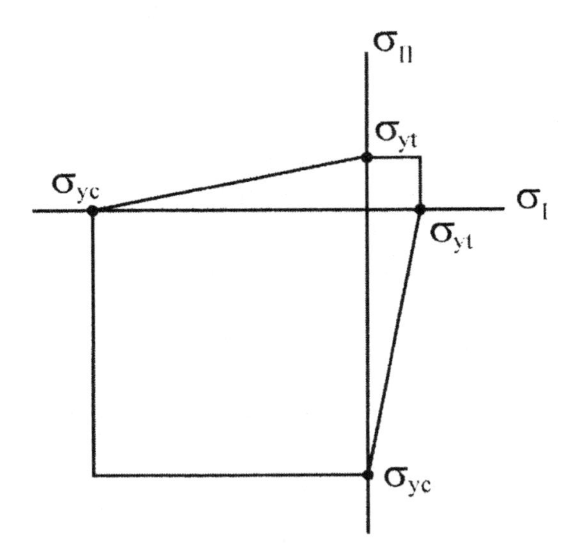

Figure 5.14 Failure envelope for the Internal Friction Theory or IFT.

Example Problem 5-3: A LNG storage vessel has a 15-inch diameter and an average wall thickness of t = 0.25 in. If the material used to construct the vessel exhibits a tensile yield strength of σ_{yt} = 40,000 psi and a compressive yield strength of σ_{yc} = 45,000 psi, determine the internal pressure required for yielding using the Internal Friction Theory.

As with the earlier examples involving pressure vessels, the applied stress states are principal in nature (with $\sigma_I > \sigma_{II}$) given the absence of applied shear stresses. As such:

$$\sigma_I = \frac{Pd/2}{t} = \frac{P(15)}{0.5} = 30P$$

$$\sigma_{II} = \frac{Pd/2}{2t} = 15P$$

Because both stresses are tensile, not to mention the fact that the yield strength in tension is lower than compression, our focus will be on the tensile part ala Case IV. Accordingly, yielding occurs when:

$$\sigma_I = \sigma_{yt}$$

such that

$$30P = 40{,}000 \quad \rightarrow \quad P = 1{,}333 \text{ psi}$$

Example Problem 5-4: A submarine propeller shaft with a 10-inch diameter has a tensile yield strength of σ_{yt} = 90,000 psi and a compressive yield strength of σ_{yc} = 120,000 psi. Determine the twisting moment required to produce yielding based on the Maximum Shear-Stress Theory (MSST), the Distortion Energy Theory (DET), and the Internal Friction Theory (IFC).

The analysis will start with a few preliminaries given the circular shaft under study:

$$J = \frac{\pi d^4}{32} = \frac{\pi(10)^4}{32} = 982 \text{ in}^4$$

$$\tau_{max} = \frac{Tr_o}{J} = \frac{T(5)}{982} = 5.1\text{x}10^{-3}T$$

MSST: If no real difference between tension and compression is noted by the theory, then the lower value (tensile in this case) must be used. As such, yielding will occur under tension first and:

$$\sigma_I - \sigma_{II} = \sigma_y = 2\tau_{max} = 90,000 \text{ psi} \quad \rightarrow \quad T = 8.84\,\text{x}\,10^6\text{in-lb}$$

DET: Again, only the lower threshold tensile yield will be considered such that:

$$\sigma_I^2 + \sigma_{II}^2 - \sigma_I\sigma_{II} = \sigma_y^2$$

such that

$$\sqrt{3}\tau_{max} = 90,000 \text{ psi} \quad \rightarrow \quad T{=}10.1\text{x}10^6\text{in-lb}$$

IFT: Unlike the first two theories, both tensile and yield behaviors must be considered, at least at the start of the analysis. Nevertheless, one does not have to travel too far down the analytical road to see that the lower value of the tensile yield will still dictate the critical load path and analysis as before. Hence, for the lower yield strength in tension:

$$\sigma_I - \left(\frac{\sigma_{yt}}{\sigma_{yc}}\right)\sigma_{II} = \sigma_{yt}$$

such that:

$$\frac{T(5)}{982} + \frac{0.75T(5)}{982} = 90,000 \quad \rightarrow \quad T = 10.1\text{x}10^6\text{in-lb}$$

Not surprisingly, the theory produces essentially the same result as the DET, but for different reasons.

5.7 THEORY VALIDITY

While the IFT presents a means for evaluating yield anisotropies for tension and compression, it is somewhat limited in realism and is rarely used. On the other hand, most experimentally determined yield-points tend to be closer to the Distortion Energy Criterion with the data usually falling between the theoretical loci of the DET and MSST. For instance, consider the case of pure shear discussed earlier in this chapter. In the absence of applied normal stresses, the principal stresses in pure shear are always equal in magnitude, but opposite in sign, or $\sigma_I = -\sigma_{II}$ as indicated in Figure 5.15. Clearly, the DET will predict a somewhat (but not overwhelmingly) larger value of σ_I (or σ_{II}) at yielding than the MSST using the loading path shown.

For a specific numerical example, it is useful to look at a 6061-T6 aluminum where a tensile yield strength of σ_y = 40,000 psi has been measured. Using both MSST and DET, the following theoretical predictions can be made:

$$\text{MSST: } \sigma_I = \tau_y = \frac{\sigma_y}{2} = 20,000 \text{ psi}$$

$$\text{DET: } \sigma_I = \tau_y = \frac{\sigma_y}{\sqrt{3}} = 23,080 \text{ psi}$$

Actual measurement taken from the data shown in Figure 5.4 indicates that τ_y = 22,900 psi for this alloy. Given the closeness of the MSST and DET predictions to the measurements, the question may arise as to which of the two theories is "better." Extensive experiments suggest that the Distortion Energy Theory or DET tends to predict experimental results more closely. Moreover, it is algebraically simpler to use and projects a smooth (realistic) envelope for design. On the other hand, given the uncertainties that may be inherent in most design analyses, not to mention materials testing, the more conservative predictions of the Maximum Shear Stress Theory may render its use well-advised. In addition, the Maximum Shear Stress Theory identifies a specific plane upon which the "failure" stress is a maximum, a feature that is lost in the use of the distortion energy criterion. In the final analysis, "better" or "best" represents a value judgment that cannot be unambiguously specified for all design situations.

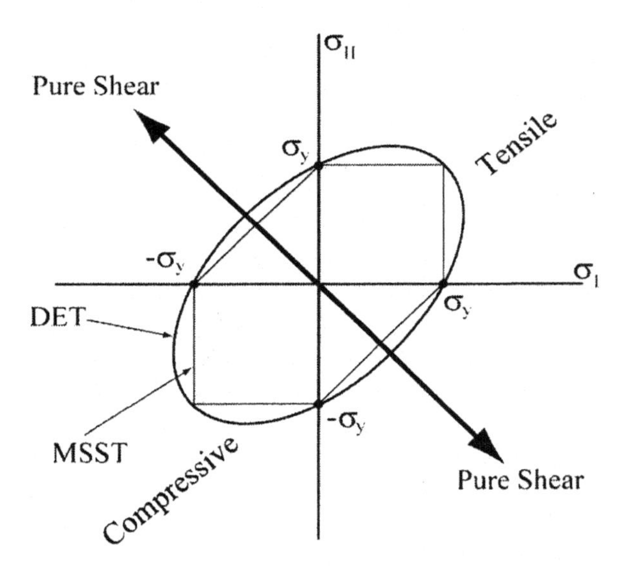

Figure 5.15. Yield envelopes for DET and MSST and a pure shear load path.

PRACTICE EXERCISES

1-2. The stress states given below were both determined to be critical in a proposed design of the water-brake system for an aircraft-carrier catapult launch system. Find the minimum tensile yield strength required for the fabrication of each element using both the Maximum Shear Stress and Distortion Energy Criteria.

$$1 \quad \sigma_{ij} = \begin{vmatrix} 25 & 15 & 0 \\ 15 & -5 & 0 \\ 0 & 0 & 5 \end{vmatrix} \text{ ksi} \qquad 2 \quad \sigma_{ij} = \begin{vmatrix} 150 & 90 & 0 \\ 90 & -30 & 0 \\ 0 & 0 & 90 \end{vmatrix} \text{ MPa}$$

3. In the movie Avatar, the greedy corporation masquerading as a human being, aka Parker Selfridge wants to rid the planet Pandora of all leaf-bearing pests and the "tree huggers" that call them home. Unfortunately, Parker may have poorly designed the closed-end, cylindrical pressure vessel for his monster "leveler" that will do the work. Help spread Human Glory by using **MSST** to determine if the vessel will yield based on for internal pressures of: (a) 1,500 psi, (b) 2,000 psi, and (c) 2,500 psi. Assume a mean diameter of 8.0 inches, a wall thickness of 0.2 inches, and a tensile yield strength $\sigma_y = 40{,}000$ psi (A1S1 303 stainless steel).

4. A critical component of the "Death Copter" is a solid, circular synchronizing shaft that is 4.0 inches in diameter and subjected to a bending moment of M=120,000 in-lb. Unfortunately, a twisting moment, T of unknown magnitude was experienced by the shaft, causing it to permanently deform beyond allowable tolerances, malfunction, and the crew to learn that gravity is an ever-present force in their lives. Using this lovely imagery, determine the twisting moment, T that will cause the shaft to yield, according to: (a) the Maximum Shear Stress Theory and (b) the Distortion Energy Theory? Note: the yield strength in tension for the shaft material is $\sigma_y = 40{,}000$ psi.

5. Even though a closed-end, tubular vessel for propane storage was fabricated from an alloy steel with a yield strength of σ_y=600 MPa, it was found to have permanently deformed while sitting at Hank Hill's store. If the mean diameter of the vessel is 10.0 cm, what wall thickness (t) would have prevented yielding if the vessel is internally pressurized to 45 MPa?

6. A thin-walled pipe of circular cross-section used for hydronic heating systems in newer homes has an internal diameter of 5.0 inches and a wall thickness of 0.2 inches. The pipe is made of a low-carbon steel with a yield strength of $\sigma_y = 30{,}000$ psi and carries a flowing fluid at 500 psi pressure. Using the Distortion Energy Theory, determine the magnitude of the twisting moment that would cause the pipe to yield.

7. Draw the load path in a two-dimensional principal stress space for the states of plane-stress given below. Determine the unknown stress parameter σ_{II} in each case as predicted by the Maximum Shear Stress Theory for a material with a tensile yield strength of σ_y=700 MPa. Note: In both cases (a) and (b), the unknown σ_{II} is positive.

$$\text{a.} \quad \begin{vmatrix} 3\sigma_{II} & 0 & 0 \\ 0 & \sigma_{II} & 0 \\ 0 & 0 & 0 \end{vmatrix} \text{ MPa} \qquad \text{b.} \quad \begin{vmatrix} -250 & 0 & 0 \\ 0 & \sigma_{II} & 0 \\ 0 & 0 & 0 \end{vmatrix} \text{ MPa}$$

8. The following strain state was determined by direct measurement at a critical point in a cam shaft using a 3-gauge, 45-degree electric resistance strain gauge rosette. If the material to be used for final construction is one of several steel alloys, all having E = 30 x 10^6 psi and ν = 0.32, find the minimum tensile yield strength required of the steel based on the Maximum Shear Stress Theory.

$$\varepsilon_{ij} = \begin{vmatrix} 1.8 & 0.6 & 0 \\ 0.6 & -0.3 & 0 \\ 0 & 0 & 0 \end{vmatrix} \text{x}10^{-3}\ \text{Psi}$$

9. The Porsche brake pedal of Problem 2.9 is to conform to the sketch below, and is to be cast of an aluminum alloy. For design purposes, assume that a maximum load of 300 lbs can be applied to the center of the pedal pad by the operator, and that the shank is the weakest part of the total pedal structure. The shank has a circular cross-section, whose radius varies linearly as one moves from the pedal pad end towards the hub end. What must the minimum yield strength of the cast alloy be, according to the Distortion Energy Theory (DET) if the design must observe a safety factor of 2.0? Hint: examine yielding only at the two cross-sections indicated and apply the safety factor after predicting yield strength via the DET.

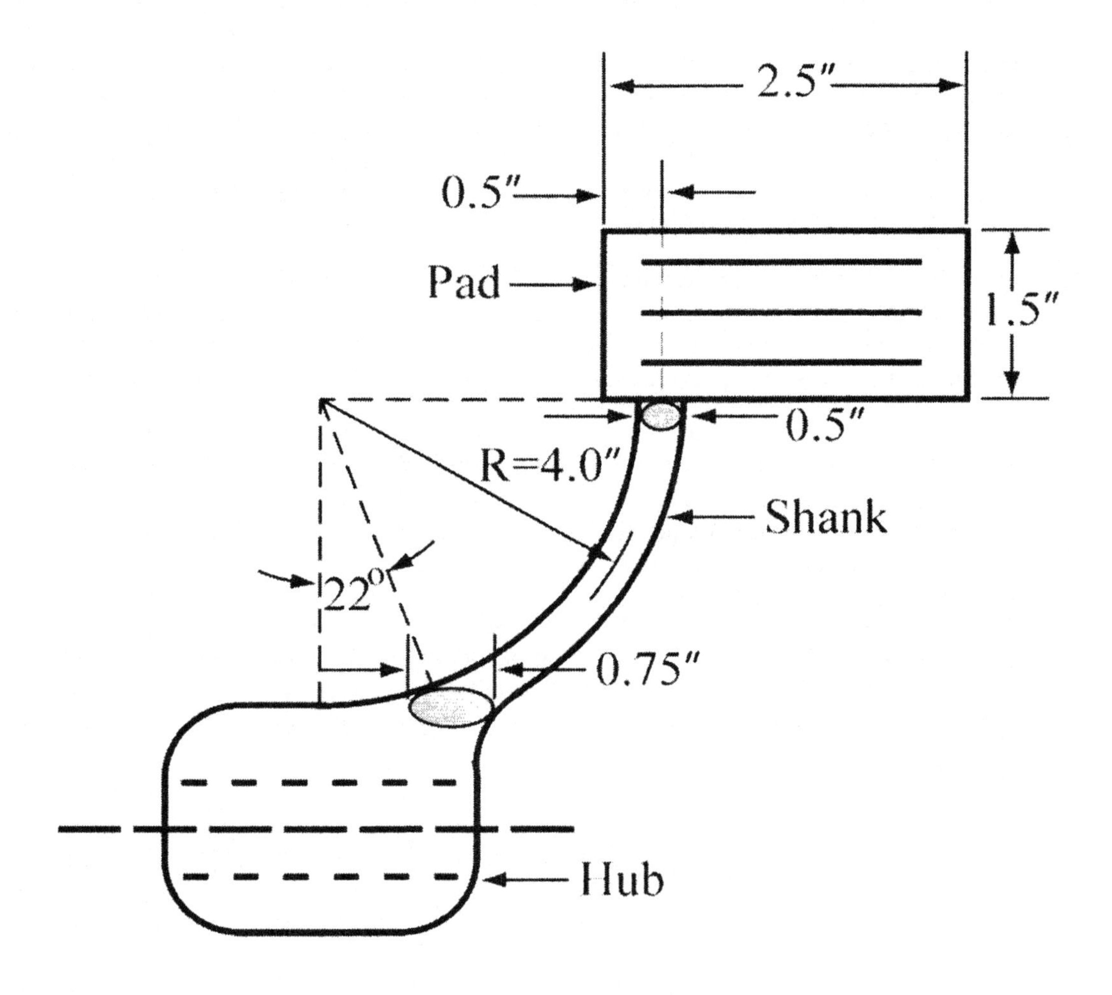

CHAPTER SIX

PLASTIC DEFORMATION AND DUCTILE FAILURE

6.1 INTRODUCTION

For engineering and its natural extension to design, plastic deformations and the structural loadings that lead to them are clearly matters of concern in both good and bad contexts. For instance, the forming of useful shapes through the plastic deformation of cast and wrought metallic ingots is an ancient technology that still is used today. Only now, modern analysis techniques allow the process engineer to accurately predict the necessary forming stresses, to design and power the required equipment, and to predict the mechanical response of finished items. In turn, the finished products may be stock material to be utilized in further deformation processing, such as sheet metal stamped into automobile bodies, or finished items such as structural I-beams. Moreover, designing crashworthy vehicles demands a thorough knowledge of the plastic response and energy absorption of various materials including metals, polymers, composites, and their utilization into foams or similar structures. On the other hand, excessive plastic deformation and the loss of critical tolerances, not to mention ductile failures are always design concerns; their prediction and avoidance deserves to be handled in a rational, rigorous, and repeatable manner. Given these important needs, the purpose of this chapter is to introduce analysis methodologies relevant to plastic design and related issues of ductile failure control.

6.2 INSTANTANEOUS (TRUE) STRAIN

Before we delve into the analysis techniques involving plastic deformation, it is useful to take another look at our definitions for stress and strain since some adjustments are warranted. For example, consider the following application of nominal strain measures in describing the large plastic deformation of a potentially failing bridge bolt. At the time of observation, the bolt with an original "*gauge*" length of L_o=2.0 inches, has now stretched to 2.8 inches. As shown in earlier chapters, the nominal strain is easily computed as:

$$\varepsilon_{nom} \equiv \frac{\Delta L}{L_o} = \frac{0.8}{2.0} = 0.40 \tag{6.1}$$

Imagine now that the deformation is considered in two steps instead of one, with the total nominal strain calculated in a piecewise fashion from two distinct observations. If each step adds 0.4 inches (for a total of 0.8 inches), the calculated strain would then be:

$$\varepsilon_{nom} = \frac{\Delta L_1}{L_1} + \frac{\Delta L_2}{L_2} = \frac{0.40}{2.0} + \frac{0.40}{2.40} = 0.367 \tag{6.2}$$

Although the final length of the strained member is still 2.8 inches in both cases, the use of the *nominal* measure, with a fixed gauge-length leads to a slightly different answer. Clearly, this disparity would not be encountered in the analysis if the strains were small ($\varepsilon < 10^{-2}$) as they would be strictly elastic. However, for the larger strains typical of *plastic* deformation, a more consistent strain measure might prove to be one that adjusts the gauge length instantaneously as deformation progresses. A definition of an *instantaneous* strain ε_i, in differential form would then be:

$$d\varepsilon_i \equiv \frac{dL}{L} \tag{6.3}$$

Here, L is now the instantaneous strain gauge length to which an increment of deformation dL is added, producing a strain increment $d\varepsilon_i$. If deformation proceeds from an initial or original gauge length, L_o to an arbitrary instantaneous length, L_i that is less than or equal to the gauge length at fracture, the total instantaneous strain or ε_t can then be described as:

$$\varepsilon_t = \int d\varepsilon_i = \int_{L_o}^{L_i} \frac{dL}{L} = \ln L_i - \ln L_o = \ln \frac{L_i}{L_o} \tag{6.4}$$

Returning to the numerical experiment described by Equations (6.1 and 6.2), the computed true strain becomes:

$$\varepsilon_t = \ln \frac{2.8}{2.0} = 0.336 \tag{6.5a}$$

For the two-step deformation process used earlier, the calculation is:

$$\varepsilon_t = \ln \frac{2.4}{2.0} + \ln \frac{2.8}{2.4} = 0.182 + 0.154 = 0.336 \tag{6.5b}$$

In either case, the final or end-point strain now has the same numerical value, thus removing the difficulty encountered when using nominal strains. However, it should be noted that it is just a coincidence that only two steps were required to converge to the answer in this example.

Coincidences aside, it is also important that the reader does not assume that there is anything "untrue" about nominal strain values. In an engineering sense, the distinction is that "*true*" measures can be employed in the entire elastic-to-plastic range, whereas *nominal* measures are only good in the elastic range of stressing. If this seems somewhat contradictory, try comparing true versus nominal values in the elastic range and you will find that there is no appreciable difference. On the other hand and as shown by the example above, significant differences will develop once the deformation is inelastic; the emerging difference between the two values can be seen in Equation (6.6):

$$\varepsilon_t = \ln \frac{L_i}{L_o} = \ln\left(\frac{L_i}{L_o} - \frac{L_o}{L_o} + 1\right) = \ln\left(\frac{L_i - L_o}{L_o} + 1\right) = \ln\left(\varepsilon_{nom} + 1\right) \tag{6.6}$$

Although somewhat more difficult to do experimentally, true strain values may also be computed from the changes observed in cross-sectional area during plastic straining (tensile or compressive). Serendipitously, it has been experimentally observed that plastic straining generates no volume change during such testing. Thus, in the plastic region at any instantaneous value of length (L_i) and cross-sectional area (A_i), one can always assume that the volume of the gauge section is unchanged relative to the original value. Mathematically, this observations leads to a simple, but very useful relationship in that:

$$L_o A_o = L_i A_i \tag{6.7}$$

where the subscripts "o" and "i" still refer to original (or nominal) and instantaneous, respectively.

In the plastic range, the true strain as espoused by Equation (6.4) can thus be alternatively computed as:

$$\varepsilon_t = \ln \frac{A_o}{A_i} \tag{6.8}$$

Taking the differential of both sides of Equation (6.8) as will prove useful later, one can also obtain:

$$d\varepsilon_t = -\frac{dA}{A} \quad \text{or} \quad d\varepsilon_t = \frac{dL}{L} \tag{6.9}$$

6.3 INSTANTANEOUS OR TRUE STRESS

During plastic straining where significant changes in the original dimensions occur, similar arguments can be made about the nature of the stresses because they must also reflect the true area of the supporting cross-section. Therefore, an instantaneous σ_i, or *true*, σ_t, normal stress may be simply defined as:

$$\sigma_i = \sigma_t = \frac{P}{A_i} \tag{6.10}$$

Here, A_i is the *instantaneous area* corresponding to the applied load *P*. In the elastic range of stressing, the true stress will be the same as that calculated using the nominal stress definition. However, in the plastic range, the true stress will always be greater than its nominal counterpart.

6.4 STRAIN HARDENING IN TENSION

While the definitions just discussed are all well and good, they are not necessary practical when investigating the elastic-plastic response of an engineering material. Fortunately, the tensile response of most ductile metals can be shown to follow a relatively simple parabolic law relating true-stress and true-strain values:

$$\sigma_t = \sigma_o \varepsilon_t^n \tag{6.11}$$

However, it should be noted that the accuracy of Equation (6.11) is considered adequate only in the plastic range of deformation; as mentioned earlier, the elastic range is accurately (and more conveniently) correlated by Hooke's Law employing nominal stress and strain measures. In Equation (6.11), the material constant, σ_o is referred to as the *strength coefficient.* Although its algebraic position is Equation (6.11) is analogous to the Young's modulus in Hooke's Law complete with units of stress, it is definitely not the same and may in fact be orders of magnitude smaller than *E*. On the other hand, the constant, *n* is known as the *Strain Hardening Exponent* and simply represents the rate of rise of the parabolic stress strain curve. Hypothetical parabolic response curves are shown in Figure 6.1 for different (and assumed) values of the exponent, *n*.

Clearly, the situation in Figure 6.1 where $n < 1.0$ is the only sensible case depicted, as it will lead to monotonically increasing response curves. Taking the Base-Ten logarithm of both sides of Equation (6.11), one obtains:

$$\log_{10} \sigma_t = \log_{10} \sigma_o + n \log_{10} \varepsilon_t \tag{6.12}$$

In logarithmic space, Equation (6.12) is linear in form; hence, the slope of the logarithmic response line is *n*, and σ_o is the true stress at an admittedly arbitrary determined $\varepsilon_t = 1.0$.

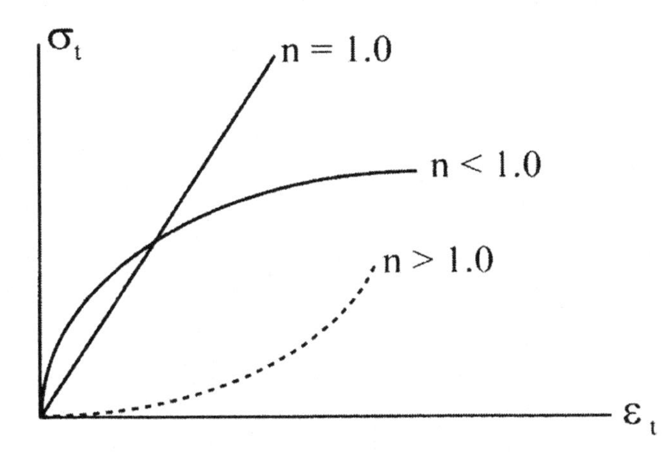

Figure 6.1 True stress and strain tensile response for differing values of n.

Example Problem 6.1: Figure 6.2 shows true-stress/strain data for a 6061-T6 aluminum alloy destined for use as light-weight tachometer gears. Using this data, determine the strain hardening exponent and the strength coefficient.

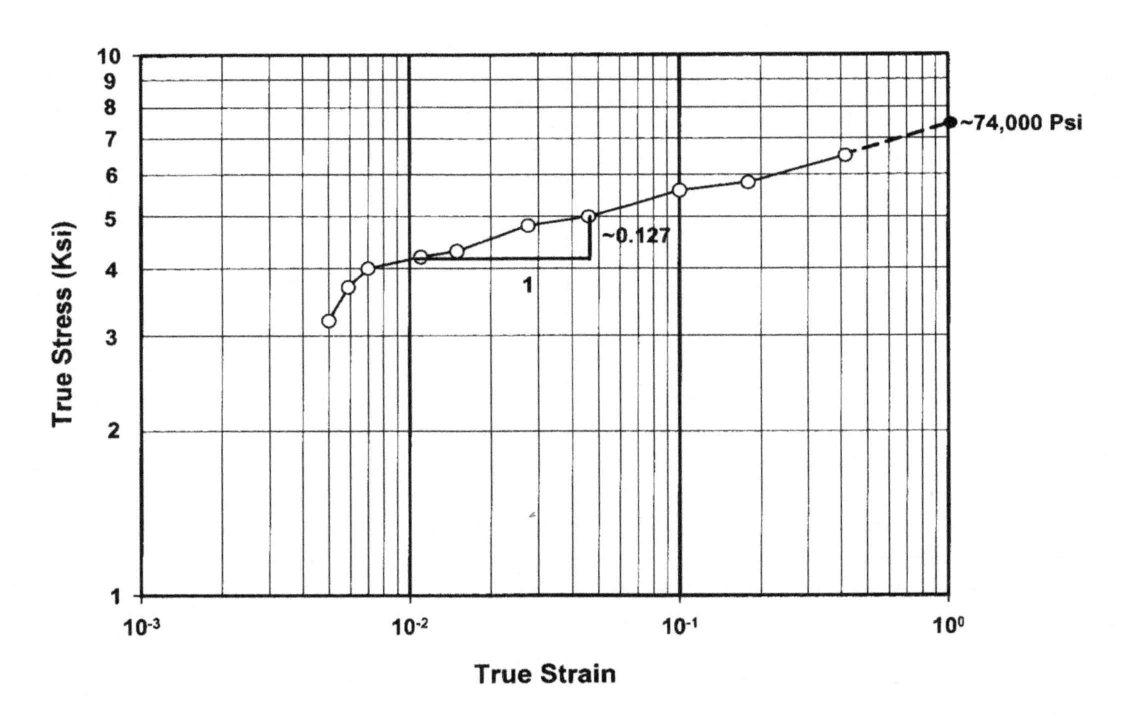

Figure 6.2 True-stress and -strain tensile response data for 6061-T6 aluminum alloy.

A review of the curve indicates that the parabolic law appears to be satisfied (linear in log space) for strains in excess of 7×10^{-3}; below that value, the data tend toward the elastic slope of n = 1.0 and the assumption of a parabolic response is no longer invalid. The fitted line above strains of 7×10^{-3} has a measured slope of n = 0.127. Things now get a little interesting in that the alloy apparently fractured at a true strain of approximately 0.43, a value well below the $\varepsilon_t = 1.0$ required to obtain σ_o. Fortunately, the parabolic law does not recognize fracture in any real mathematical sense, so the straight-line response can be simply extrapolated to a strain value of 1.0. Using this extrapolation, it can be shown that σ_o= 74,000 psi. Thus for the sample of 6061-T6 aluminum alloy tested, the following stress-strain relationship is valid: $\sigma_t = 74,000\varepsilon_t^{0.127}$

6.5 LOAD INSTABILITY IN TENSION

Although Equation (6.11) successfully correlates true-stress and -strain, it contains no information on the failure of the tensile member. For a tensile response described in nominal stress and strain measures, specimen instability occurs at the maximum load point usually denoted as the *tensile strength* or *ultimate tensile strength*. Physically, ductile materials do not really exhibit load instabilities in tension. Instead, the true-stress continues to increase even as the post instability load decreases because the cross-sectional area is shrinking at a faster rate. A measure of instability based on true stress and strain concepts can therefore be found, by examining changes in stress as the maximum load is reached. For brevity in the following discussion, all stress, strain, and geometrical quantities will be starred (e.g., $P_{max} = P^*$) at the *maximum load point.*

To begin with, we will first look at the load stress relationships in differential form:

$$dP = d(\sigma_t A_i) = \sigma_t dA_i + A_i d\sigma_t \tag{6.13}$$

However, at $P = P_{max} = P^*$, the slope goes to zero as it is a maxima and $dP \equiv 0$. Using Equation (6.13):

$$\frac{d\sigma_t^*}{\sigma_t^*} = -\frac{dA_i^*}{A_i^*} \tag{6.14}$$

From the area-correlated definition of true strain:

$$d\varepsilon_t^* = -\frac{dA_i^*}{A_i^*} \tag{6.15}$$

Combining Equations (6.14) and (6.15), one obtains:

$$\frac{d\sigma_t^*}{d\varepsilon_t^*} = \sigma_t^* \tag{6.16a}$$

At the maximum load point, $\sigma_t^* = \sigma_o \varepsilon_t^{*n}$ so when substituted into Equation (6.16a), one gets:

$$\frac{d\sigma_t^*}{d\varepsilon_t^*} = n\,\sigma_o \varepsilon_t^{*(n-1)} = \sigma_o \varepsilon_t^{*n} \tag{6.16b}$$

Finally, after canceling like terms, the true strain corresponding to load instability is shown to be nothing more than the exponent, *n*.

$$\varepsilon_t^* = n \tag{6.17}$$

However, and this is another "biggie," Equation (6.17) is only valid at the maximum load or point of instability. The true stress at load instability is then easily found via Equation (6.11) to be:

$$\sigma_t^* = \sigma_o \varepsilon_t^{*n} = \sigma_o n^n \tag{6.18}$$

Example Problem 6.2: Calculate the (nominal) tensile strength of the aluminum alloy 6061-T6 found in Example Problem 6.1, using the parabolic strain hardening law. You may fondly recall that from Example Problem 6.1, it was found that:

$$\sigma = 74,000\varepsilon^{0.127}$$

At the maximum load point or ultimate tensile strength (instability), Equation (6.17) tells us that:

$$\varepsilon_t = n = 0.127$$

Substituting this relationship back into the stress-strain equation:

$$\sigma_t^* = 74,000(0.127)^{0.127} = 56,940 \text{ psi}$$

This value represents the true tensile or ultimate strength of the material. If the nominal tensile strength, σ_{uts} is required, it should reflect the original area, A_o that will always be larger than the true area, A_i, at least when tension is applied. Although we do not know the original or true areas, the product of the stress and area ($\sigma_t A_i$ or $\sigma_{nom} A_o$) will always equal the load, P at any point during the test including P^* at the ultimate strength. As such,

$$56,940A_i^* = \sigma_{uts}A_o \quad \rightarrow \quad \sigma_{uts} = 56,940\frac{A_i^*}{A_o}$$

While we still do not know the areas, we do know something about the strain at the highest load, P^* because it is the point of instability and the strain is simply the exponent, n:

$$\varepsilon_t^* = \ln\frac{A_o}{A_i^*} = n$$

Hence,

$$\frac{A_i^*}{A_o} = e^{-n} = e^{-0.127} = 0.88$$

Finally,

$$\sigma_{uts} = 0.88 \text{ x } 56,940 = 50,107 \text{ psi}$$

As would be expected, the nominal tensile strength is "lower" than the true value since the nominal or original area is greater or $A_o \geq A_i$.

PRACTICE EXERCISES

1. Load, elongation, and the resulting instantaneous diameter data are given below for two ductile alloys used for ammunition (A) and mountain bike rims (B). Using this data, please do the following:

   i. Plot three stress-strain curves for each material showing nominal and true values using both ln(L/Lo) and ln(A_o/A) true strain measures. Why are the three curves different for each material?

   ii. Plot true stress-true strain data on log-log paper, and evaluate the material parameters σ_o and n for each alloy. Use both true strain measures.

   iii. Does ε_t = n, at P = P_{max}, compare well with the actual observed value? Does it matter if you use the length or area based true strain relationship and if so, which one is better?

   A. Cartridge Brass (70% Cu – 30% Zn), Lo = 1.0 inch, Do = 0.249 inches.

P (lb)	ΔL (inch)	D (inch)
2,250	0.015	0.247
2,370	0.035	0.244
2,450	0.055	0.242
2,500	0.08	0.239
2,550	0.10	0.236
2,570	0.12	0.234
2,350	0.20	0.195
2,250	0.21	0.185

   B. 304 Stainless Steel, Lo = 1.0 inch, Do = 0.249 inches.

P (lb)	ΔL (inch)	D (inch)
4,930	0.02	0.245
5,060	0.045	0.241
5,180	0.08	0.236
5,300	0.125	0.231
5,380	0.175	0.225
5,450	0.245	0.217
5,460	0.320	0.205
5,350	0.360	0.191
4,750	0.405	0.165

2. A forensic failure analysis has revealed that the mild structural steel under scrutiny plastically deformed and therefore exceeded design tolerances. In a tension test of the steel using a circular cross-section specimen and an original diameter of D = 0.36 inches, the following data points were found in the plastic range.

Load (lb)	Diameter (inch)
4,900	0.342
6,400	0.328

   Assuming that both data points fall on the straight line in a log-log plot of all true stress and strain values, compute σ_o and *n*.

3. Its Nerd/I-love-math time so embrace the fun and determine the relationship between the percent reduction in area (a nominal ductility measure) and the true strain at fracture?

4. A tensile member of 2024-0 aluminum alloy used for aircraft fuselages had an original cross-sectional area of 4.0 square inches. If the strain hardening exponent, n = 0.21 and the strength coefficient σ_o = 56,000 psi, determine the following:

   a. The maximum load that can be carried by the member.
   b. The change in length of the member assuming an original length of 4.0 feet.

5. An iron-nickel alloy used for high temperature turbine blades has the following plastic deformation parameters shown below.

$$\sigma_o = 90{,}000 \text{ psi and } n = 0.15$$

   Using this data, find

   a. The ultimate tensile strength.
   b. The percent reduction in area if the true strain at fracture is 0.25.

CHAPTER SEVEN

BRITTLE AND LOW DUCTILITY FAILURE ANALYSIS

7.1 INTRODUCTION

Up to this point, we have primarily discussed the behavior of ductile materials capable of relatively large deformations and plasticity. Thus, if service flaws (dislocations, cracks, scratches, pitting, etc.) were to be inherent and/or develop due to the fabrication or usage, the high stresses near these irregularities can be relieved by plastic flow. However, many applications require materials possessing little or no apparent ductility, especially now with energy efficiency driving operational temperatures up. In fact, brittle materials often possess many desirable properties, even though they lack the "forgiving" ductility of common metallic alloys. For instance, ceramics may exhibit high hardness, oxidation resistance, desirable elevated temperature capabilities, and/or passivity toward aggressive environments that make them very attractive to high temperature, wear, and/or corrosion situations. On the other hand, extremely brittle glassy solids may be required due to their transparency to visible light.

While the discussion has so far been centered on the "classic" brittle materials such as ceramics and glass, there is also a trend toward using high-strength forms of metallic alloys in applications where load-bearing efficiency or high-temperature capabilities is desirable. As would be expected, there is a penalty in that raising the yield-strength of metals usually results in a loss of ductility. In these materials including the so-called ceramets, the operational stresses that become concentrated near any flaws will not be relieved by plastic deformation and catastrophic failure may result. Hence, the high-strength materials will ultimately fracture and not have the same leeway as their more ductile counterparts in tolerating damage before failing. Given this tendency to fail in a less graceful way (a nice way of saying catastrophic), it is essential that the designer be aware of the differences between ductile and brittle behaviors, and more importantly, be able to ensure safe designs.

7.2 FRACTURE OF PERFECTLY BRITTLE MATERIALS

The concept of a perfectly brittle solid is in reality, an engineering approximation where the solid fails while stressed exclusively in the elastically region. For such materials subjected to uniaxial tension, the *ultimate tensile strength* (σ_{uts}, *UTS*, or *TS*) occurs as the final achievable point on the elastic stress-strain curve as already discussed in previous chapters. Solids such as window glass or household ceramics including the venerable toilet bowl come to mind as obvious examples of these materials. Yet (and in spite of our common perceptions), most of these materials do actually exhibit some form of local irreversible deformation that might be termed plastic, at least at the microstructural level. For instance, this is particularly true for compressive loads applied over a very small portion of the material's surface as in the hardness indentations discussed in Chapter Four. Nonetheless, attempts to measure permanent deformations on a larger scale, or even invent a failure theory for brittle materials akin to DET or MSST have not proven successful. As a result, or perhaps merely a surrender to practical reality, the engineering approximation of these materials as acting in a purely brittle fashion is clearly reasonable from a design perspective.

Because the behaviors of brittle materials are clearly different from their ductile counterparts, there was obviously a need for some form of comprehensive approach to fracture. The ancients who built very large structures from brittle stone no doubt recognized this importance, especially since failure would definitely tend to ruin the day of the buildings occupants. From a more modern perspective, it was not until around 1920, that Alan Griffith provided the first successful analysis to predict failure stresses in brittle materials. In his analysis, Griffith assumed that brittle materials contain flaws that although not necessarily visible, can concentrate stresses locally and catastrophically; for any given material, the flaws will propagate when the stresses reach some critical strength in a region adjacent to the flaw tip.

While certainly a reasonable assumption, evidence for the existence of these flaws is often indirect. For example, size-effect experiments have shown that flaws are indeed present in the material, even if not easily observed. One such exercise involved glass fibers, with both length and diameter (and thus surface area and volume) separately controlled. During these tests, it was consistently observed that the failure loads increased as specimen size decreased. As indicated by the data, the strength appeared to be inversely proportional to the size (area and/or volume) of the material. Based on these observations, it was astutely hypothesized that the smaller the material sample, the less the statistical likelihood that a large, strength-degrading flaw was present. To make things even more complicated, it was also observed that the strength would tend to degrade with time and/or after touching the samples with bare hands. Unbeknownst to Griffith and others at the time, atmospheric water vapor, as well as oil and dust from ones hands could attack fresh glass surfaces and produce or worsen surface flaws.

Although Griffith probably looked at a number of approaches to the problem, his analysis eventually considered an energy balance between competing processes occurring at the tip of a stressed crack or flaw. As shown in Figure 7.1, a surface crack of length, *a* has a volume element *dV* immediately adjacent to its tip. Letting σ symbolically represent all stress components that may be present and acting on the element, the elastic strain energy stored is:

$$dU = \frac{\sigma^2}{2E} dV \tag{7.1}$$

If the crack extends uniformly through the thickness an amount *da*, then the volume element cannot support the stress σ (it is cracked), and the energy of the structure is reduced by the amount ($\sigma^2/2E$ dV). Interestingly, this result would seem to indicate that all stressed cracks should spontaneously extend because the energy stored from any cracked structure would tend to be reduced. Because we do not usually observe structures crumbling into dust (a very good thing indeed), Equation (7.1) must not represent a complete accounting of the energy changes that take place as the crack extends. It was therefore necessary to invent an energy term that represents the ability of any materials to resist crack extension. Denoted herein as γ, this resistance is defined as the reversible, adiabatic energy necessary to create a unit area of new (crack) surface in the solid.

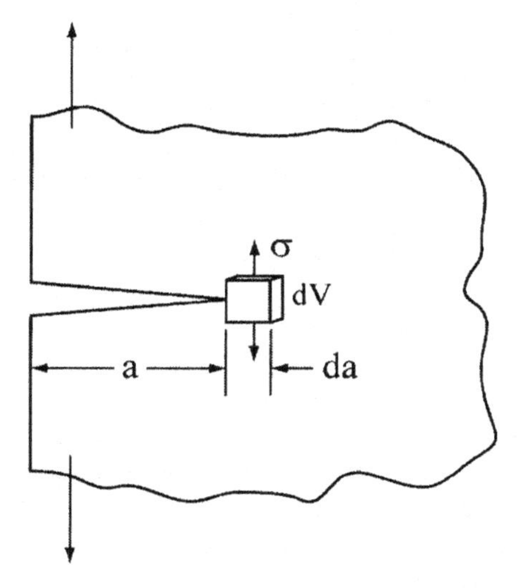

Figure 7.1 Stressed volume element adjacent to a crack tip.

As the crack lengthens by da, thus creating new surface area dA, the energy of the solid is increased by dU:

$$dU = \gamma \, dA \tag{7.2}$$

If the other boundaries of the structure carrying the applied forces and moments do not move (often referred to as the "fixed grip" condition), the total change in energy of the structural system becomes:

$$dU_{total} = \frac{\sigma^2}{2E} dV + \gamma dA \tag{7.3}$$

In the above relationship, the first term is the driving force for crack extension while the second is the resistance of the material to that extension. If $dU_{total} \leq 0$, the energy of this structure will be reduced by the crack extension such that:

$$\frac{\sigma^2}{2E} dV + \gamma dA \leq 0 \tag{7.4a}$$

Hence, Equation (7.4) represents the instability condition for the structure. In terms of analyzing a cracked structure for load-bearing safety, Equation (7.4) must be used to determine if the crack will propagate and lead to catastrophic failure. However, it is worth noting that the balance given by the equations does not allow for other forms of energy dissipation such as acoustic emissions, heat generation, and/or plastic flow.

Example Problem 7.1: An armor plate carries a uniform tensile stress, σ as shown in Figure 7.2. If an internal, penny-shaped flaw of radius, r and thickness, t is introduced into the plate as shown in the figure, under what conditions will the flaw extend?

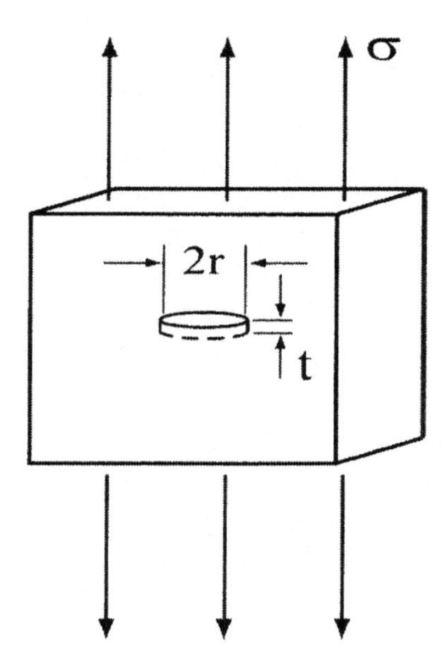

Figure 7.2 "Penny"-shaped crack or flaw in a stressed plate.

Before starting the fracture analysis, we must first do some preliminary calculations pertaining to the flaw volume and surface area. The volume, V of the stressed material "lost" by flaw will be simply $V = -\pi r^2 t$ while the surface Area, A of the flaw (top and bottom surfaces, but neglecting periphery) is $A = 2\ \pi r^2$. As the crack extends radially, the differential forms of both surface area and volume become: $dA = (4\pi r)dr$ and $V = (-2\pi rt)dr$, respectively. Applying Equation (7.4) with the minimum condition for instability being a net energy change of zero and the critical stress denoted as σ_c:

$$\frac{\sigma_c^2}{2E}(-2\pi r t dr) + \gamma(4\pi r dr) = 0 \qquad (7.4b)$$

Solving for the critical stress σ_c:

$$\sigma_c = \sqrt{\frac{4E\gamma}{t}} \qquad (7.5a)$$

In reality, the stress-state surrounding the flaw is in neither simple tension as assumed, nor the analysis/result rigorous in any sense. Nonetheless, it does present a relatively straightforward (and first of its kind) way for estimating a design stress when fracture is considered. In a similar vein, Griffith applied the energy balance via Equation (7.4) to the case of an elliptical through crack centered in a tensile plate (crack length 2a), as well as (equivalently) to an edge crack in a tensile plate (crack length a). For a crack geometry that is a sharp ellipse and relatively small compared to the plate, Griffith found:

$$\sigma_c = \sqrt{\frac{2E\gamma}{\pi a}} \quad a \approx \text{ plate thickness} \qquad (7.5b)$$

$$\sigma_c = \sqrt{\frac{2E\gamma}{\pi a(1-\nu^2)}} \quad a << \text{plate thickness} \qquad (7.5c)$$

where ν is Poisson's ratio as defined in earlier chapters. From a practical perspective, both equations (7.5 a or b) are virtually numerically indistinguishable, differing only by about 5% for typical Poisson ratio values such as $\nu = 0.3$. Given this similarity in results, Equation (7.5b) will be primarily used herein for all cases as long as the flaw or crack is small compared to the cross-section. Finally, it is worth noting that one way to determine properties such as γ would be to use Equation (7.5b) along with a tensile test of a plate (to fracture) after a sharp edge crack of known length was artificially introduced. More accurate and convenient methods for determining fracture resistance are discussed later in this chapter.

7.3 THEORETICAL COHESIVE STRENGTH

The *theoretical cohesive strength* of a material is defined as the tensile strength in a completely flaw-free condition. Using this definition, the theoretical cohesive strength can be readily estimated by applying Equation (7.5) in a limiting case, such as where the flaw is simply a missing atom. Strictly speaking, Equation (7.5) is derived for a featureless continuum, and its use in analyzing discrete atomic arrays will result in approximate answers at best.

Example Problem 7.2: Find the theoretical cohesive strength of fused silica (amorphous or glassy quartz) used for laboratory equipment and compare it to experimental data. Approximate numerical values of fused silica properties include E = 75 GPa, γ =1.91J/m^2, with the flaw size on the order of one atom or a = 3 x 10^{-10} m.

Using Equation (7.5a), the theoretical cohesive strength can then be determined:

$$TCS \cong \sqrt{\frac{2E\gamma}{\pi a}} \cong \sqrt{\frac{150\,x\,10^9\,(1.9)}{\pi\left(3\,x\,10^{-10}\right)}} = 17\ GPa$$

Experimentally, the tensile strength for fine silica fibers has been measured to be as high as 26 GPa, so the estimate is reasonable. Presumably, these specimens were free of non-atomistic size flaws, or flaws of such small magnitude are not as influential as suggested by the relationships. It is also important to note the closeness of the predicted and experimental cohesive-strength values relative to the elastic modulus, *E*. More precise theoretical cohesive strength predictions indicate that it is usually about one-tenth of the *Young's modulus* or ~ E/10 for many materials. Given these very high-strength numbers, it is easy to see the natural beauty and strength of spider's silk, not to mention the justification for utilizing fine fibers and filaments of high modulus materials in composites.

7.4 BRITTLE FAILURE UNDER COMBINED STRESS

In reality, and unfortunately for those of you who seek a simple life, tensile members are a rather small minority among all stressed components that one will experience in engineering practice. Therefore, it is necessary to apply the general technique of Equation (7.4) to cracked components subjected to multi-axial stresses. Griffith clearly recognized this need, and pursued the topic way back in 1924 by examining typical bi-axial elements not unlike Figure 7.3. As indicated by the figure, the sharp crack was positioned relative to the axes of any choice of principal stresses, such that a minimum magnitude would cause catastrophic crack extension. In other words, there might indeed be orientations that would not tend to extend the crack. Because it is also reasonable to assume natural flaw populations that are random in both size and orientation, such special cases will not be considered at this time. Thus, the most extreme and therefore, strength-robbing flaw orientations will be expected to be present. Using this as a basis for the model, the loci of brittle fracture points are indicated in Figure 7.4. As with the yield theories espoused in Chapter Five, the failure loci envelope of Figure 7.4 is used in the same fashion, except that failure via fracture is the result when the boundary is breached.

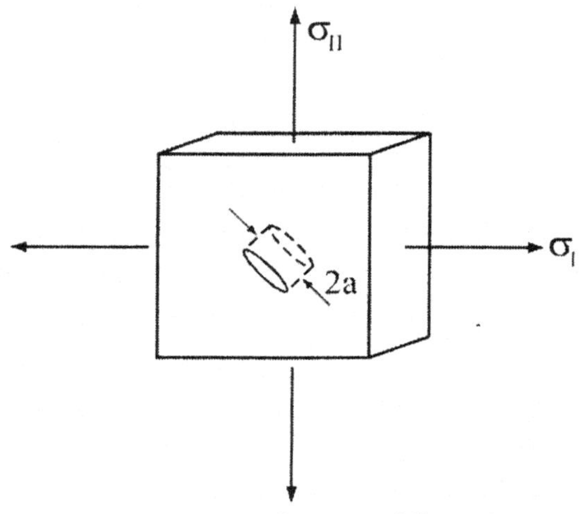

Figure 7.3 Plane stress element of flawed material.

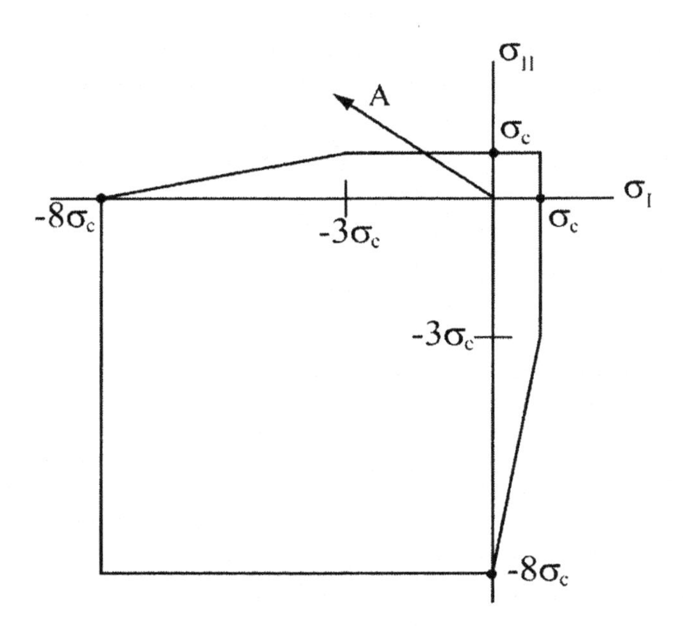

Figure 7.4 Plane stress brittle failure envelope (Griffith).

The intersection of the failure locus with the tensile axes, σ_c is the value predicted by Equation (7.5) with the crack length determined from the natural flaw population; the critical value, σ_c can be found by simply determining the Tensile Strength of the material. As shown in Figure (7.4), the fracture locus is a bit more fragmented than those seen earlier for cases involving yielding. For load paths such as the one designated as A in Figure (7.4) where $\sigma_I > 0$ and $-3\sigma_c \leq \sigma_I \leq \sigma_c$, the presence of σ_I has no effect and $\sigma_{II} = \sigma_c$ at failure; this corresponds to a maximum tensile stress theory of fracture. If so inclined, the reader can further examine the role tensile stresses play in fracture, as well as renew their love affair with Mohr's circle by twisting a piece of chalk and observing the fracture surface. When this torsional load, path "A" is rotated 45 degrees counterclockwise from the σ_{II}-axis, the resulting spiraled plane of fracture is the maximum tensile plane. For the curved region of the fracture locus or $-8\sigma_c \leq \sigma_I \leq -3\sigma_c$, both stress components contribute strain energy release to propagate the crack with failure predicted by:

$$(\sigma_I - \sigma_{II})^2 + 8\sigma_c(\sigma_I + \sigma_{II}) = 0 \tag{7.6}$$

Interestingly, the first term in Equation (7.6) with its difference in the principal stresses appears to indicate that shear is a controlling factor in fracture. Appearances aside, this is definitely not the case, especially since the second term involves a principal stress sum that has no simple or similar physical interpretation. On the other hand, the compressive strength influence via $-8\sigma_c$, is eight times as great as its tensile counterpart. This does indeed have significance, because it is a common observation that brittle solids are much stronger in compression than in tension. Great care must therefore be used in the design process to avoid confusing the two for reasons of experimental convenience. Finally, it should be noted that the Griffith failure criterion as depicted in Figure 7.4 does not hold for all brittle solids. While Figure 7.4 does accurately describes a good number of ceramics and glass, many materials fall outside its domain. In all likelihood, brittle materials not accurately described by its boundaries probably contain flaw distributions that cannot be approximated as sharp and flat ellipses.

Clearly, the shortcomings of Griffith's multi-axial failure criterion are at least three in number. Firstly, many brittle materials simply do not agree with its predictions, probably due to more complex flaw systems as just mentioned. In fact, many if not most strength controlling flaws and the maximum stress-states that drive them reside on the surface. Secondly, the approach does not lend itself readily to describe most structures containing service-induced flaws that are far more complex than the tensile plate example. The third and perhaps the most serious shortcoming is the constraint that the material in question must fail in a perfectly brittle manner; the inclusion of plastic energy dissipation in a balance such as Equation (7.4) is impractical and ultimately limits its usefulness. Given these shortcomings, a more practical theory of brittle behavior and fracture was required. Unfortunately, it was not until the 1950s that a more viable approach to the mechanics of fracture was developed.

7.5 MODES OF CRACK EXTENSION

Despite the apparent complexities of fracture, that there are only three basic and independent ways or modes that a crack can grow when under the influence of forces (internal and/or external). Referring to Figure 7.5, the crack can be extended either by displacing the crack faces perpendicular to the crack plane (*Mode I*), opening the crack faces perpendicular to the leading edge of the crack in shear (*Mode II*), or displacing the crack faces parallel to the leading edge of the crack (*Mode III*).

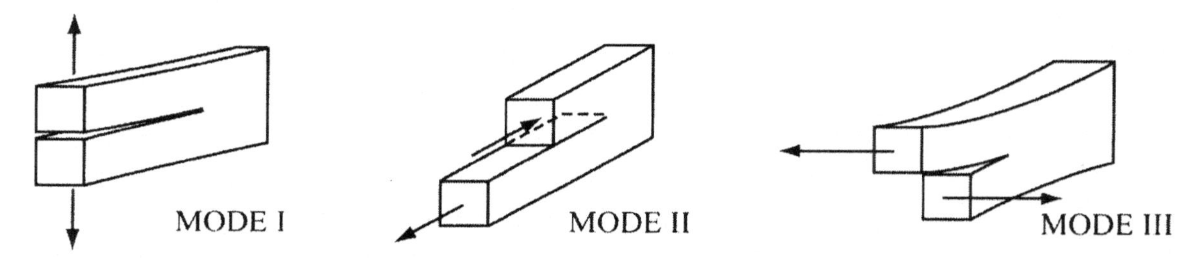

Figure 7.5 Three crack extension modes.

In reality, any given crack may have displacements imposed on its faces that are combinations of the three modes rather than the pure cases illustrated in Figure 7.5. Nevertheless, such structural problems are more easily solved by considering the three modes separately, and then superimposing the individual results to complete the analysis.

Most flaws or cracks of engineering concern reside on the surface and are either inherent, developed during manufacture (grinding, etc.), and/or enhanced during service via stresses or corrosion. It is quite common (in fact, the norm) for the application of cyclic stresses or deformations to directly lead to the development of cracks in what is known as the *fatigue degradation* of the structure (see Chapter Eight). Moreover, even static stresses imposed on structural members in the presence of aggressive environments such as water vapor or caustic pollutants may lead to a type of crack formation known collectively as *static fatigue*, *stress-corrosion cracking*, or *environment-aided cracking*. Conversely, but no less deleteriously, cyclic stresses combined with aggressive environments often lead to *stress-corrosion Fatigue Cracking*; this complex flaw initiation and propagation process may produce cracks of intolerable size faster than either of the two processes taken separately.

Regardless of the means for their generation and/or propagation, cracks tend to seek out planes normal to the directions of the maximum (principal) tensile stress, so most engineering studies (including the one you are presently enjoying) have been devoted to Mode I behaviors. Nevertheless, the general analysis methods and procedures, not to mention their application to the design process will be similar for the remaining modes; the reader is encouraged to check the fracture mechanics literature for guidance in dealing with such cases.

In order to begin an analysis of cracked structures using a more realistic approach, we must first envision a sharp crack of length, a entering a structure from its surface as shown in Figure 7.6. In this analysis, the crack surfaces are assumed to be elastically displaced in a Mode I fashion by whatever structural loadings are present in the system (not shown).

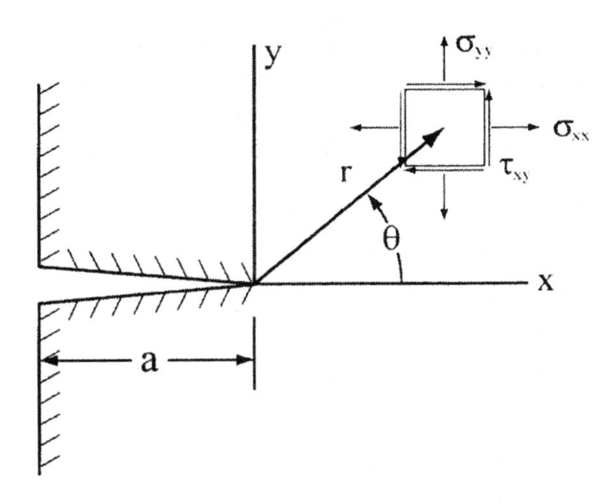

Figure 7.6 Stresses near the tip of a Mode I crack.

With reference to Figure 7.6, and following a very sophisticated analysis thankfully beyond the scope if this text, the components of stress in the plane surrounding the crack tip are defined as follows:

$$\sigma_{xx} = \frac{K_I}{\sqrt{2\pi r}} \cos\frac{\theta}{2}\left[1 + \sin\frac{\theta}{2}\sin\frac{3\theta}{2}\right] \tag{7.7a}$$

$$\sigma_{yy} = \frac{K_I}{\sqrt{2\pi r}} \cos\frac{\theta}{2}\left[1 - \sin\frac{\theta}{2}\sin\frac{3\theta}{2}\right] \tag{7.7b}$$

$$\tau_{xy} = \frac{K_I}{\sqrt{2\pi r}} \left[\sin\frac{\theta}{2}\cos\frac{\theta}{2}\cos\frac{3\theta}{2}\right] \tag{7.7c}$$

It can be seen in Equations (7.7) that the stresses predicted by the equations above are clearly singular (go to infinity) at the crack tip as $r \to 0$. Mathematically, our problems get worse in that the assumptions of stresses and strains being linearly related are physically unrealistic. Practically speaking, nonlinear behavior of one sort or another such as plastic deformation for some materials, must begin as $r \to 0$ because the local stresses can never achieve indefinitely large magnitudes.

Examining the y-component of normal stress, σ_{yy} important for a Mode I opening on the crack plane, one finds from Equation (7.7) that when $\theta = 0$:

$$\sigma_{yy} = \frac{K_I}{\sqrt{2\pi r}} \tag{7.8}$$

In this form, it becomes clear that σ_{yy} will become less and less significant as r becomes very large compared the crack length. Thus, Equations (7.7) only remains valid in a region near the crack tip.

Although relatively simple as relationships go, Equation (7.7) must hold for all Mode I loading. That is, the relationships must be valid regardless of the shape of the structure, the particulars of its loading, and/or the geometry of the crack save that it be flat and sharp. Given these constraints, the parameter, K_I must contain all of the missing details of loading and geometry. Furthermore, since K_I sets the actual magnitude of the stresses at any point r and θ, it has come to be known as the *stress intensity factor* or SIF. However, and this is another in the "biggie" class, the *stress intensity factor* should not be confused with the *stress concentration factors* discussed in Chapter Two, even though they unfortunately share very similar nomenclature. The SIF details the intensification of local stresses in the vicinity of a crack while K_σ relates the magnification of far-field stresses by some geometric change.

The derivation of the *stress intensity factors* for any given cracked and loaded structural member is a complex task requiring any number of the methods including, but certainly not limited to the mathematical theory of elasticity, numerical methods, as well as experimental stress-analyses techniques including photoelasticty. Given these difficulties, all examples used herein will simply be quoted from other readily available sources as the solutions demand a relatively sophisticated competence in mechanics beyond the scope of this book.

Example Problem 7.3: A historically significant stress intensity factor was obtained for a large tensile-plate with a sharp through crack in its center as shown in Figure 7.7a. By using the terms "large" to describe the plate, it must be orders of magnitude larger than the crack in all directions.

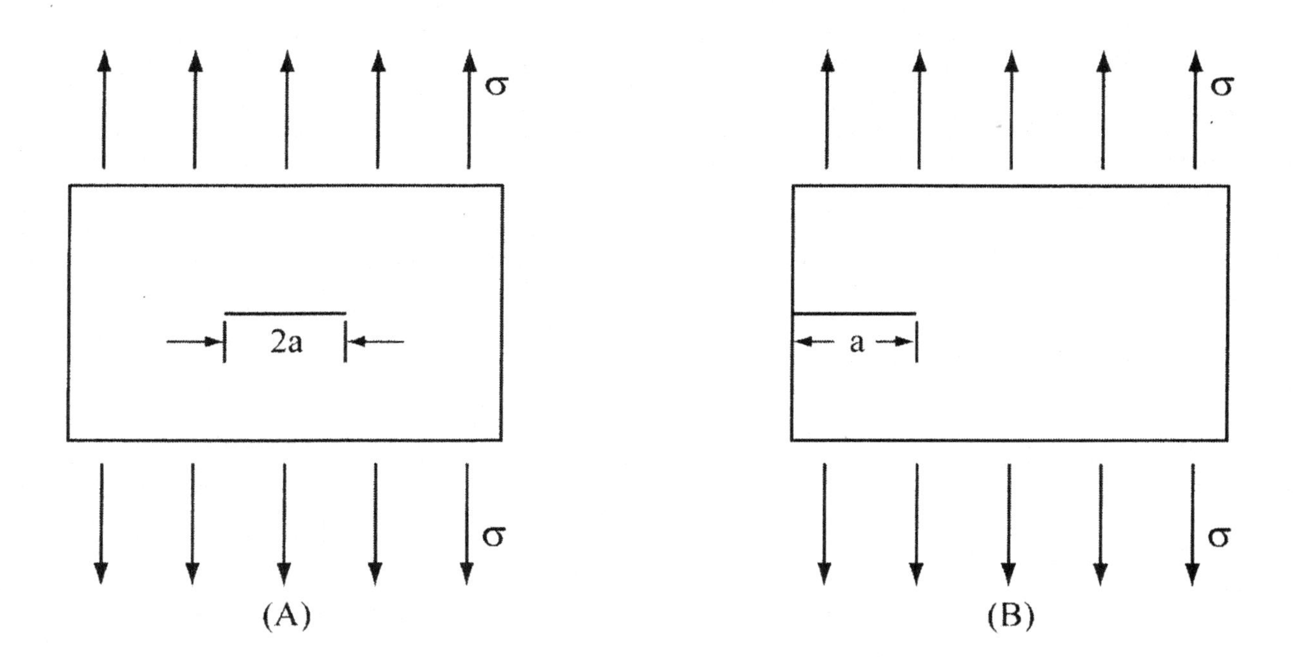

Figure 7.7 Central crack (A) and (B) edge crack geometries.

For the configuration shown in Figure (7.7A), the Mode I stress intensity factor or SIF has been shown to simply be:

$$K_I = \sigma\sqrt{\pi a} \tag{7.9a}$$

Clearly, this equation simply relates the stresses and flaw size without any additional terms save for π. Given this simplicity, Equation (7.9a) shall be considered herein the "Mother" of all SIF equations since it is the basis or starting point for every other configuration. Case in point: consider the same plate, but now with an edge-cracked through its thickness as shown in Figure (7.7b). For this geometry, the SIF takes on the following form:

$$K_I = 1.1\sigma\sqrt{\pi a} \tag{7.9b}$$

As indicated by the factor of 1.1, the SIF is 10 percent greater due to the increased compliance of the free-edge relative to the central crack that has the sides of the plate restricting the deformation.

Equation (7.9a) applies to the same structural configuration as studied by Griffith, and eventually leading to Equation (7.5). Due to the historical significance of Equation (7.9a), those who have truly "cracked up" and study this sort of thing often write the solution to K_I in a more general way:

$$K_I = Y\sigma\sqrt{\pi a} \tag{7.10}$$

Here, Y is a "correction factor" that renders the Equation (7.9) specific to a particular structural configuration; in some solutions, it is not uncommon to find the square-root of π included in the term Y. For the edge cracked plate discussed above, Y = 1.1 and is a relatively simple numerical correction. More generally, the correction factor Y (also called F, F_1, Φ, etc.) will prove to be very complex as it is dependent on crack size, loading, and/or structural dimensions. Fortunately, these solutions have been derived and tabulated for just about all useful configurations known to engineering, with a few selected examples and their inherent complexity given below.

Example Problem 7.4: The stress intensity factor solution for the circumferentially cracked, round tensile bar shown in Figure 7.8 is:

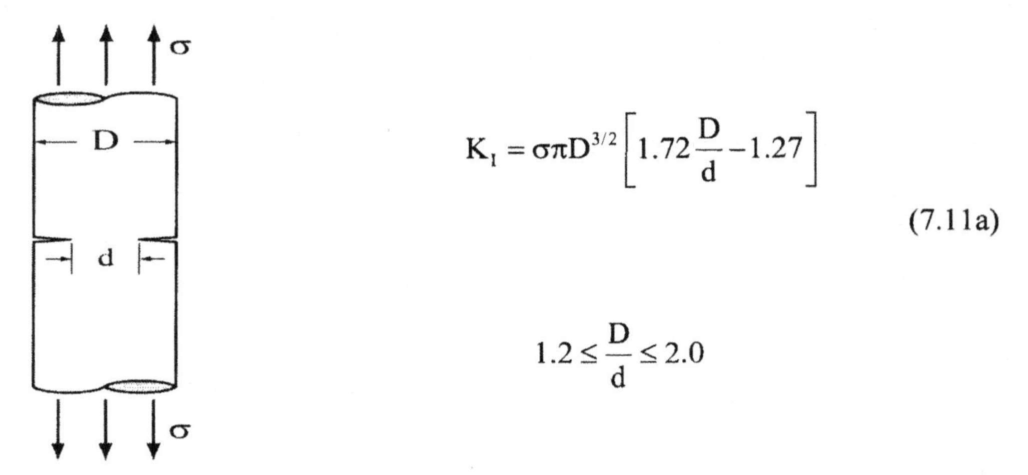

$$K_I = \sigma\pi D^{3/2}\left[1.72\frac{D}{d} - 1.27\right] \tag{7.11a}$$

$$1.2 \leq \frac{D}{d} \leq 2.0$$

Figure 7.8 Circumferentially cracked, round tensile bar.

Example Problem 7.5: Consider the stress intensity factor for the edge cracked beam (Figure 7.9):

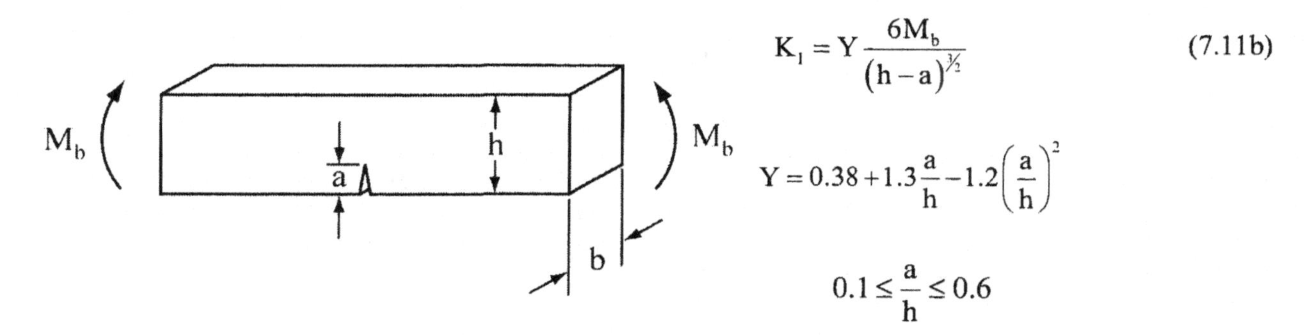

$$K_I = Y\frac{6M_b}{(h-a)^{3/2}} \tag{7.11b}$$

$$Y = 0.38 + 1.3\frac{a}{h} - 1.2\left(\frac{a}{h}\right)^2$$

$$0.1 \le \frac{a}{h} \le 0.6$$

Figure 7.9 Edge cracked beam under bending.

7.6 MATERIAL FRACTURE RESISTANCE

In addition to providing information on the magnitude of the stress field near sharp cracks, Equation (7.7) can also be used to describe the onset of crack Mode I instability and failure under increasing loads. It is not too difficult to visualize the rapid build up of the stresses at the crack-tip as increasing larger loads are applied to the flawed structure. Eventually, the material will not be able to tolerate further increases in the total state of stress around the crack and a catastrophic failure will ensue. As discussed earlier, the localized process leading up to failure at the crack tip can be monitored by the SIF, K_I; if the SIF helps describe the state of stress at the crack tip, then it must have an upper bound that is representative of the resistance of the material crack extension. Thus, a useful fracture-free criteria can emerge if:

$$K_I < K_{IC} \tag{7.12}$$

where K_{IC} represents the critical observed magnitude of K_I for which the crack is unstable. The use of K_I and K_{IC} is analogous to the use of stress and yield strength in that both K_I and the stresses driving it are parameters derived from applied loads and structural geometries while K_{IC} and yield strength are the material-dependent upper bound of the parameters. Although the just-defined critical stress identity factor or K_{IC} may sound like a new concept, it is commonly referred to as the *fracture toughness,* a name that may seem somewhat more familiar.

For perfectly brittle materials, all that is necessary is to test a cracked structural element such as a beam or plate for which a K_I solution is available. When the critical load is reached and failure at hand, $K_I = K_{IC}$. Although this is certainly the most commonly used method, alternative measures of the fracture resistance of perfectly brittle materials are also available. Consider the Griffith structural problem, with K_I given by Equation (7.9) repeated here for convenience.

$$K_I = \sigma\sqrt{\pi a} \tag{7.9}$$

Crack instability dictates that fracture will occur when the SIF reaches a critical value:

$$K_I \rightarrow K_{IC} = \sigma_C\sqrt{\pi a} \tag{7.13}$$

However, the energy-driven process Griffith originally analyzed leading up to Equation (7.5) indicates that:

$$\sigma_c\sqrt{\pi a} = \sqrt{2E\gamma} \qquad \text{(Plane Stress)} \qquad (7.14a)$$

$$\sigma_c\sqrt{\pi a} = \sqrt{\frac{2E\gamma}{1-\nu^2}} \qquad \text{(Plane Strain)} \qquad (7.14b)$$

The relationships between crack size and specimen thickness (plate) has been relabeled *plane stress* or *plane strain*; the thicker the specimen, the more the internal restraints set up a situation of triaxial stress and the accompanying strain limited to a single plane. By combining Equations (7.13) and (7.14), the fracture resistance parameters can then be related by the elastic constants:

$$K_{IC}^2 = 2E\gamma \qquad \text{(Plane Stress)} \qquad (7.15a)$$

$$K_{IC}^2 = \frac{2E\gamma}{1-\nu^2} \qquad \text{(Plane Strain)} \qquad (7.15b)$$

Considering typical values for Poisson's ratio, there is in fact little numerical difference between the two forms of Equations (7.15). To complicate matters further, another related fracture resistance parameter may be considered for perfectly brittle materials by reexamining the original Griffith crack instability condition:

$$\sigma_C = \sqrt{\frac{EG_C}{\pi a}} \qquad \text{(Plane Stress)} \qquad (7.16)$$

where G_C is defined as the *strain energy release rate* for the material; in this form, G_C refers to the released elastic strain-energy that will create unit-area of new crack surface. The factor 2.0 in Equations (7.5) has been absorbed in G_C because extension of the crack generates two surfaces per unit-area of crack extension. Comparing Equations (7.15) and (7.16), it can finally be surmised that:

$$K_{IC}^2 = EG_C \qquad \text{(Plane Stress)} \qquad (7.17a)$$

$$K_{IC}^2 = \frac{EG_C}{1-\nu^2} \qquad \text{(Plane Strain)} \qquad (7.17b)$$

7.7 HIGH-STRENGTH ALLOYS AND PLASTICITY ASPECTS

Metallurgists have clearly made significant strides over the past several decades in producing alloys with greatly increased yield strengths. Unfortunately, these gains in yield strength are usually accompanied by significant losses in ductility. In the presence of sharp cracks, high-strength alloys will often behave in a *quasi-brittle* fashion, that is, they appear ductile when un-notched, but fail while seeming to be elastically deformed when cracks are present.

As demonstrated earlier, the theoretical or maximum strength a material can exhibit is close in magnitude to E/10, with *E* still the elastic modulus. Using this order-of-magnitude estimate, it is therefore reasonable to scale a "high-strength" metallic alloy such that $\sigma_y > E/150$. Regardless of the strength of a material (brittle or ductile), it is likely that the yield strength will eventually be exceeded near the roots of any sharp flaws or cracks as working stress levels are increased. Under this scenario, and provided the *plastic zone radius* or r_p measured from the crack tip is small compared to the remaining elastic ligament, the stress field given by Equations (7.7) can be assumed valid outside of the plastic zone. The plastic zone in turn, can be located by moving through the elastic stress field toward the leading edge of the crack until the yield criterion is satisfied. As discussed in Chapter Five, the *distortion energy criterion* in nonprincipal form can be stated as:

$$(\sigma_x - \sigma_y)^2 + (\sigma_x - \sigma_z)^2 + (\sigma_y - \sigma_z)^2 + 6(\tau_{xy}^2 + \tau_{xz}^2 + \tau_{zy}^2) = 2\sigma_y^2 \qquad (5.21b)$$

By substituting the elastic stress field via Equations (7.7) for σ_{xx}, σ_{yy}, and τ_{xy}, recognizing that the remaining shear stresses and σ_{zz} are all zero, and finally solving for the *plastic zone radius*, r_p, one obtains:

$$r_p = \left(\frac{K_I^2}{2\pi\sigma_y^2}\right)\cos^2\frac{\theta}{2}\left(1 + 3\sin^2\frac{\theta}{2}\right) \qquad \text{(Plane Stress)} \qquad (7.18)$$

The size of the plastic zone immediately ahead of the crack in the xz-plane for $\theta = 0°$ then becomes:

$$r_p = \frac{K_I^2}{2\pi\sigma_y^2} \qquad \text{(Plane Stress)} \qquad (7.19)$$

For a plane strain condition where $\varepsilon_{zz} = 0$, there are stresses generated in the z-direction. As a result, there will also be a z-directed Poisson strain ahead of the crack for $\theta \approx 0^0$. However, as θ approaches 180 degrees, the normal stress components go to zero so the crack faces are essentially unstressed. Without σ_{xx} or σ_{yy}, there is no Poisson strain or displacement in the z-direction and the unstressed crack faces act as a displacement constraint near the crack tip, thus providing a physical basis for a plane strain state. Under this scenario with $\varepsilon_{zz} = 0$, then the stress in the z-direction becomes:

$$\sigma_{zz} = \nu(\sigma_{xx} + \sigma_{yy}) \qquad (7.20)$$

Resolving Equation (7.17) for the plastic zone radius under plane strain conditions, the following relationship evolves:

$$r_p = \frac{K_I^2}{2\pi\sigma_y^2}\cos^2\frac{\theta}{2}\left[(1-2\nu)^2 + 3\sin^2\frac{\theta}{2}\right] \qquad \text{(Plane Strain)} \qquad (7.21)$$

For a typical Poisson ratio of $\nu = 0.3$, the radius of the plastic zone ahead of the crack tip simplifies to:

$$r_p = 0.16\frac{K_I^2}{2\pi\sigma_y^2} \qquad \text{(Plane Strain)} \qquad (7.22)$$

A quick comparison between Equations (7.19 and 7.22) reveals that the size of the plane strain plastic zone is only 16 percent of the same quantity under plane stress conditions. As engineering fortunes would have it, this difference in plastic zone radii is truly critical in assessing the fracture resistance of a material.

In a ductile material, an about-to-propagate flaw must expend its strain energy not only to create new surface area, but also to induce plastic flow ahead of the crack. The greater the amount of plastic work that must be done in the solid, the more strain energy must also be released to propagate the crack. Thus, plane stress conditions imply more fracture resistance than plane strain for the same material. The million dollar question then becomes: how does one know which condition to choose since both conditions seem reasonably sensible? Fortunately, the answer is relatively easy (and yes, sensible) even though it comes from a somewhat indirect route. As one increases the thickness of any given fracture-testing specimen, the thicker crack faces become more of a constraint and tend towards plane strain conditions and a reduced fracture resistance. Figure 7.10 poignantly illustrates these tendencies as a function of specimen thickness for the fracture resistance of 7075-T6 aluminum alloy.

Interestingly, if one examines the fracture planes as the specimen thickness is increased and K_{IC} decreases, the appearance of the fracture surface changes. For relatively small thicknesses and high K_{IC} values, the fracture plane is slanted or beveled in a chisel-shape fashion at approximately 45 degrees to the specimen faces as shown in Figure (7.11A). At the other extreme, that is, thick sections and minimal K_{IC} values, the fracture plane becomes flat intersecting the specimen sides at 90 degrees as demonstrated in Figure (7.11B). Intermediate thicknesses are partly flat in the center of the fracture plane and slanted in opposite directions on the specimen sides as can be seen in Figure (7.11C).

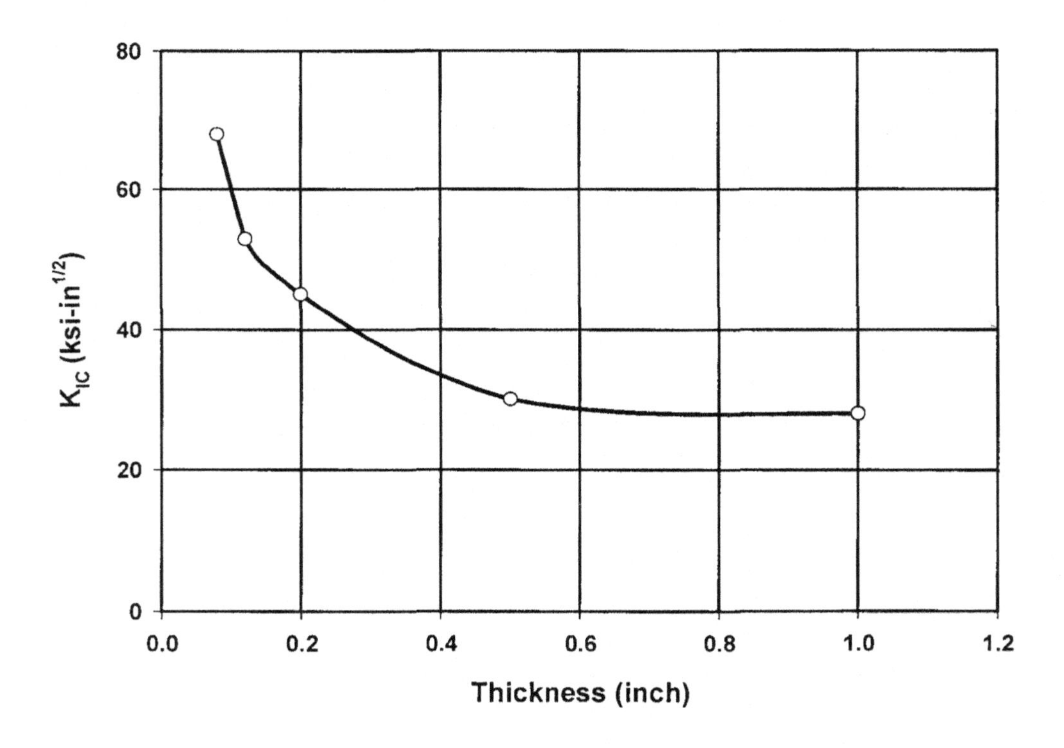

Figure 7.10 Fracture toughness of 7075-T6 aluminum as a function of specimen thickness.

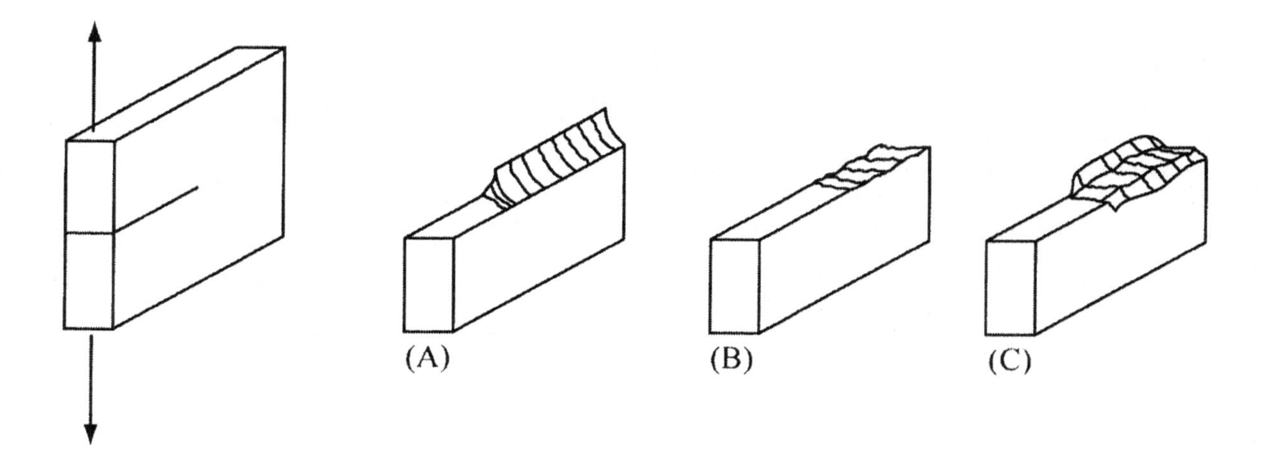

Figure 7.11 Fracture plane appearance for (A) plane stress, (B) plane strain, and (C) intermediate cases of specimen thickness, t.

Flat fracture planes have long been recognized to be indicative (and equivalent) to plane strain conditions; when measured in plane strain conditions, K_{IC}, will be the minimum expected value of fracture resistance. As a minimum value, its conservative use in design ensures that the ability to resist fracture will not be overestimated. Over the years, experimental observations on many materials have led to the following "rules of thumb" for determining whether an actual fracture is the result of plane stress or plane strain deformation.

The first step is to not prejudge the situation and calculate the plane stress radius via Equation (7.19):

$$r_p = r_{pc} = \frac{K_{IC}^2}{2\pi\sigma_y^2} \tag{7.19}$$

Once calculated, this value must be compared to the specimen thickness, t using the following criteria:

$$\text{If } r_{pc}/t > 1.0 \qquad \text{Plane Stress Deformation} \tag{7.23a}$$

$$\text{If } r_{pc}/t < 1/5 \qquad \text{Plane Strain Deformation} \tag{7.23b}$$

If the ratio $r_{pc}/t < 1/5$ holds, then plain strain conditions prevail and Equation (7.22) must be used to calculate the actual size of the plastic zone if required. A comparison of the plastic zone radii for both deformation modes is shown in Figure 7.12.

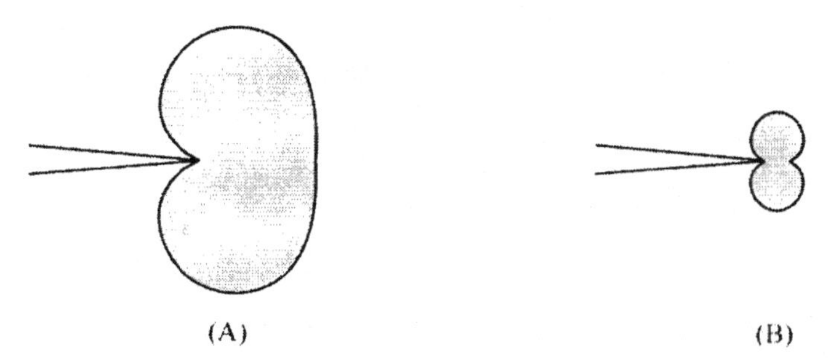

Figure 7.12 Comparison of plastic zone sizes for (A) plane stress and (B) plane strain.

7.8 PLANE STRAIN FRACTURE TOUGHNESS TESTING

Since the plane-strain fracture toughness conservatively represents the critical stress intensity factor, it is essentially a design property. Due to the criticality of fracture resistance in design, a detailed procedure for its measurement has been developed by the American Society for Testing and Materials (ASTM) and is outlined in Standard E-399. One of the test specimen configurations (three-point bend) is shown below in Figure (7.13). For this configuration, the stress intensity factor is:

$$K_I = \frac{PL}{th^{3/2}}\left[\left(2.9\left(\frac{a}{h}\right)^{1/2} - 4.6\left(\frac{a}{h}\right)^{3/2} + 21.8\left(\frac{a}{h}\right)^{5/2} - 37.6\left(\frac{a}{h}\right)^{7/2} + 38.7\left(\frac{a}{h}\right)^{9/2}\right)\right] \tag{7.24}$$

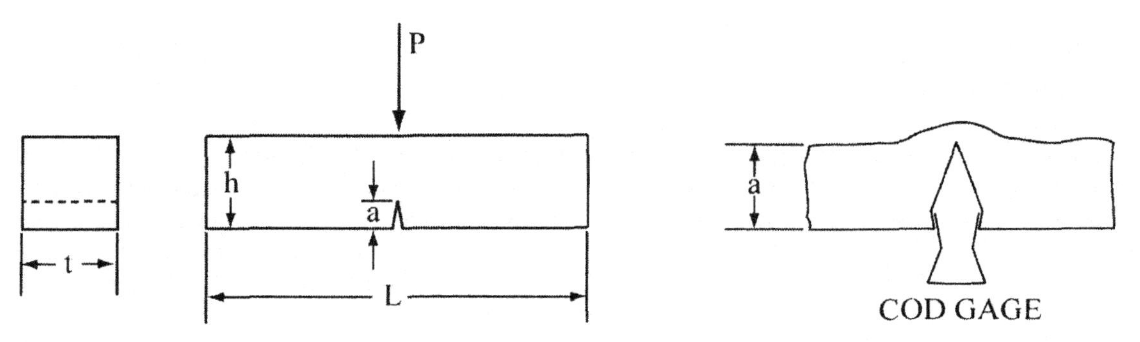

Figure 7.13 Standard test specimen for K_{IC} determination.

During testing, the crack opening displacement or COD is recorded, leading to one of the three possibilities shown in Figure 7.14. Graphical construction lines labeled OQ are drawn such that their slope is 5 percent less than the tangent to the COD record; the parameter P_Q is the load where the graphical construction line OQ intersects the test record.

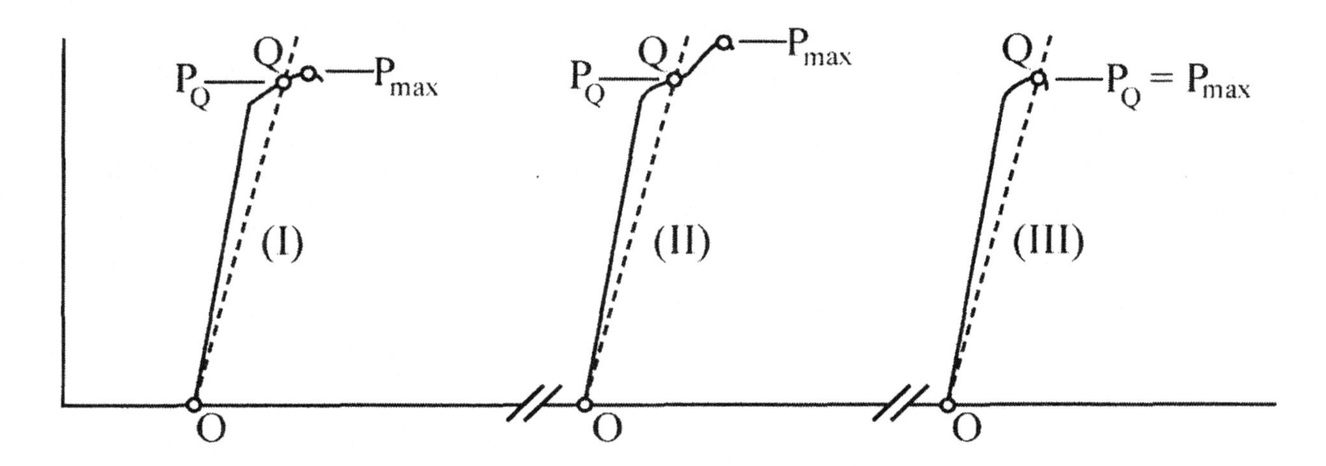

Figure 7.14 Three possible loads, COD curves for high-strength alloys.

Whichever of the three load records is generated by the test, the following assessments are then made:

Case I: Here, every load preceding P_Q is less than P_Q; if $P_{max}/P_Q < 1.1$, then P_Q is used in Equation (7.24), leading to a K_I value denoted K_{IQ}.

Case II: In this case, a load value preceding P_Q exceeds P_Q; use this load as P_Q, not the value found by the graphical line. Again, if $P_{max}/P_Q < 1.1$, proceed to calculate K_{IQ}.

Case III: Here, $P_Q = P_{max}$; calculate K_{IQ} directly.

Using the value of K_{IQ} determined from the appropriate case (I, II, or III), calculate r_p from Equation (7.21). If this zone radius then satisfies the plane strain criterion described by Equation (7.23b), then K_{IQ} is simply K_{IC} and all is well. If not, back to the drawing board since the specimen simply did not generate enough constraint at the crack tip! If one does not succeed the first time, try again by increasing the specimen thickness and/or the crack depth. If frustration continues indefinitely, check to see whether the yield strength reaches or exceeds E/150. It may be the material in question is too ductile for the current fracture mechanics analysis. Although other analytical techniques do exist for such ductile materials, they are not addressed in this book.

7.9 FRACTURE MECHANICS DESIGN CONCEPTS

As discussed earlier in this chapter, fracture-safe design procedures are centered around the stress intensity factor or SIF and its magnitude relative to the fracture toughness:

$$K_I \rightarrow K_{IC} \quad \text{where} \quad K_I = Y\sigma\sqrt{\pi a} \tag{7.25}$$

While relatively straightforward as equations go, its practical usage for a safe design is a little more complicated as described in the five scenarios below.

(a) The existence of undetected flaws: It is not always feasible or even desirable to section large structural members and search for flaws; indeed, there is no guarantee that you would not miss them in the sectioning process anyway. Perhaps the only practical solution would be to utilize nondestructive techniques such as x-radiography, ultrasound, thermography, and/or any other method sensitive to physical discontinuities such as a crack. Unfortunately, all existing techniques have distinct flaw-size thresholds usually denoted as a_0, below which they do not possess sufficient resolution to detect a defect. Although the threshold certainly depends upon the method employed, other factors such as the component geometry, site inaccessibility, and/or service environment may act to increase the minimum flaw size that is detectible. In such cases, the design procedure must start with the assumption that flaws of size, a_0 may be present, but are undetected. If the material has a known resistance to fracture as defined by K_{IC}, then the design stress can be specified by:

$$K_{IC} \geq \sigma_{des} Y \sqrt{\pi a_0} \tag{7.26a}$$

(b) Assessment of critical flaw sizes: If a design is already completed such is the case for all existing structures and components, then the section stresses and all material response measures are inherently fixed; in this situation, both K_{IC} and σ_{des} are no longer variable. Presumably, the stresses were determined assuming a flaw-free structure, especially if the structure was designed back in the "good old days" before modern fracture mechanics techniques were available.

If flaws are indeed present (certainly a reasonable assumption), their critical size, a_c is then found by solving the following form of the SIF equation:

$$K_{IC} = \sigma_{des} Y \sqrt{\pi a_c} \tag{7.26b}$$

(c) Rational development of proof tests: In the past, designer ignorance could be circumvented by over-sizing the components (that is why things used to last and last) or using some form of "proof test" to assess reliability and performance. In this scenario, a structural member that is designed to carry a specified load (including any safety factors), would be tested at the same or higher values. Medieval armorers customarily supplied breastplates with clear evidence of a successful projectile proof test because the warriors demanded resistance to arrows shot by English longbows. With respect to proof tests of structures made of crack-sensitive materials, the fact that some of them survive while others do not may be a reflection of the distribution of flaws present in the material. However, there is also the very distinct risk that the proof test will not cause failure, but will inflict some form of damage (see the next chapter) that renders the structure more vulnerable to destruction while in service.

(d) Failure analysis: If a member fails in service from fracture, the actual stress states can be accurately reconstructed if the critical flaw can be located and measured.

Example Problem 7.6: The grooved pump shaft shown in Figure (7.15) is made of 4340 steel with a yield strength of $\sigma_y = 220{,}000$ psi, a plane strain fracture toughness, $K_{IC} = 55{,}000$ psi-in$^{1/2}$ and is subjected to a bending moment, M_b. Moreover, the groove results in a stress concentration factor of 4.0 for the larger diameter section. Determine: (a) what bending moment induces yielding at the root of the groove, (b) will this moment lead to catastrophic failure if a very small crack develops at the groove root, and (c) what size crack in the groove root will cause failure when the moment in part (a) is applied?

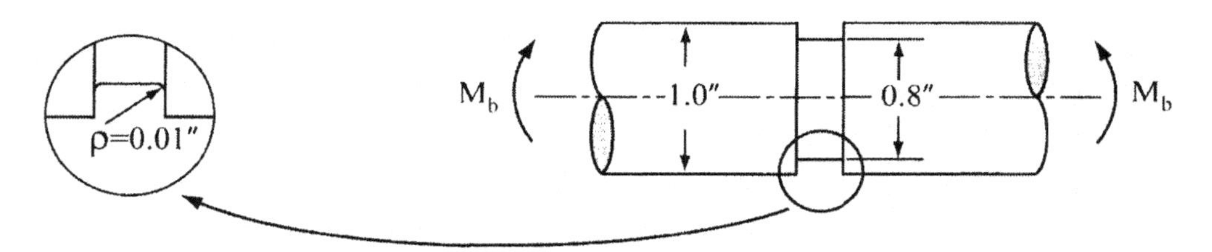

Figure 7.15 Grooved shaft under bending moment, M_b.

In order to solve part (a), we must first determine the nominal flexural stresses in the 1.0-inch-diameter section in terms of the unknown Moment, M_b:

$$\sigma_{nom} = \frac{M_b C}{I} = \frac{M_b (0.5)}{\pi/4 (0.5)^4}$$

Since only bending is applied, the stress state is uniaxial and therefore, inherently principal in nature. Using the MSST and the stated stress concentration factor of k_σ=4.0

$$\sigma_y = 220{,}000 \text{ psi} = 4\sigma_{nom} \quad \rightarrow \quad M_{b,\,yield} = 5390 \text{ in-lb}$$

For the solution of part (b), it is first necessary to make a few assumptions, the first being that a crack developing in the groove root is circumferentially symmetric. In addition, the groove plus crack is equivalent to a radial crack of the same total depth, or in more precise and highly technical terms, the groove depth plus a "tad." For this configuration, the stress intensity factor is given by:

$$K_I = \sigma\sqrt{\pi a}\ \sqrt{1-a/b}\ \frac{3}{8}\left[1+\frac{1}{2}\left(\frac{a}{b}\right)+\frac{3}{8}\left(\frac{a}{b}\right)^2+\frac{5}{16}\left(\frac{a}{b}\right)^3+\frac{0.25}{128}\left(\frac{a}{b}\right)^4+0.054\left(\frac{a}{b}\right)^5\right] \quad (7.27a)$$

where

$$\sigma = \frac{4M_b}{\pi a^3} \quad (7.27b)$$

Letting a = 0.4", b = 0.5", and M_b =5390 in-lb, a direct calculation revels that K_I = 41,900 psi-in$^{1/2}$. Because $K_I < K_{IC}$, a slight crack as just imagined will not cause failure.

For part (c), the critical crack size under a moment of 5390 in-lb will correspond to $K_I = K_{IC}$ = 55,000 psi-in$^{1/2}$. Unfortunately, one cannot directly solve for the critical crack size given the nature of the SIF solution. In this case, it best to iteratively or graphically solve for the unknown crack size, a_{cr}. Solving K_I by numerical iteration for the minor diameter 2a:

$$2a_{cr} \cong 0.74 \text{ inches}$$

Since the original diameter was 0.80 inches, a radial crack must grow inward (or radially) from the groove a value of 0.03 inches to induce fast fracture under the applied moment of M_b =5390 in-lb.

7.10 FAILURE ANALYSIS AND FRACTOGRAPHY

Although the analysis so far has focused on the analytical side of fracture mechanics, much can be learned from the remnants of such an event. The art of studying such post-fracture features is appropriately named Fractography. Although few like to admit it, Fractography is an "interpretive" skill in that one must be able to translate markings on the surface in the same way that trackers read a trail. Fortunately, most fracture events have sufficient similarities such that some general rules can be developed as described in the following sections.

Crystalline materials such as many ceramics and alloys will usually fail in one of two ways: a crack will propagate either through the crystalline grains in a *transgranular* fashion, or between the grains in an *Intergranular* way. Generally, the crack path is the one of least resistance for a given material, with some fractures observed to be alternately transgranular and intergranular; this implies that local conditions and principal stress directions all interact to determine the crack path. However, it is often the case that the fracture surface appears undeviating to the unaided eye so microscopy (optical and/or scanning electron) is required to detect the salient features.

Intergranular failure is relatively easy to identify as can be seen in Figure 7.16. When viewed on a microscale level, the faceted fracture surface is a reflection of the morphology of the grain polyhedra. Often times in metals and ceramics, an intergranular failure occurs when the grain boundary have been weakened by the deposition of segregating phases or aggressive environmental contaminants. Grain boundaries may also be weakened by the development of porosity at elevated temperatures or when grains lacking sufficient plasticity generate localized boundary micro-cracks. Such behaviors are often observed in polycrystalline ceramics.

Figure 7.16 Scanning electron micrograph showing intergranular fracture of alumina (Courtesy of G. Quinn, Guide to Fractography of Ceramics and Glasses, NIST Special Publication 960-16, May, 2007).

Transgranular failure can occur in either a brittle or a ductile fashion. Brittle transgranular fracture takes place by *cleavage* of the grains so the planes are usually crystallographically defined. Figure 7.17 shows a typical transgranular fracture. If one were to observe the actual cleavage plane, they would observe parallel lines often referred to as *river patterns*. Such patterns are actually ledges on the cleavage plane; smaller ledges come together and combine to form a higher ledge, so in appearance, small "streams" combine to form larger "rivers." Because the crack is known to propagate in the same direction as the "flow" of the stream (from smaller ledge regions toward higher), the patterns literally point back to the failure origin.

Ductile transgranular fracture occurs due to the generation and coalescence of micro-voids within the solid. Typically, such behaviors are observed for materials where the ductility is low, but not negligible as in high strength metals of the face-centered cubic crystal habit. A vast majority of these alloys derive their strength from the presence of intermetallic phases that are harder and more brittle than the metallic matrix. However, strain incompatibilities arise between small particulate phases and the ductile matrix as the material is strained beyond the elastic limit. Such incompatibilities lead to phase/matrix separation and void formation. Voids also form from dislocation interactions when second-phase particles are not present. At any rate, the micro-voids will coalesce as plastic strain continues, and fracture eventually ensues as depicted in Figure 7.18. The resulting fracture plane is covered with depressions that appear to be round. Often referred to as *equiaxed dimples,* depressions such as these are characteristic of failures where only one principal stress is appreciable, and of course, tensile.

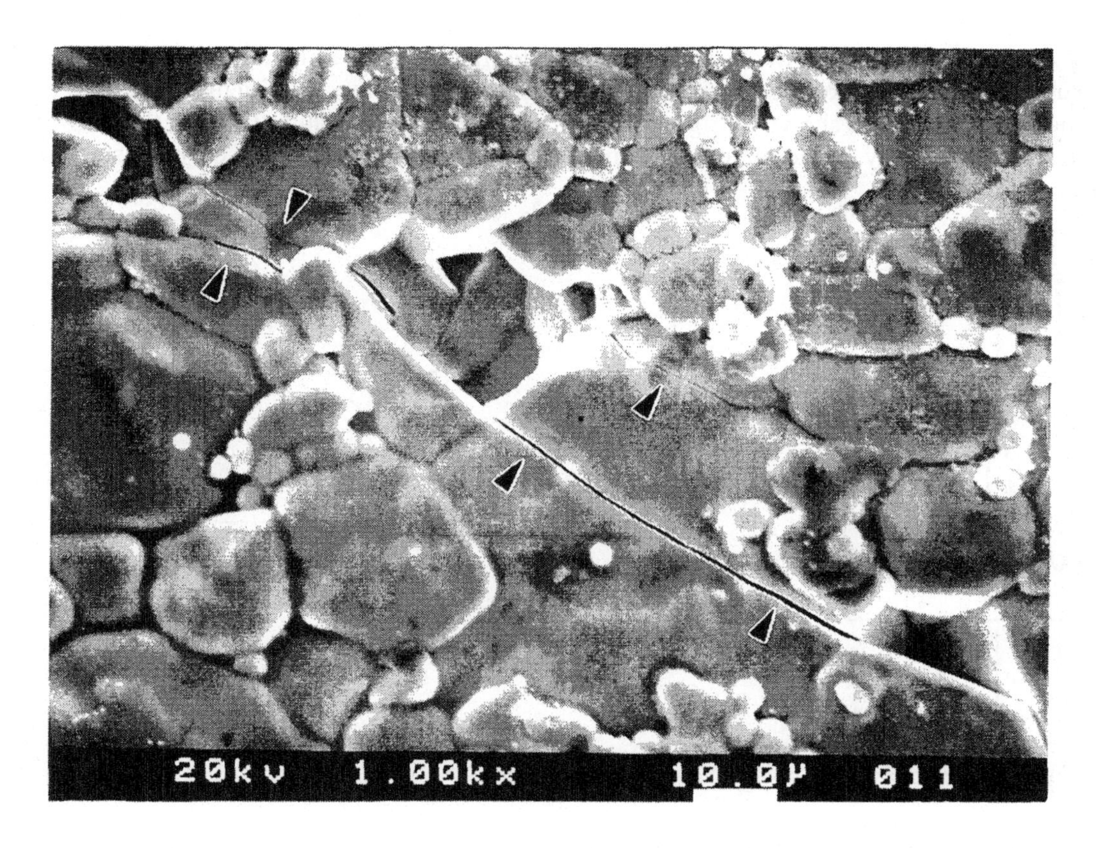

Figure 7.17 Scanning electron micrograph of transgranular fracture in alumina after thermal shock.

Microvoid coalescence under the influence of shear stresses leads to an elongation of the dimples on the fracture plane. Figure 7.19 illustrates the generation of the elongated dimples, as well as the parabolic depressions on the fracture faces that point in opposite directions.

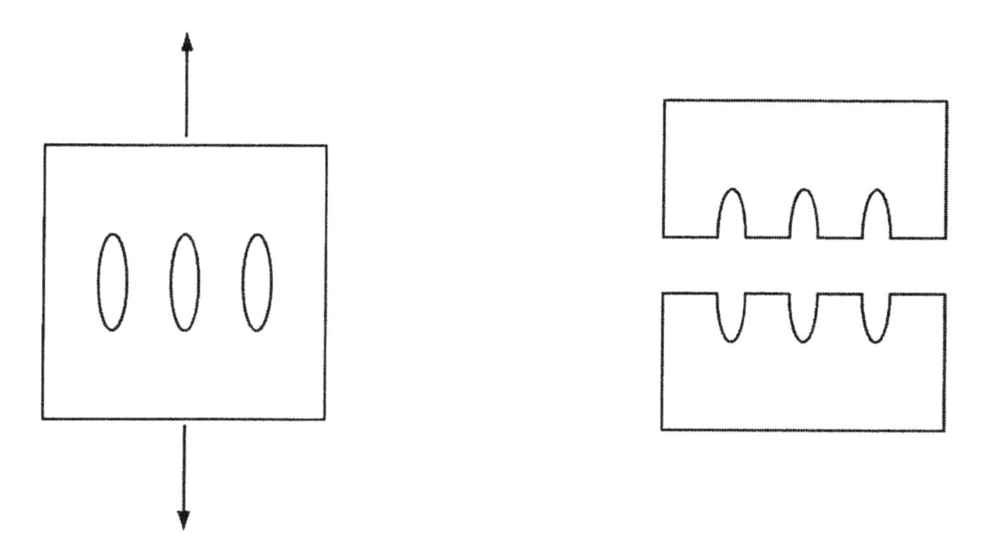

Figure 7.18 Generation of fracture surface dimples due to microvoid coalescence in tension.

Finally, for non-uniform states of normal stress that generate fracture such as bending, elongated dimples may also result. However, unlike the shear-driven case mentioned above, the fracture faces are mirror images and the planes appear as shown in Figure 7.20. Reconstructing the nature (but not the magnitude) of the failure stress state may be possible by examining the dimple configuration on both sides of the fracture surfaces.

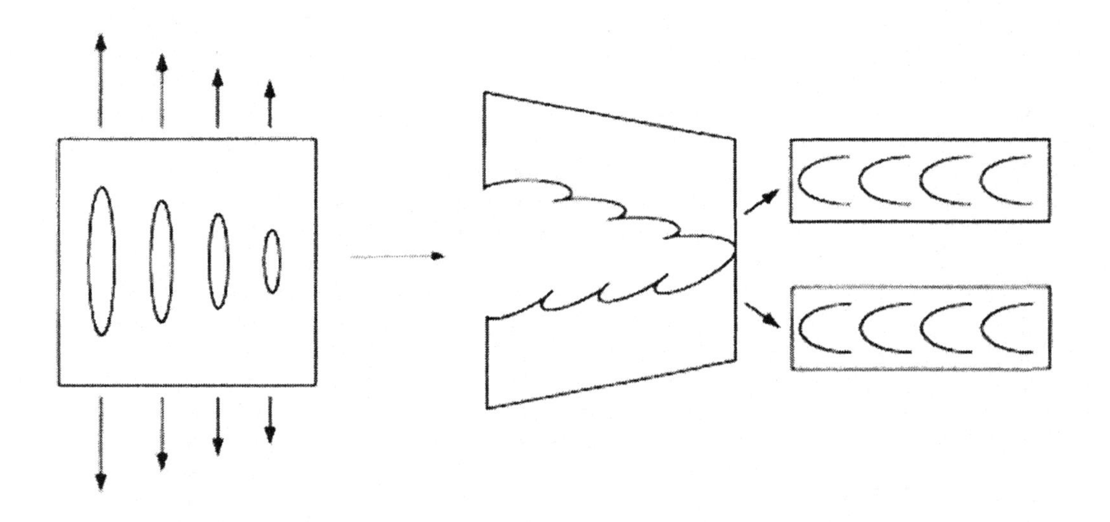

Figure 7.19 Elongated fracture dimples generated by non-uniform tensile stress caused by bending.

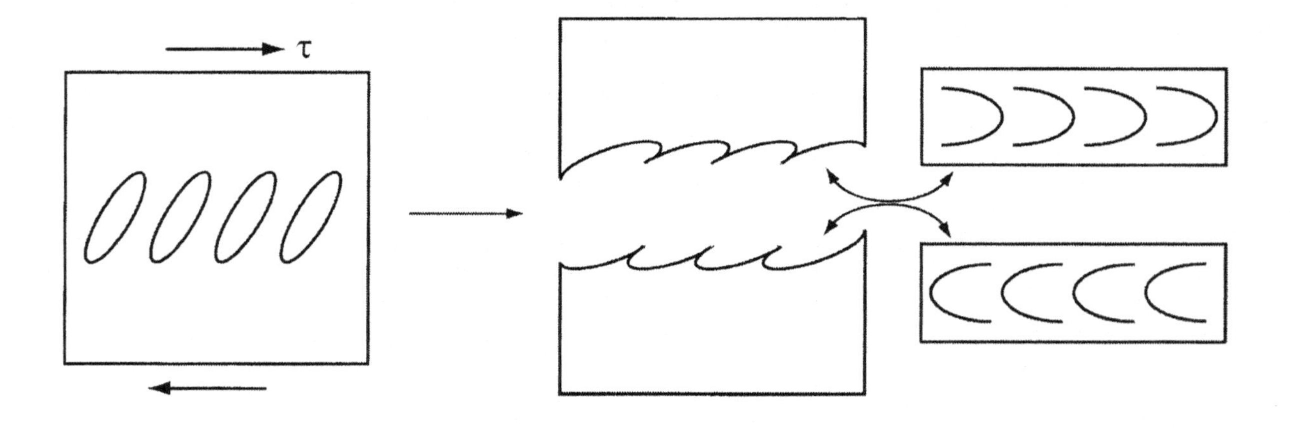

Figure 7.20 Elongated fracture dimples generated by dominant shear stress components.

PRACTICE EXERCISES

1. A brittle glass used for standard issue, collegiate beer mugs is known to have a measured surface energy of 0.3 Joules/m^2 and an elastic modulus of 70 GPa.

    a. Determine the theoretical cohesive strength of the glass, assuming flaws no longer than 3 Angstroms long. How does this strength compare with the elastic modulus?

    b. If large samples of the glass were observed to fail in tension at 25 MPa, what would be the expected size of the largest flaws present?

2. In a clear case of poetic justice, the glass of Problem 1 is also used for 3 cm (diameter) cylindrical laser rods to zap annoying professors.

    a. According to the Griffith failure theory, what torque will cause the rods to fail?

    b. If the rod were loaded in pure compression, what load will cause failure?

3. Your first exciting assignment at the ACME Engineering Company is to determine the dependence of plane-strain fracture toughness on the tensile strength of heat-treatable steel (0.45C-Ni-Cr-Mo Steel). Assuming a design stress equal to 90 percent of the tensile strength, using the stress intensity factor expression suitable for small flaws in large tensile plates and the collected data graphically shown below:

$$K_I = \sigma_{design}\sqrt{\pi a}$$

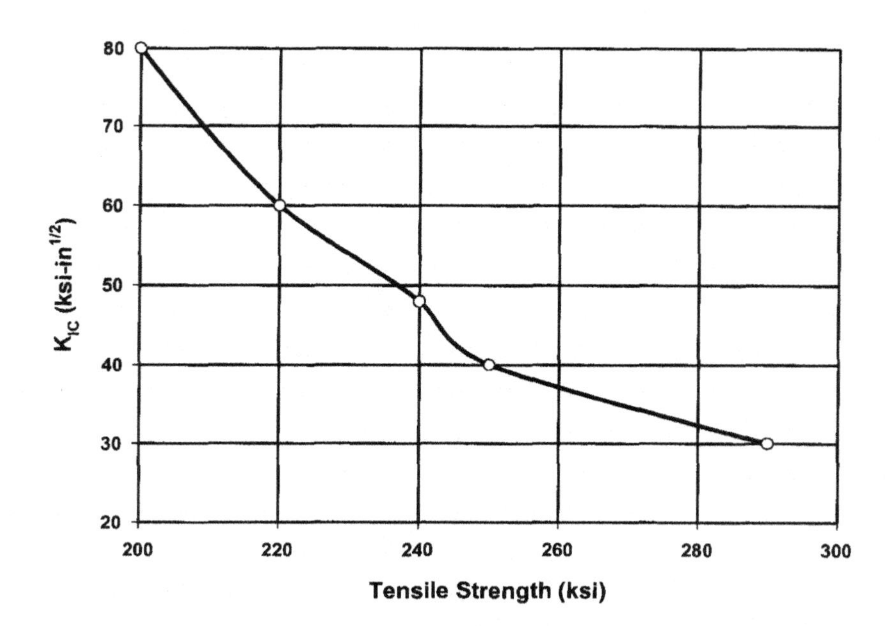

What are the critical flaw sizes, a_c, for the steel if its tensile strength is metallurgically manipulated to be: (a) 300 ksi, (b) 250 ksi, and (c) 220 ksi?

4. The mechanical controls for an aircraft, rear stabilizer requires the tensile plate shown below with a width of 2.0 inches, a center hole 0.25 inches in diameter, and a thickness of 0.25 inches. When fabricated from a durable heat-treatable alloy steel, this materials boasts a tensile yield strength of σ_y = 180,000 psi and a fracture toughness of K_{IC} = 60,000 psi-in$^{1/2}$

   a. What design stress corresponds to the onset of yielding at the "equator" of the hole?

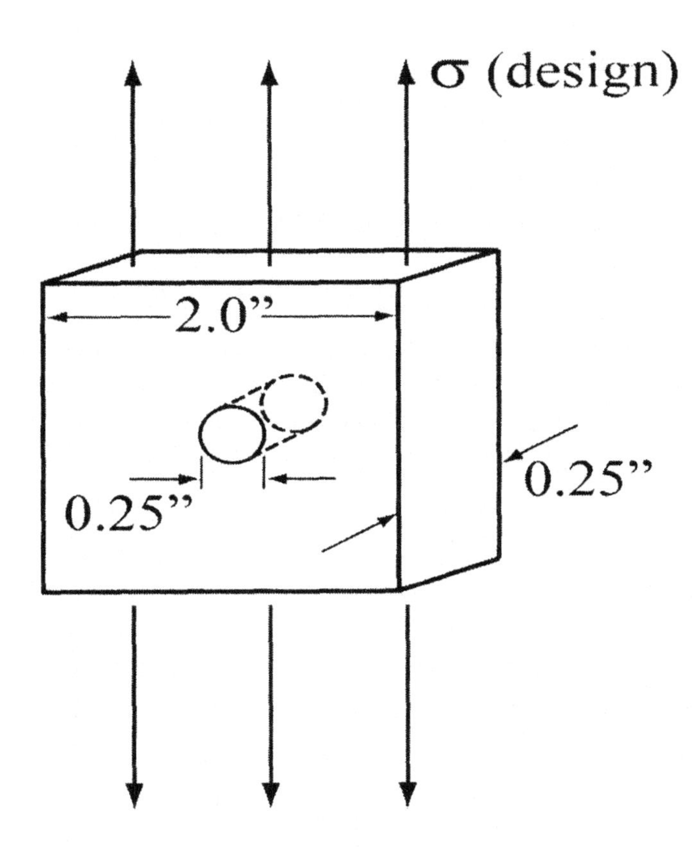

   b. Based on the stress intensity factor information given below (and next page), will the structure be fracture-safe for the stresses of part-a, if a slight (say 0.001") cracks develop at the equator?

$$K_I = \sigma\sqrt{\pi a} \quad Y\left(\frac{r}{b}, \frac{r+a}{b}\right)$$

   c. If the answer to part of b is affirmative, how long can the cracks be before they reach a critical length?

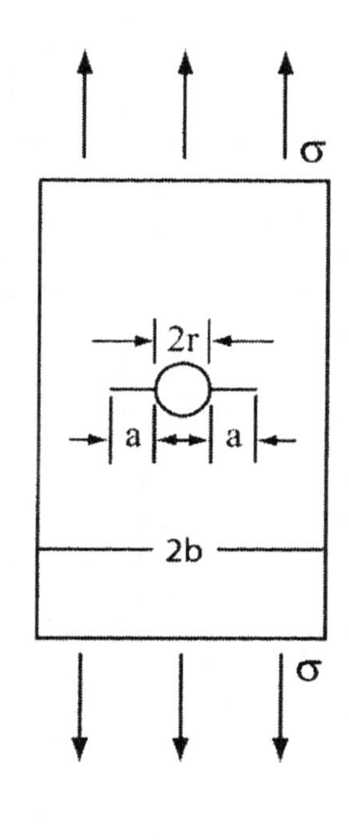

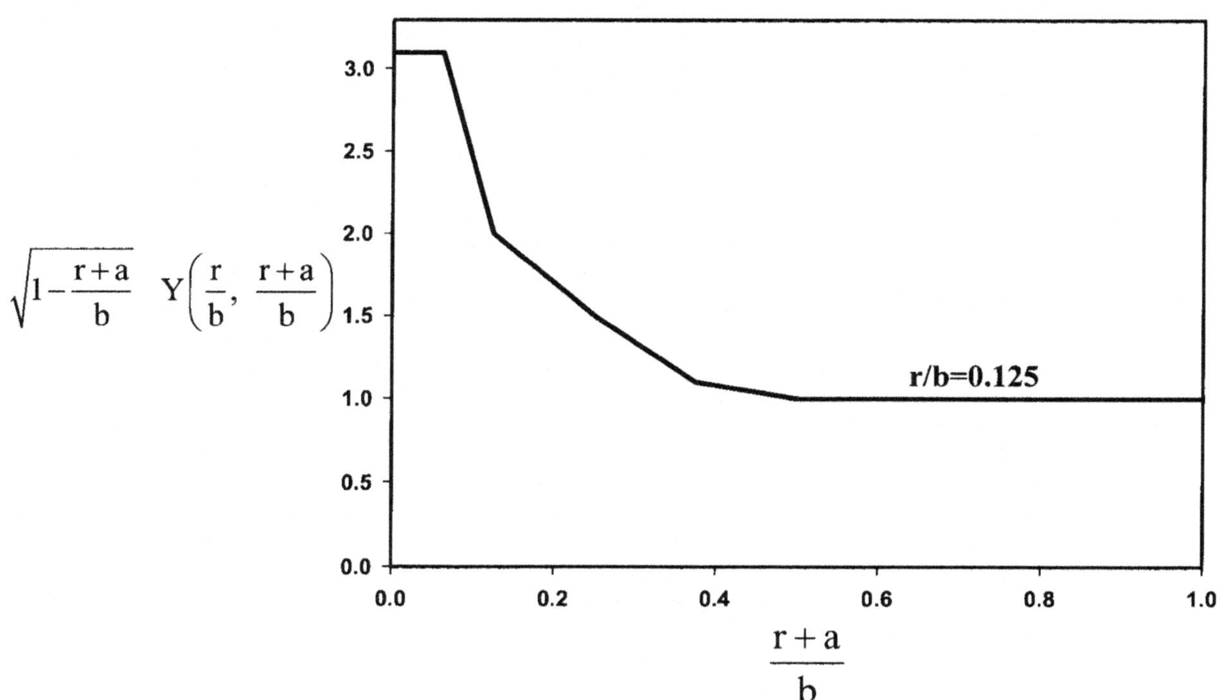

5. For the steel in Problem 3, what structural thickness is required to achieve plane-strain fracture conditions if the yield strength is approximately 85 percent of the tensile strength, and the achievable tensile strength of the material is: (a) 200 ksi, (b) 250 ksi and (c) 300 ksi?

6. Shown below is the relationship between the stress-intensity factor, K_I, crack length and internal pressure for a radially cracked cylindrical pressure vessel used for the Space Shuttle air filtration system. If the vessel is fabricated from 2024-T351 aluminum alloy with a measured fracture toughness of 44 MPa-$m^{1/2}$, r_i=8 cm, and r_o=10 cm, what internal pressure is indicated for:

   a. Crack initiation (assume $a = 10^{-2}$ mm) as source of fracture.

   b. Leak before break (crack is through wall but no catastrophic failure).

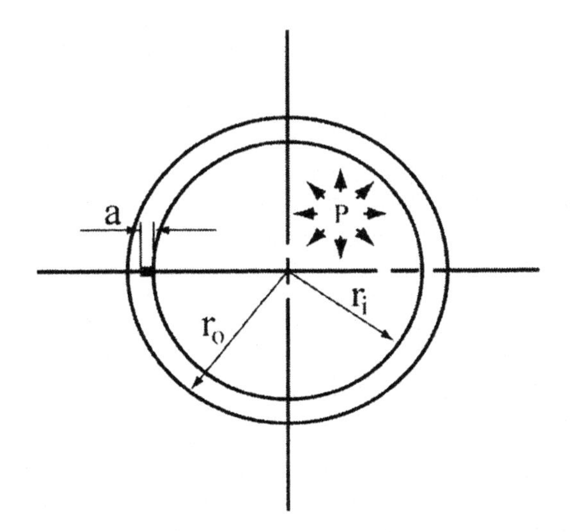

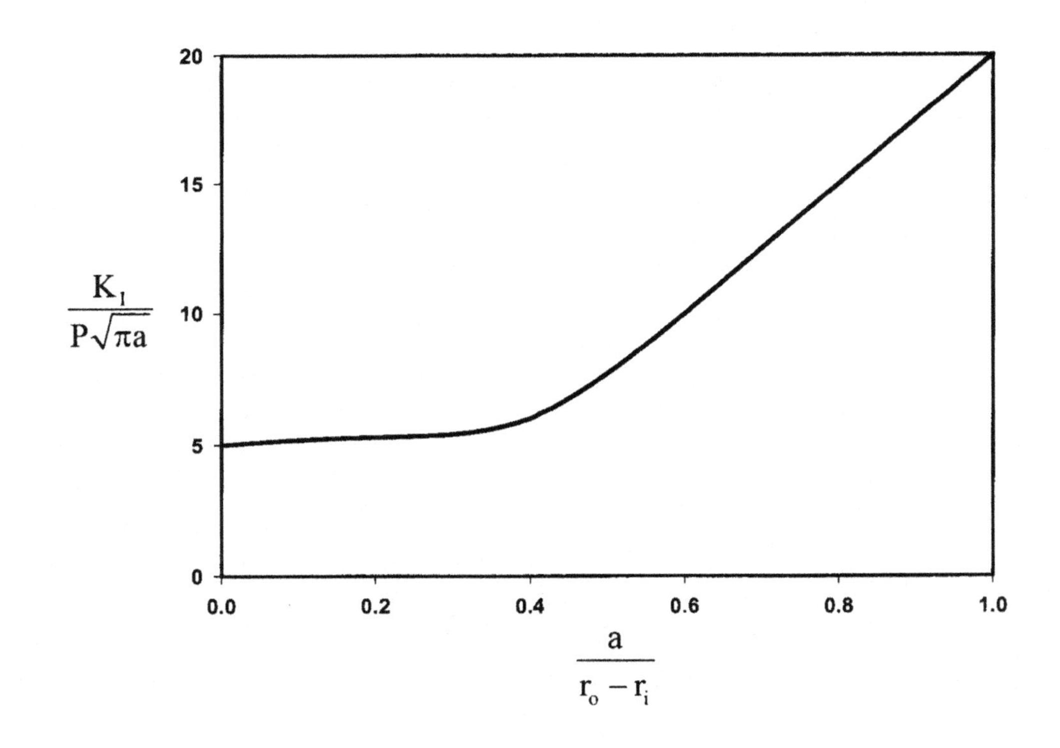

7. The relatively thick eyeglasses of a legendary football coach residing at an undisclosed location somewhere in central Pennsylvania can be modeled as a wide beam that carries a bending moment, M_b and has an elliptical surface crack as shown below. After extensive testing that rightfully annoyed the cantankerous coach, the material was found to exhibit a yield strength of $\sigma_y = 150{,}000$ psi and a fracture toughness $K_{IC} = 50{,}000$ psi-in$^{1/2}$. The stress intensity factor, K_I at point A is given by the following expression:

$$K_I = 0.8\,\sigma\sqrt{\pi a/Q} \qquad Q = 1 + 1.464\left(\frac{a}{c}\right)^{1.65} \tag{7.28}$$

where σ is the maximum tensile bending stress and Q is the shape factor for the crack as already defined. Given this information, determine:

a. What magnitude bending moment would cause fracture?

b. What is the size of the plastic zone?

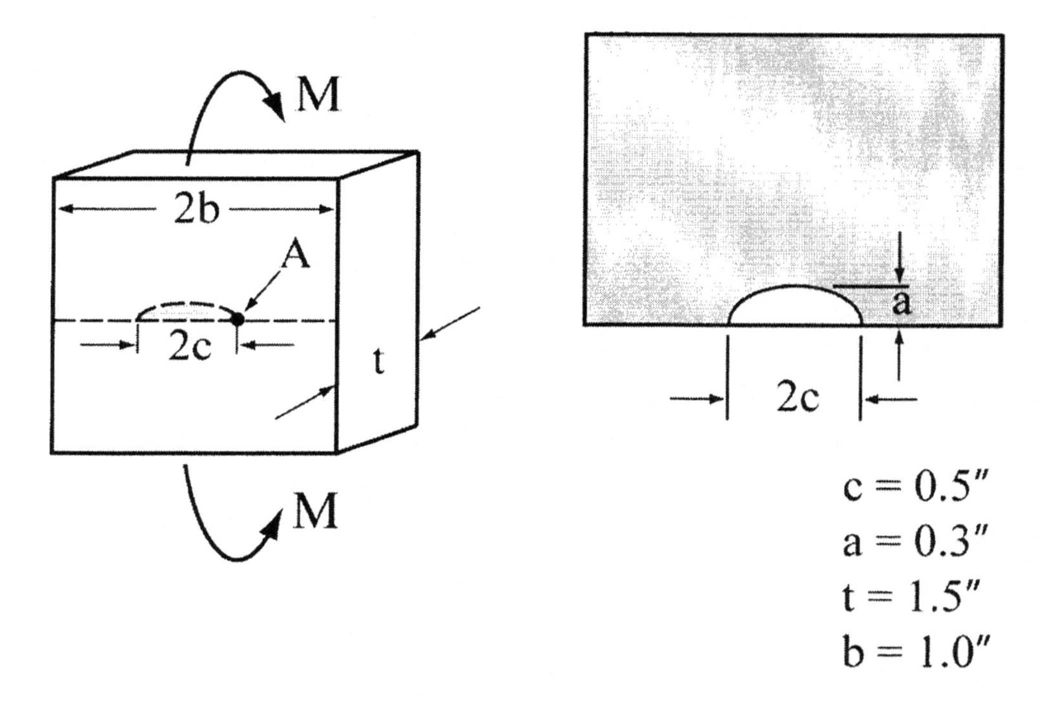

CHAPTER EIGHT

FATIGUE FAILURE

8.1 INTRODUCTION

Your life and wallet are both constantly stricken by what amounts to severe fatigue. No, I am not talking about being tired of paying excessive amounts for tuition and taxes (who isn't?) In this expensive case, *fatigue* is the name given to the eventual loss of load-bearing capacity of a structural member subjected to stresses of varying cyclical amplitude; although the application of a single load may not cause failure, applying the same magnitude load repeatedly can. Unfortunately, this scenario applies to just about anything that costs you money, going from light bulbs to the entire range of car or electronic parts that mysteriously fail just as the warranty expires! Failures of this type were first observed systematically when rotating railway axles began failing in Europe in the mid-1800s leading to loss of life, public unease over the newly introduced technology, and (most importantly to the capitalists of the day) economic distress. Although the axles in question were understood to be safely loaded as far as static levels were concerned, they still failed (fractured) when the same loads/stresses were applied over tens or hundreds of thousands of cycles. In reality, these behaviors should not have been a surprise. After all, humans seem to have an intuitive sense of this behavior even when young, as who has not broken a piece of wire or paper clip by simply bending it back and forth until fracture? Although you probably did not ponder it at the time, there appears to be a deleterious combination of changing stress levels that somehow leads to the type of fracture discussed in the previous chapter.

8.2 FATIGUE ENDURANCE

As far as fatigue is concerned, the question of engineering relevance is how long a structure or component will last when subjected to cyclically applied loads? In this context, the historical example of the rotating railway axle is still relevant since it is merely a beam where the outer surface experiences alternating tensile/compressive stresses. As the axle rotates, any given point on the outer diameter goes through states of maximum compression, followed by zero stress at the neutral axis, then onto a maximum tensile stress, back to zero, and so on as shown in Figure 8.1. For the loading case just described, the tensile and compressive stress magnitudes are equal, and the *completely reversed stress amplitude* or σ_{ao}, is the only parameter of interest. From a design standpoint, the material response of concern is the number of cycles to structural failure, or N at the completely reversed stress amplitude just discussed.

In engineering parlance, the number of cycles to failure, N is usually referred to as the *endurance*, whereas the stress amplitude for that particular life expectancy, σ_{ao} is the *endurance strength* or *endurance limit*. A graphical comparison of the endurance of a material as a function of the magnitude of the reversed, stress-amplitude is referred to as an *S-N diagram*, so named in a simpler day (but definitely not predating the Greeks) when stress was known as "S" and not "σ." Two typical forms of S-N curves are shown in Figure 8.2. Because component lifetimes may range from tens of thousands to tens of millions of cycles, it is usually necessary to use a logarithmic scale for the abscissa (x-axis).

While the practice of bending a wire back and forth usually resulted in the desired effect, namely fracture, the noted success required sufficient deformation and stress levels. Whether you realized it or not at the time, the stress magnitude was key to ensuring failure; slight bends and the correspondingly lower stress levels would have literally taken forever and a day for failure to occur. In fact, materials with endurance response curves such as the one labeled "A" in Figure 8.2 exhibit a range of stress amplitudes from zero to σ_e such that failure never occurs. Obviously, this is a design stress of great interest and is often referred to as the *endurance limit*, *fatigue limit*, or simply σ_e. Ferritic steels with body-centered cubic crystal structures, as well as some titanium alloys all possess endurance limit strengths. It is also not unusual that the endurance limit may be a large fraction of the yield strength, further enhancing the applicability of these materials in fatigue-prone service applications. Unfortunately, most materials have endurance responses like that labeled "Material B" in Figure 8.2. Here, fatigue failure appears to occur eventually, no matter how low the stress amplitude; aluminum alloys are notable for this type of response. Despite this inevitable tendency towards a fatigue failure, *fatigue* or *endurance strengths* are still quoted; in this case, the cited stress amplitude usually implies failure at an admittedly arbitrary 10^6 to 10^8 cycles. Given this wide range, caution is certainly warranted as there is no universal or even clear standard on the number of cycles to failure for endurance strength.

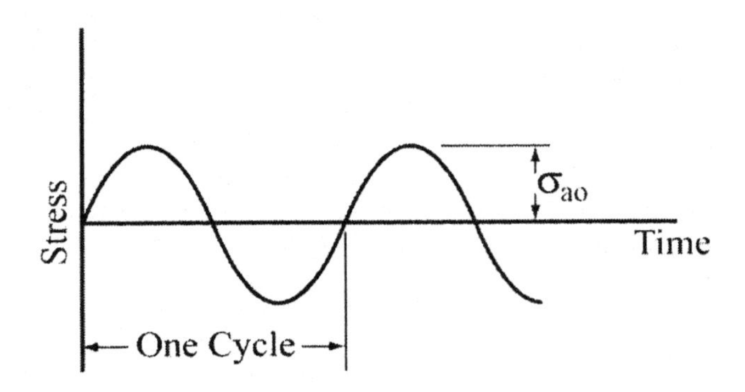

Figure 8.1 Load (or stress) history for a completely reversed, uniaxial stress.

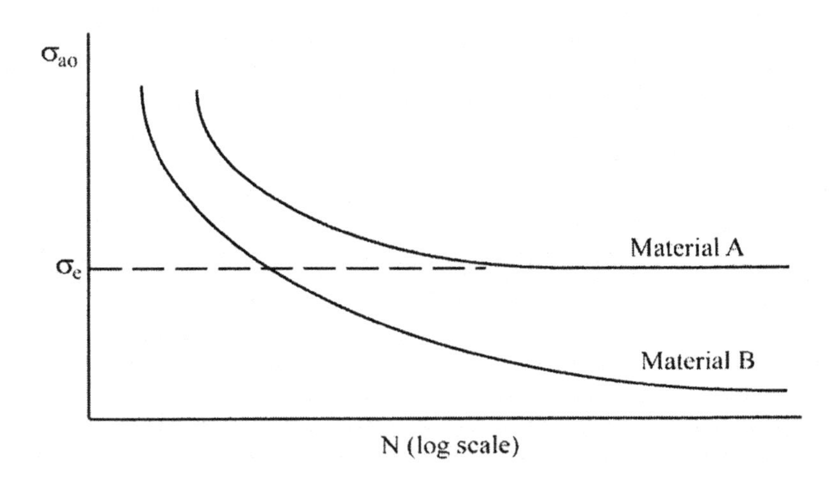

Figure 8.2 Typical S-N endurance curves for two metallic alloys.

8.3 THE STATISTICAL NATURE OF S-N DIAGRAMS

To make matters worse, statistical variations in the endurance data cannot be blissfully ignored in the same way in which disparities in yield or tensile strength are often wished away. To help compensate, most endurance data are obtained from specimens that are fabricated with a degree of care not typical of ordinary manufacturing processes. In addition, the loading of these specimens is relatively precise compared to actual service stressing, the laboratory environment is more fully understood, and so forth. Even in this relatively well-defined laboratory setting, a great deal of scatter is noted in the actual cycles to failure for a given stress amplitude. Unfortunately, it is the physical nature of the fatigue process including the underlying flaws and their stochastic distributions that ultimately leads to such variations.

Consequently, the certainty or accuracy of any failure prediction is greatly reduced for actual structural elements in a service environment. Given these potential problems, a more candid approach to the gathering and utilization of endurance data might be the following.

Imagine a relatively large sample of experimental specimens to be tested, all at the same reversed stress amplitude, say σ_{ao}'. All are tested, and the cycles to failure noted for each. A histogram of the results is then constructed as speculatively shown in Figure 8.3. Although the scatter band can be tightened by carefully preparing and testing the specimens, the observed variations may be far more extreme for actual structural members; some may fail almost immediately while others may survive indefinitely. In either case, a single line on an S-N plot implies an average, or mean value of cycles to failure where 50 percent of the members will either have failed or survived; the N-value corresponding to 50% survival is indicated in the figure. Accordingly, it might be more instructive for the designer to know the failure rate, or, conversely, the survival rate rather than to blindly apply a safety factor to the 50% (survival) data commonly catalogued. Similar failure/survival rates could be determined from endurance histograms at other amplitudes. Finally, a more candid S-N diagram with survival rates explicitly indicated could then be constructed as shown in Figure 8.4. Unfortunately, the great cost and time involved in generating such (complete) survival rate data usually precludes the idea.

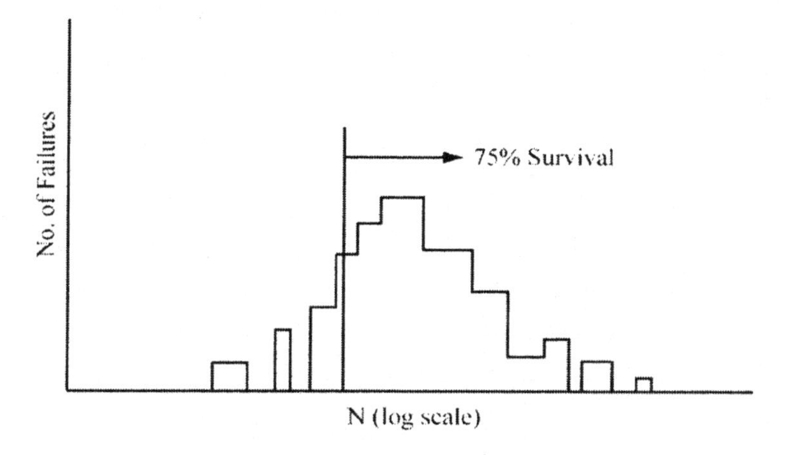

Figure 8.3 Histogram of cycles to failure for similar fatigue specimens loaded to amplitude σ_{ao}'.

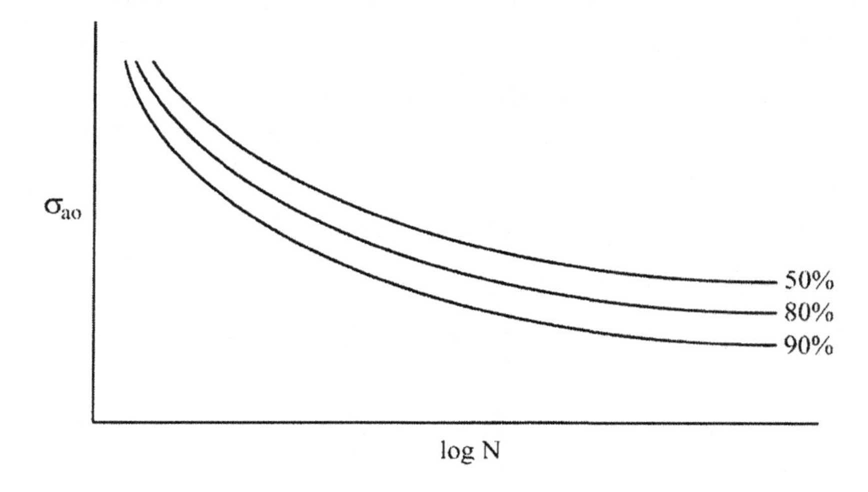

Figure 8.4 S-N curves depicting expected survival rates.

8.4 MEAN STRESS EFFECTS

In addition to the statistical issues just discussed, many if not most load histories are not as simple as the completely reversed case depicted in Figure 8.1. Indeed, countless structural elements carry a mean, or what is essentially a static stress level, upon which the fluctuating cyclic load is superimposed. Consider the two load histories labeled "History A" and "History B" indicated in Figure 8.5. "History A" is completely reversed, with a stress amplitude σ_a while "History B" has the same amplitude superimposed on a static or mean-stress, denoted herein as σ_m.

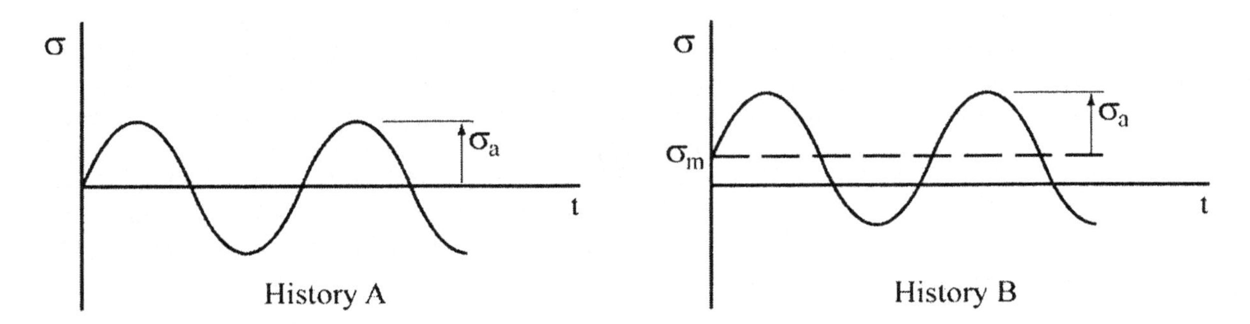

Figure 8.5 Load histories of equal stress amplitude, without and with a mean stress.

If the stress histories of both A and B are applied to identical fatigue specimens, sample B would consistently fail in fewer cycles than that the same material subjected to the loading of A; *the moral of the story being that mean stresses that are tensile tend to reduce fatigue endurance while compression will have the opposite effect!* It is not that the stresses are being "mean," only that the ever-present tensile stresses from the non-zero mean help extend or grow the crack at a greater rate. Conversely, a compressive mean-stress would tend to extend the fatigue life of a component since it is essentially "clamping down" on the crack; in fact, compressive residual stresses are routinely used to reduce or even reverse tensile mean-stresses.

Despite the obvious importance of the mean stress on fatigue behavior, it would be economically unfeasible to generate S-N data with all levels of mean or static stress that are relevant. Given this costly reality, the best approach would be to derive a predictive theory that is based on relatively simple tests (as was the yield theories discussed in Chapter Five), yet still sufficiently sophisticated to encompass a wide range of stress levels and behaviors.

Any fatigue theory worth its weight in failed parts must include a generalized, cyclic stress-behavior as shown in Figure 8.6. Using the conceptualization of the stress history shown in Figure 8.6, the following design parameters as a function of the maximum and minimum stress levels emerge:

$$\sigma_a = \frac{\sigma_{max} - \sigma_{min}}{2} \tag{8.1a}$$

$$\sigma_m = \frac{\sigma_{max} + \sigma_{min}}{2} \tag{8.1b}$$

$$R = \frac{\sigma_{min}}{\sigma_{max}} \tag{8.1c}$$

where σ_a is the stress amplitude, σ_m is the mean stress, and R represents the stress ratio between the minimum and maximum values.

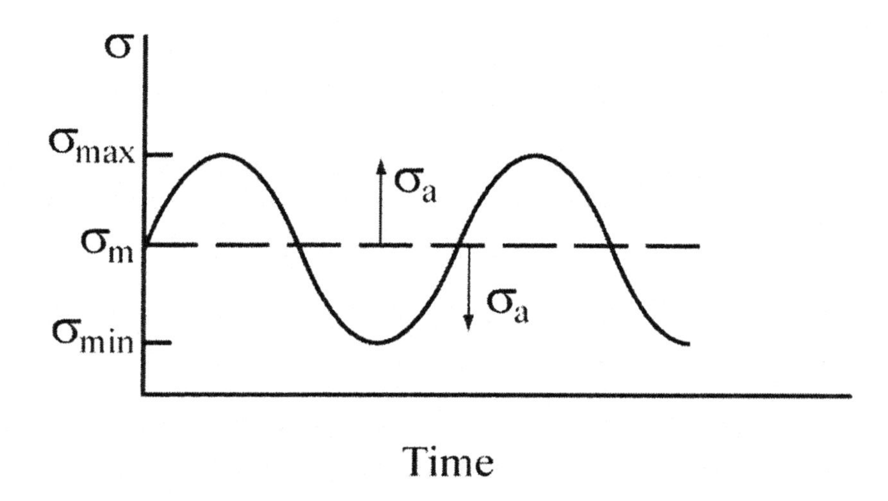

Figure 8.6 General stress history parameters for cyclic loading.

Note that the completely reversed case of R = -1.0 emerges when $\sigma_m = 0$ such that $\sigma_a = \sigma_{ao}$. With regard to the deleterious effect of the mean stress, one possibility could be to conduct a series of endurance tests using various combinations of σ_a and σ_m pairs such that failure occurs in the exact same number of cycles, N'. Of course, the previously mentioned statistical variations make it practically impossible to achieve such a discrete value of N. Ignoring the latter reality, one could then arrange the data points in a space with mean stress and stress amplitude axes as shown in Figure 8.7.

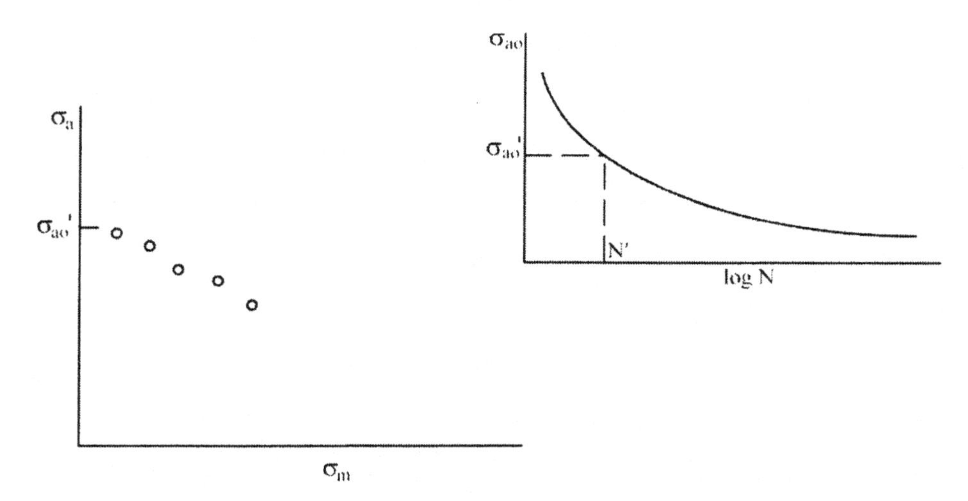

Figure 8.7 Values of σ_a and σ_m leading to failure after N′ cycles.

Note that, for $\sigma_m = 0$, $\sigma_a = \sigma_{ao}'$ and is therefore correlated with an actual endurance value, N' from the S-N data set. At the other extreme, the ultimate strength will dictate another "natural" boundary because failure is all but guaranteed. Given these boundaries, the next step is to determine an empirical correlation of the data in between that has some practicality to it. Although several correlations of the desired type have been proposed over the years, the one attributed to Goodman will get the honors. Goodman's correlation may be written as follows:

$$\frac{\sigma_a}{\sigma_{ao}} + \frac{\sigma_m}{\sigma_{uts}} = 1.0 \tag{8.2a}$$

where σ_{uts} is the *tensile strength* for the alloy in question. In terms of determining the equivalent, completely reversed alternating-stress, σ_{ao}, the relationship can be rearranged such that:

$$\sigma_{ao} = \frac{\sigma_a}{\left(1 - \frac{\sigma_m}{\sigma_{uts}}\right)} \tag{8.2b}$$

The triangular region in Figure 8.8, between the axes origin and the Goodman failure line consists of values of σ_a and σ_m for which the member will survive N' cycles. Conversely, points to the right of the Goodman line fail in less than N' cycles. When the position of (hypothetical) data points is outside of the Goodman failure line as shown in Figure 8.8 then Equation (8.2) is conservative in nature. Finally, it should be assumed that the corresponding Goodman failure prediction will work for stresses within the elastic behavior region that are well short of the tensile yield strength of any given material.

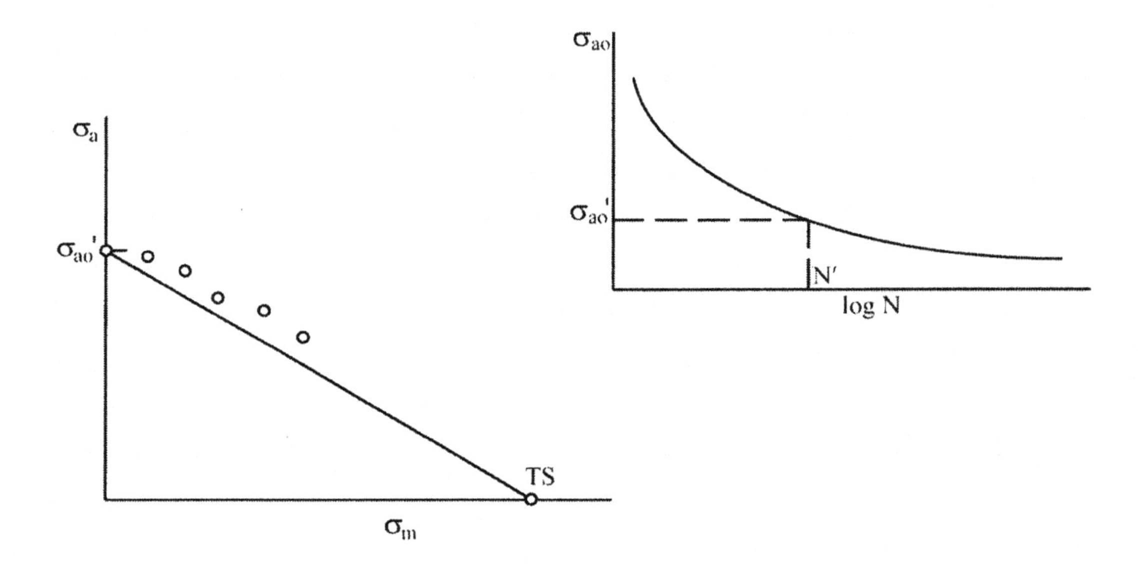

Figure 8.8 Goodman failure criterion compared to hypothetical data points.

8.5 CUMULATIVE DAMAGE CONCEPTS

While the idealized load history depicted in Figure 8.5 is useful for design purposes, it is not sufficiently general for many if not most applications. Consider just about any components in an automobile engine that could see widely varying loads as the car is cruising along a highway, climbing a steep mountain, or even traveling through Alaska in winter or the Mohave Desert at noon. Another example could be an aircraft wing whose flexural stresses will vary due to altitude, turbulence, maneuvers, or an ill-matched collision with a hapless bird. Actually, the most general stress histories might closely resemble those depicted in Figure 8.9, where the stress levels have odd cyclic patterns or alternatively, are persistently changing over time.

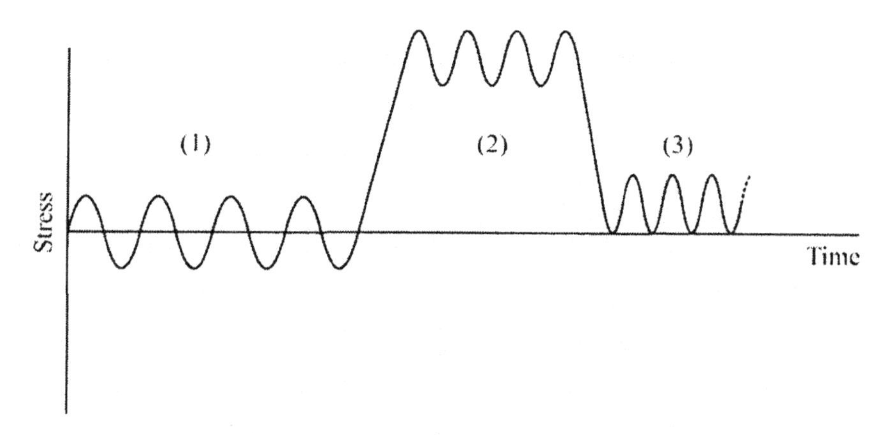

Figure 8.9 General stress history with variations in cyclic amplitude.

Taken alone, the endurance for any given cycle depicted in Figure 8.9 could be ascertained from S-N data. Moreover, the Goodman equation could also be used to calculate the equivalent, completely reversed amplitude for any cyclic stress level when a mean stress is present. However, the real question to be addressed is to what manner the different cyclic histories combine to contribute to eventual failure? To answer this question, it is first necessary to invent a concept of *damage,* that will describe the contribution of each ith stress state to failure. In order to begin this type of analysis, we must first define the associated damage, d_i as follows for each corresponding stress state:

$$d_i = \frac{n_i}{N_i} \tag{8.3}$$

In Equation (8.3), n_i is the actual number of cycles accumulated by the member under cyclic stress condition "i," whereas N_i is the number of cycles to failure if only the condition "i" were applied in a completely reversed sense. Obviously, failure occurs when the damage has reached 100 percent or simply, $d_i = 1.0$. If many separate conditions are applied sequentially, the fatigue damage from each condition can be expected to accumulate, and failure may be postulated to occur when:

$$\sum_{i=1,2,3...} d_i = \sum_{i=1,2,3...} \frac{n_i}{N_i} = 1.0 \tag{8.4}$$

Equation (8.4) is commonly referred to as *Miner's rule*, after the engineer who first applied it during the 1950s. Note that the cumulative damage rule given by Equation (8.4) is linear in nature and is therefore, not entirely accurate on that basis alone. An even more serious criticism of Miner's rule is that the sequence of application of the load conditions i = 1, 2, 3 ... and so on is not recognized, where in practice it is known to influence the final failure. For example, a few large amplitude cycles applied at the beginning might shorten a components life. Although this damage is included, the extent is usually greater than predicted by Equation (8.4). To further cloud the picture, initial or occasional overstressing is sometimes purposely practiced because it often increases endurance through a process know as *crack retardation*. Intentional overstressing is believed to cause plastic flow at the tips of unseen flaws, thereby "blunting" the defects that are the sites for fatigue crack initiation. It is somewhat ironic that those of us who live in Pennsylvania (or similar states) with its many potholes may actually benefit from that all too common bump. However, and geographical advantages notwithstanding, Miner's rule cannot account for anything but decreased endurance from varying stress levels.

Example Problem 8.1: The daily, multi-stress loading cycling encountered by car struts in Pennsylvania fabricated from SAE/AISI 2330 Steel is summarized in Table 8.1. S-N data for the hardened steel is presented below in Figure 8.10. In addition to the data shown, the tensile strength was found to be σ_{uts}=250 ksi. Using this loading history and S-N data, determine how many days the struts will survive.

Table 8.1 General stress history with variations in cyclic amplitude.

Loading Condition	σ_{max} (ksi)	σ_{min} (ksi)	n_i (cycles)
1	50	-50	1,000,000
2	175	75	60
3	196	4	4
4	190	-190	0.5

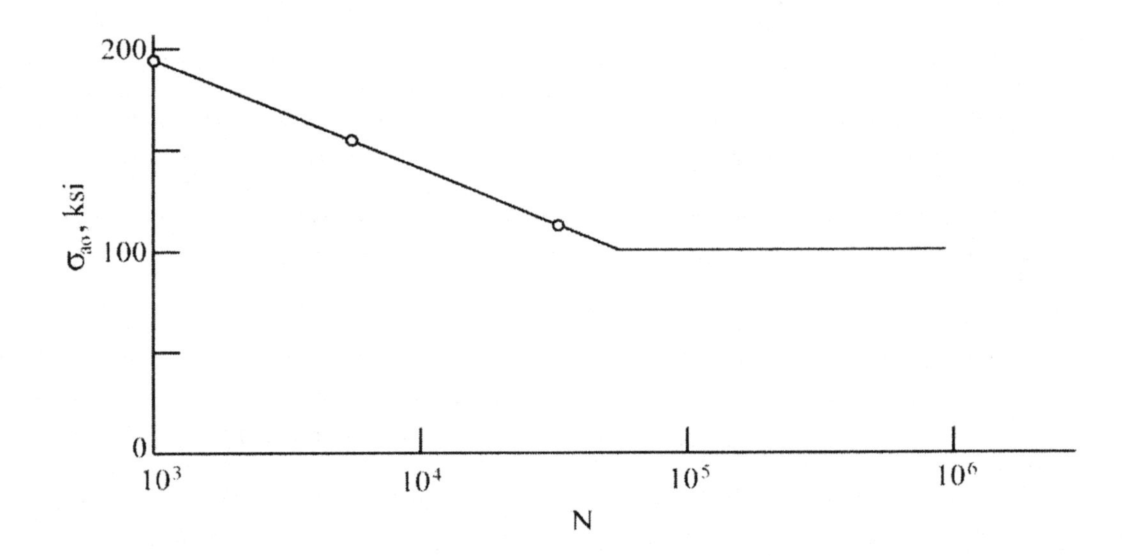

Figure 8.10 S-N data for SAE/AISI 2330 hardened steel.

In order to determine the component life it is first necessary to find N_i for each condition or i = 1, 2, 3, and 4. The following table summarizes the results of the necessary calculations and the taking of endurance values, N_i, from the S-N curve.

Table 8.2 Summary of stresses and associated cycles to failure.

Loading Condition	σ_m (ksi)	σ_a (ksi)	σ_{ao} (ksi)	N_i (from S-N)	$d_i=n_i/N_i$
1	0	50	50	∞	0
2	125	50	100	6×10^4	10^{-3}
3	100	96	160	4×10^3	10^{-3}
4	0	190	190	10^3	0.5×10^{-3}

It should be noted that the completely reversed, stress amplitudes were calculated using Equation (8.2b) since the mean stress, $\sigma_m \neq 0$. Per Miner's rule, the damage accumulated in one day of wonderful PA roads will be the ratio of the individual (daily) cycles, n_i to the total number of cycles, N_i for a given stress level. Using this construct, the accumulated damage is:

$$D = 0+0.001+0.001+0.0005 = 0.0025$$

Since damage must equal unity at failure, the total number of days to failure times the daily damage must equal the same or:

$$D\,(0.0025) = 1.0.$$

Solving for the number of days until failure results in D = 400 days or a little over one year; one better have a good warranty! It should be noted that although the answer is "magically" an integer for this example, most if not all calculations will not be a nice round number. In these less than magical instances, the answer must be *rounded down* as a conservatism unless you are required (or willing) to predict a fractional life.

8.6 NOTCH EFFECTS

Although not explicitly mentioned in the preceding sections, notches or notch-like discontinuities will contribute to the fatigue endurance of a structural member in two important ways. Firstly, the notch may operate as a virtual crack, especially with a small radius of curvature so that it may be construed as "sharp." Even if the radius is relatively large, the discontinuity will certainly act as a stress riser with the elastic stresses concentrated at a local level. This is where it gets potentially complicated; even though the stress state is nominally elastic, fatigue damage is essentially an inelastic process and enhanced plasticity at or near the defect may actually result in a greater endurance than expected, even if the concentrated stress was used to predict endurance directly from S-N data.

In order to include these effects, a *fatigue stress concentration factor*, K_f, is usefully defined as the ratio of endurance limits for unnotched specimens to that of notched members, or simply:

$$K_f = \frac{\sigma_e(\text{unnotched})}{\sigma_e(\text{notched})} \tag{8.5}$$

For materials without clear endurance limits, any fatigue strength based on 10^6-10^8 cycles could be used in Equation (8.5) as suggested earlier. Clearly, K_f must be determined experimentally as it is a material response. No matter how unlikely the situation, if $K_f = 1.0$, notched and unnotched samples will exhibit the same endurance, and stress risers have no effect whatsoever. At the other (and more likely) extreme, $K_f = K_\sigma$, in which case the full value of the concentrated stress is the amplitude that determines endurance from un-notched S-N data. Intermediate cases may be scaled by the *fatigue notch sensitivity factor* that is denoted herein as q:

$$q = \frac{K_f - 1}{K_\sigma - 1} \tag{8.6}$$

As indicated by Equation (8.6), values of q = 1.0 implies a highly notch-sensitive material, whereas q = 0 implies $K_f = 1.0$ and a notch-insensitive solid. Assuming q = 1.0 in design practice would constitute a highly conservative approach and may be the only sensible path if reliable values of K_f are not known.

8.7 THE PHYSICAL NATURE OF FATIGUE FAILURE

The fatigue failure process is composed of three distinct, physical phenomena that are ultimately related. Assuming there are no crack-like defects present in the structure, a crack must initiate at some point. Usually, preexisting flaws (often at the surface) will be present from manufacturing processes such as welding, grinding or machining, and/or due to handling or aggressive environmental attack. When the structure is essentially unflawed, the initiated crack will be found in the regions of highest stress amplitude including local stress risers, material discontinuities on a micro-scale, or locations of high residual (tensile) stress.

In metallic materials, the slip deformation process may actually generate micro-cracking where none previously existed. Under this scenario, initial micro-cracks are propagated along planes of maximum cyclic shear stress in what is known as stage I growth. The termination of stage I growth is therefore and appropriately referred to as *fatigue crack initiation.* It should be noted that some materials seem to possess a cyclic threshold stress, below which initiation does not appear to take place. In S-N testing, the endurance limit may simply be the threshold stress or higher, the later implying that cracks may initiate but subsequently arrest.

During the next portion of fatigue crack growth known as stage II, the crack seeks and propagates along planes with the maximum tensile stress amplitude. If the crack encounters a compressive field such as seen in bending, it will choose the path of least resistance and go mixed mode. The final phase of crack growth known as stage III is *fast fracture*, where the crack has reached the critical size for the local stress state. Figure 8.11 schematically illustrates the various stages of crack initiation and growth on a S-N diagram. Note that the propagation phase accounts for most of the endurance limit at high stress magnitudes, whereas initiation is the dominant phase for stress levels near the endurance limit.

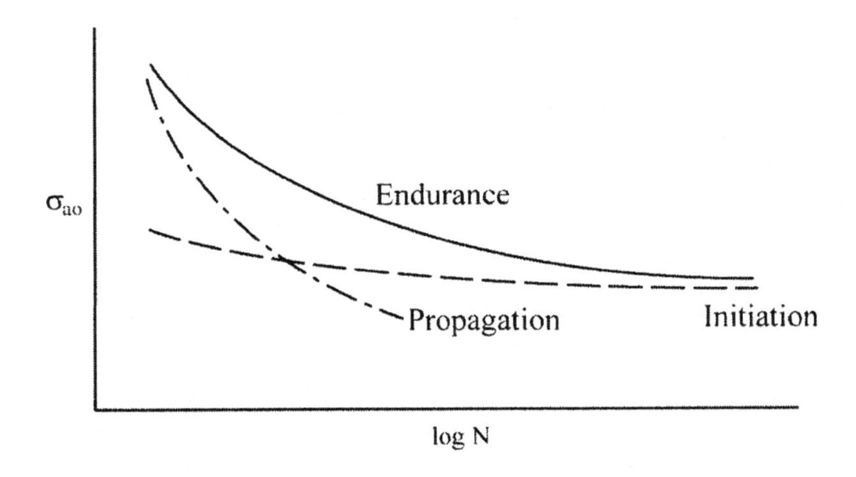

Figure 8.11 Schematic of the total fatigue endurance compared to the relative contributions crack initiation and propagation.

8.8 FATIGUE CRACK PROPAGATION

Predictions of fatigue endurance in a structural member can be accomplished in either one of two ways. The first is the traditional or macroscopic approach (Sections 8-2 through 8-6) where simple test specimens are used to observe behaviors without any concern about the underlying defects. An alternative and more microscopic approach is to predict the number of cycles that is required to propagate a specific fatigue crack to a critical size; the latter technique implies knowledge of a particular flaw, as well as its actual growth rate. Figure 8.12 illustrates such growth of a typical fatigue crack subjected to a constant amplitude cyclic stress with a_0 representing the size of some preexisting crack or flaw. Note that crack growth proceeds slowly at first, but accelerates with continued cycling as the crack lengthens and the corresponding stress intensity factor (SIF) increases with each cycle.

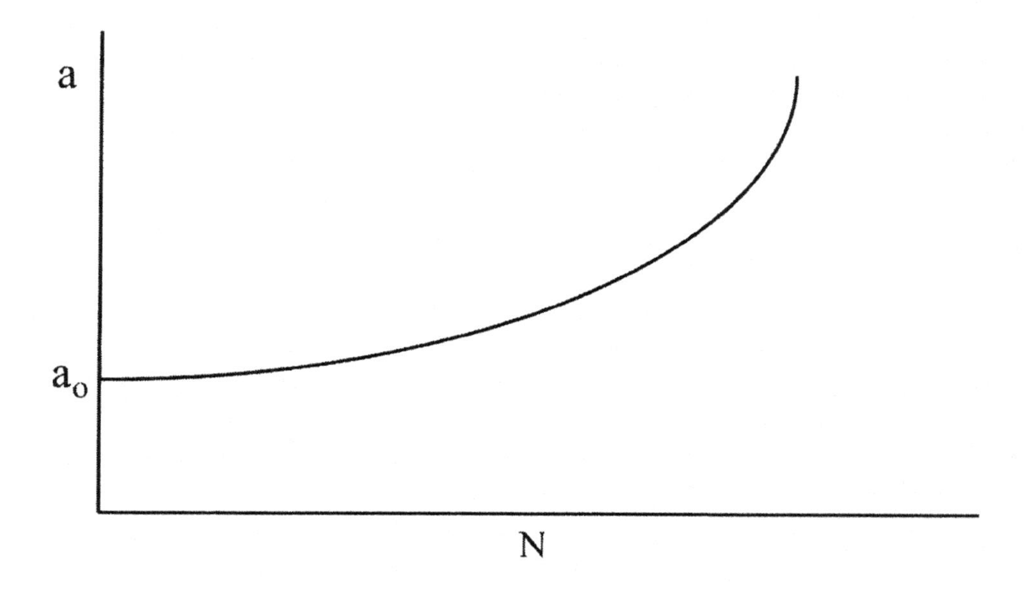

Figure 8.12 Schematic growth of a fatigue crack as a function of constant amplitude cyclic stress.

As discussed earlier, there is a distinct plastic zone surrounding the crack tip where irreversible (damage) processes are taking place; the size of this zone for Mode I cracks is proportional to:

$$r_p \propto \left(\frac{K_I}{\sigma_y}\right)^2 \tag{8.7}$$

For the case of cyclic loading, the stress range $\Delta\sigma = \sigma_{max} - \sigma_{min}$ is clearly the loading parameter of interest. There is then, a related stress-intensity factor range ΔK_I from Equation (7.10), where:

$$\Delta K_I = Y\Delta\sigma\sqrt{\pi a} \tag{8.8}$$

Given this proportionality, it is not unreasonable to expect that the fatigue crack growth rate, denoted herein as da/dN, will be proportional to ΔK_I. This supposition is in fact borne out as shown by the schematic sketch of a typical da/dN response curve for a structural alloy correlated with the logarithm of ΔK_I in Figure 8.13. Examination of Figure 8.13 reveals that there is an apparent threshold value of ΔK_{th}, below which a crack will not propagate; this threshold is somewhat (but not entirely) analogous to the endurance limit seen in the S-N response discussed earlier and has obvious design significance. However, the very existence of ΔK_{th}, as well as specific (measured) values of it remains an area of some controversy because little things like observer impatience may ultimately affect its determination.

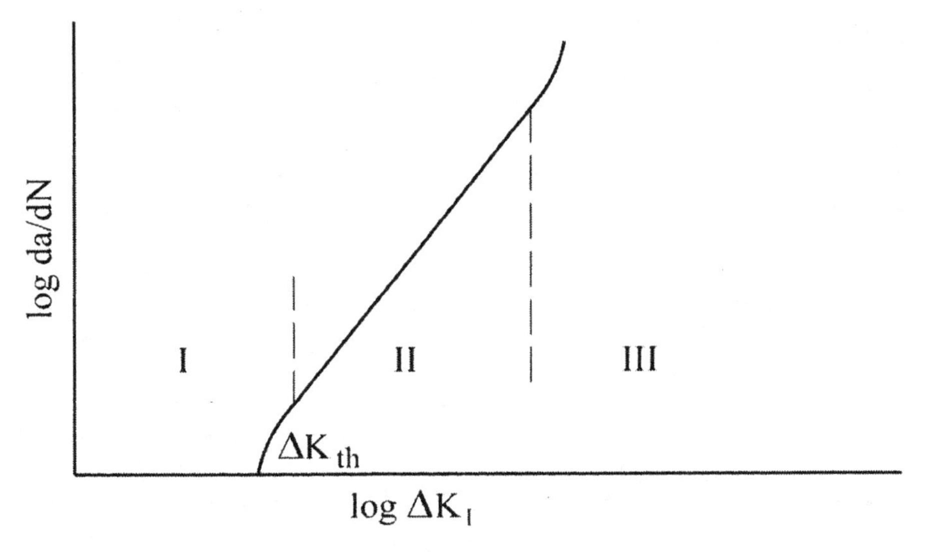

Figure 8.13 Fatigue crack growth rate, da/dN versus the stress intensity factor range ΔK_I for a typical structural alloy.

In any case, region II as shown in Figure 8.13 is the response of interest from an engineering standpoint, as it may well constitute nearly all of the useful lifetime of a member containing a propagating fatigue crack. Because Region II is approximately a straight line for a log-log correlation of *da/dN* with ΔK_I, a power law is suggested for relating the two quantities. *Viola*, the ensuing relationship given by Equation (8.9) is often referred to as the *Paris Crack Propagation Law* (after P.C. Paris) and his work in the early 1960s.

$$\frac{da}{dN} = A\left(\Delta K_I\right)^n \tag{8.9}$$

where A and n are both material dependent constants. Since it is the endurance as defined by the number of cycles to failure, N_c that is ultimately sought, the relationship can be rearranged and integrated to yield a more useful form:

$$\int_{N_o}^{N_c} dN = \int_{a_o}^{a_c} \frac{da}{A\left(\Delta K_I\right)^n} \tag{8.10}$$

Here, N_o may be the actual number of cycles accumulated as the crack has grown to an initial length a_o, or N_o can simply be set at zero because N_c is the additional endurance available. The final or critical crack length, a_c may be found from the crack instability criterion defined earlier:

$$K_{IC} = Y\sigma_{max}\sqrt{\pi a_c} \tag{8.11}$$

Thus, from a crack propagation viewpoint of fatigue, the fracture toughness serves to define fatigue endurance by fixing the critical flaw size. On the other hand, the *initial* or starting crack size, a_o is usually a more difficult parameter to specify. If an easily observable macro-crack is the flaw in question, setting a_o is relatively straightforward. However, if the flaw in question has not been observed directly, accurately specifying it is definitely far more challenging.

The concept of an initial, but unobserved fatigue crack is best understood by examining the nondestructive evaluation (NDE) techniques used to probe structural elements for flaws. For the small flaws of concern to the designer, a certain probability can (at best) be assigned to detecting the flaw due to the inherent limitation of the NDE technology employed. Currently, the thresholds for flaw detection are approximately on the order of 0.001 inch (0.03 mm) depending on the method used. Typical probability of detection (POD) curves are shown in Figure 8.14 for the most common forms of NDE probing currently employed that include x-radiography, dye penetrants, and ultrasound radiation. As shown by the plots, there is a wide range of flaws that can be missed by any given method. *Whatever the limitation of the system, a_o must be fixed at that value as a conservatism if a flaw is not detected.*

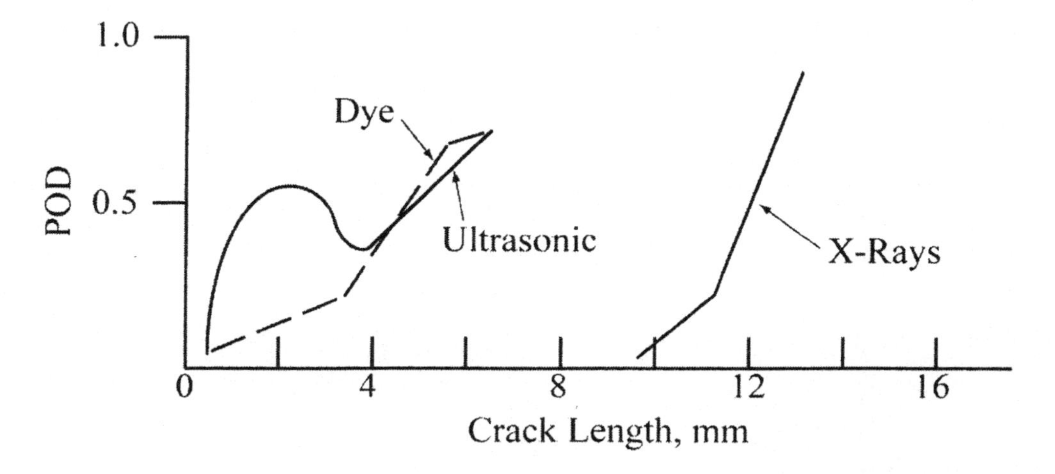

Figure 8.14 Probability of flaw detection for three NDE techniques.

While, the initial flaws may be smaller than the threshold, and N_c may therefore be an underestimate, such calculated lifetimes may serve as a rational basis for specific in-service flaw inspection schedules.

Although susceptible to the full gamut of experimental errors, the material parameters A and n in Equation (8.10) can be experimentally determined for the structural materials in question. Some average values for typical structural alloys in units of ksi and inches are:

Martensitic steels: $da/dN = 0.66 \times 10^{-8} (\Delta K_I)^{2.25}$

Ferritic (Pearlitic) steels: $da/dN = 3.6 \times 10^{-10} (\Delta K_I)^{3.0}$

Austenitic (stainless) steels: $da/dN = 3.0 \times 10^{-10} (\Delta K_I)^{3.25}$

Although the data scatter has proven to be far more extensive for aluminum and titanium alloys, some average values for A and n also in ksi and inches come in at:

Aluminum alloys: $da/dN = 8.2 \times 10^{-8} (\Delta K_I)^{2.14}$

Titanium alloys: $da/dN = 8.23 \times 10^{-13} (\Delta K_I)^{5.0}$

It must be noted that per the units of the constant, A, ΔK_I is always in units of either ksi√in or MPa√m, and da/dN in corresponding units of inches/cycle or meters/cycle.

8.9 FATIGUE FRACTOGRAPHY

Equation (8.10) implies that fatigue crack propagation occurs on an incremental basis for each loading cycle. Extensive observations on actual fatigue surfaces along with a host of physical models have established the validity of incremental growth in structural alloys. At first glance, the observer (using an unaided eye) might observe concentric lines that resemble successive crack front positions as shown in Figure 8.15 for a bicycle crank arm. These coarse lines are labeled "beach marks," since their appearance is much like the lines left on a sand beach as the tide retreats. Such lines are also referred to as "clamshell" markings with their foci in Figure 8.15 indicating the crack origin site (s). Although these markings are useful in identifying fatigue as the cause of failure, they do not represent individual crack fronts as one would expect. Indeed, when they are clearly visible as in Figure 8.15, they are indications of significant changes in the stress range (or contamination during the event), such as temporary overstressing, load removal, and so on.

If one used higher magnifications to examine the fracture surface where fatigue crack growth is suspected, a finer set of markings that are parallel to, and in between the beach marks can be observed (Figure 8.16). These markings or *striations* are indeed the true successive positions of the advancing crack front or da/dN; each line corresponds to an incremental position change of the crack front. However, the potential topography and magnifications required for fatigue fractography dictates that electron microscopy be employed for surface imaging.

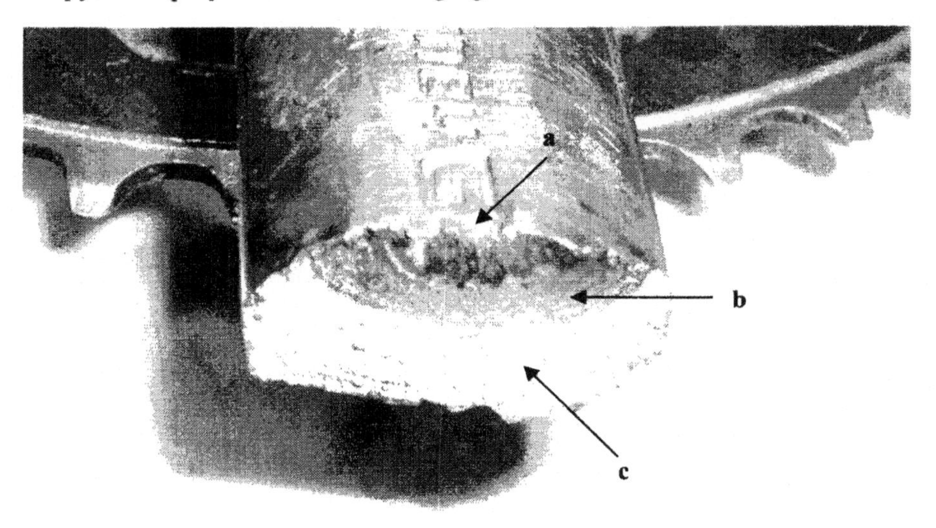

Figure 8.15 Fracture surface of a bicycle crank arm showing (a) initiation site, (b) beach marks, and (c) fast fracture surface.

For forensic studies, striation lines are truly useful as they can help determine the state of cyclic-stress present in the member during the crack propagation phase; the growth rate can be directly measured and therefore, determined from the striations seen in Figure 8.16. If the total crack length at the position of the striations is measurable, Equation (8.9) can also be used to calculate the stress intensity range, which in turn will provide the stress range and load as shown in Example 8.2.

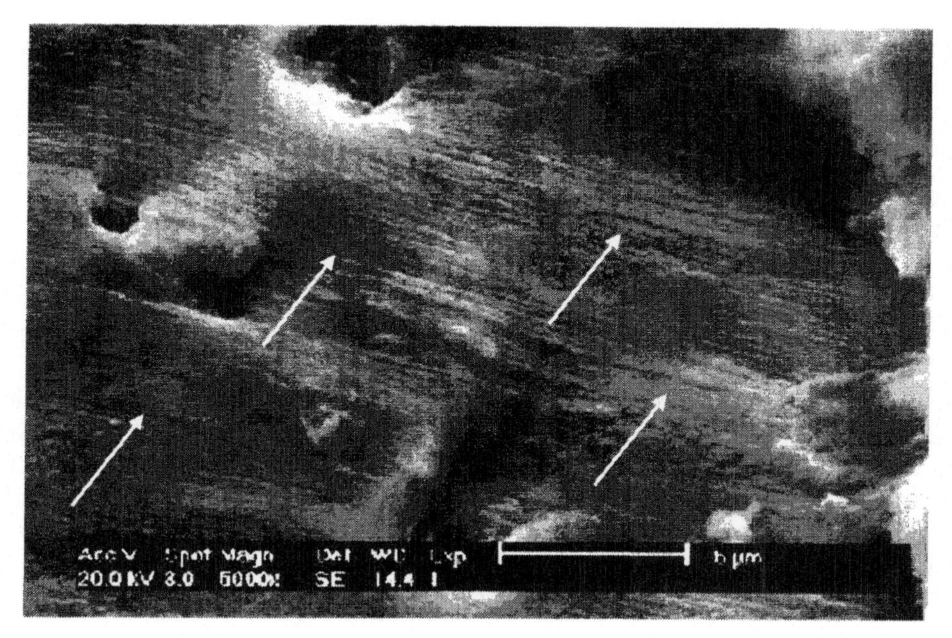

Figure 8.16 Fatigue crack striation lines (a few highlighted by arrows) representing da/dN as viewed in the scanning electron microscope from the failed crank arm of Figure 8.15.

Example Problem 8.2: It turns out that the crank arm shown in Figures 8.15 and 8.16 belonged to an adventurous person whose day was, shall we say, ruined when the crank failed. Because the ambulance chasing lawyers are smelling blood, you have been hired by the manufacturer to determine the cause of the failure and hopefully save their hides. Brinell hardness tests indicated the tensile yield stress to be around $\sigma_y = 28{,}000$ psi, an acceptable value for the aluminum casting alloy used. Note: the yield strength only serves to verify that the manufacturer had achieved reasonable or expected properties for the alloy and as a potential threshold for design stresses since they clearly should be in the elastic range.

Since striations are clearly present and the failure not a simple overload, you must now examine the possibility of fatigue and the underlying loads generated by the rider. The striation spacing measured from the photomicrograph of Figure 8.16 suggests that on average:

$$\frac{da}{dN} = 1.43 \text{ x } 10^{-5} \text{ (inches/cycle)}$$

From an analysis of the striations, their apparent end, and the start of fast fracture in Figure 8.16 indicates that:

$$a_{cr} = 0.213 \text{ inches}$$

Data for similar aluminum alloys suggest that for a given stress-intensity range, the Paris equation can be written as:

$$\frac{da}{dN} = 5.47 \text{ x } 10^{-15} \left(\Delta K_I\right)^{2.3}$$

Substituting in the measured value of da/dN and solving for the underlying stress intensity range:

$$\Delta K_I \approx 12,000 \text{ psi}\sqrt{\text{in.}}$$

In order to determine the stress level, you must first assume a crack type. Choosing a common semicircular ("thumbnail") surface crack, the stress range can then be approximated as:

$$\Delta K_I \approx \frac{2}{\pi} \Delta\sigma\sqrt{\pi a} \quad \rightarrow \quad \Delta\sigma \approx 23,000 \text{ psi}$$

Given the proximity of the predicted stress range to the yield strength of the material, the applied stresses were indeed, excessive and fatigue failure could be expected to occur at some point. For the dimensions of the crank arm under scrutiny, the determined stress-levels translate to a bending moment of 760 in-lb or a foot force of approximately 190 lbs. Sustained pedal loadings of at least 130 percent of the rider's weight have been shown to be possible for a serious cyclist as in the case. Clearly, the section size must be increased and/or a stronger and more fatigue resistant material should be specified for a successful redesign. Unfortunately for the crank manufacturer, the lawyers may actually have a case.

PRACTICE PROBLEMS

Questions 1 through 4 pertain to either a 2024-T4 aluminum alloy used for lightweight automobile components or a 130B titanium alloy used in nuclear attack submarines. The S-N endurance data (50 percent survival) for the two alloys is shown below. The tensile strengths are 70,000 psi and 155,000 psi for aluminum and titanium, respectively.

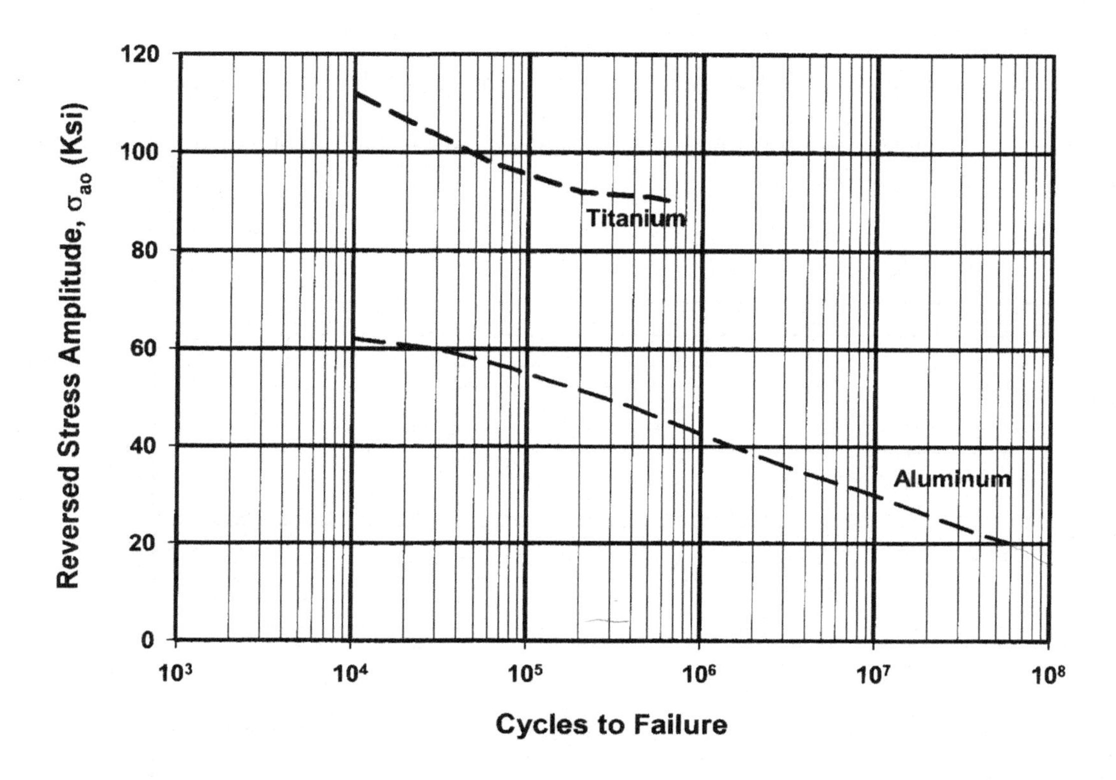

1. What is the fatigue strength for the 2024-T4 aluminum alloy used for lightweight auto components? If subjected to a cyclic stress featuring a mean stress of 15 ksi, what is the maximum allowable stress amplitude if 50 percent of all material specimens are to survive 10^8 cycles? Note: σ_{uts}=70 ksi

2. What is the endurance or fatigue limit for 130B titanium alloy used in nuclear attack submarines? If subjected to a cyclic stress amplitude of 60 ksi, what is the maximum mean stress that could be tolerated on the average, so that no failures are expected? Note: σ_{uts}=155 ksi

3. A chassis support member made from 2024-T4 aluminum accumulates the following cyclic stress histories:

Stress Condition	1	2	3
Maximum Stress	55 ksi	50 ksi	32.5 ksi
Minimum Stress	-55 ksi	20 ksi	2.5 ksi
# of Cycles	20,000	5,000,000	n_3

   Find the number of cycles n_3 for the stress condition 3 that the member can accumulate after experiencing the cyclic stress history of stress conditions 1 and 2. Base all calculations on 50 percent survival rates. Note: σ_{uts}=70 ksi

4. Ironically, the 130B titanium alloy submarine is being attacked by irate mermaids and is subjected to the daily stress history given below: Note: σ_{uts}=155 ksi

Stress Condition	1	2	3
Maximum Stress	118 ksi	95 ksi	128 ksi
Minimum Stress	38 ksi	-95 ksi	-24 ksi
# of Cycles	10,000	400	20

What is the maximum number of such days wherein 50 percent of all subs will not have failed in fatigue?

5. The crack growth rate data for a copper infiltrated, quench and temper 4600 steel prepared by powder metallurgical processing is given below.

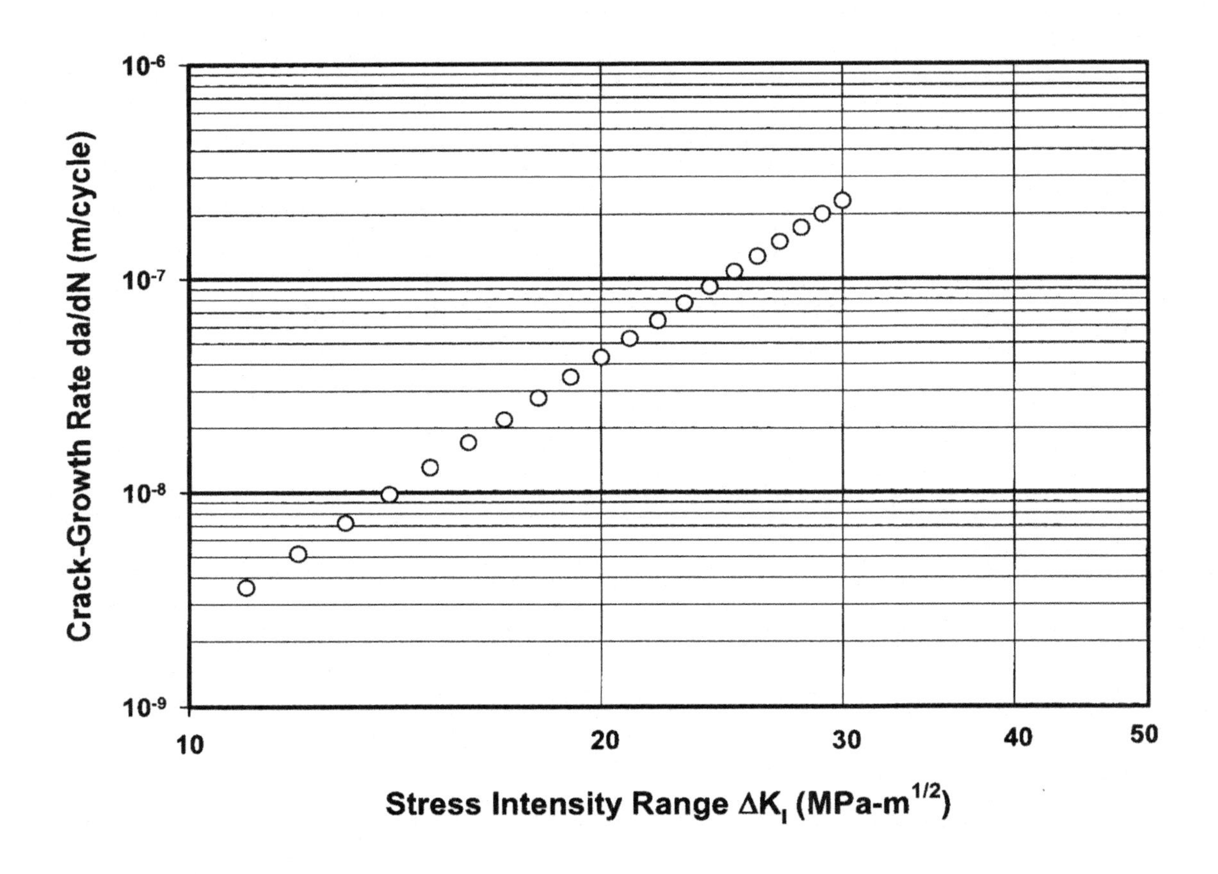

What are the values of the material parameters in the fatigue crack propagation law for this composite?

6. For a small edge or surface crack of depth "a" in a tensile stress field that ranges through $\Delta\sigma$, the stress intensity range is approximately:

$$\Delta K_I = \Delta\sigma\sqrt{\pi a}$$

Using the Paris equation, prove that the number of cycles to failure, N_c, is given by:

$$N_c = \frac{2}{(n-2)A\Delta\sigma^n \pi^{n/2}}\left[\frac{1}{a_o^{(n-2)/2}} - \frac{1}{a_c^{(n-2)/2}}\right]$$

Where a_o = initial crack size, a_c = is the critical or final crack size at instability.

7. An initial surface flaw with a depth of 0.001 inches is detected in the ferritic/pearlitic steel matter/anti-matter flux capacitor that makes temporal travel at warp speeds possible. If the fracture toughness of the steel is 60,000 psi$\sqrt{in}$, how many cycles will lead to failure for stresses ranging from zero to 20,000 psi? Note: assume $A=3.6x10^{-10}$, $n=3.0$, and $Y=1$.

8. The main instrumentation port for a high temperature coal gasifier consists of a large plate with a relatively small circular hole subjected to completely reversed stress amplitudes. For this port, it is safe to assume that the hole diameter is significantly smaller than the plate width. In addition, the steel is known to have an unnotched endurance limit of 50,000 psi. What is the maximum reversed stress amplitude that can be applied, if failure is never to occur, for (a) $q = 0$, and (b) $q = 1.0$.

CHAPTER NINE

VISCOELASTICITY

9.1 INTRODUCTION

Although not considered up to this point, time can affect both stress and strain distributions in structures. No doubt, any owner of eyeglasses with plastic frames gradually feels a loss of gripping power as the frame and arms slightly deform over time. Clothes tend to develop more wrinkles, even if you are able to avoid sleeping in them. Asphalt pavements often move/shift during the summer under the heat and loads from the cars passing by. On the other hand, securely tightened bolts (especially the plastic kind used in toy erector sets) may loosen with time, even in the absence of movement, vibrations, or any environmental effects that could explain the change. Although the above examples are primarily drawn from observations of organic materials, metals and ceramics can experience similar phenomena. However, in these instances, it is usually necessary to elevate the service temperature to values far in excess of everyday ambient temperatures. The present chapter addresses the interrelation of stress, strain and time for those service situations where temperatures are sufficient (but not necessarily high) to produce time-related effects. Higher temperature behaviors of various engineering materials are considered with more detail in Chapter Ten.

9.2 VISCOELASTIC MODEL ELEMENTS

In order to investigate such time-related affects, we must first look at how a material (such as common plastics) might respond under load. Usually, the instantaneous response or deformation is no different from the materials already discussed (alloys, ceramics, etc.). However, the deformation (or conversely, the stress) will usually begin to change over time, even if the load or imposed deformation remains constant. Hence, there appears to be both an elastic and time dependent response that must be considered.

As discussed in earlier chapters, elastic behavior may be idealized to correlate stress and strain linearly. Such a Hookean response for fully elastic materials is the same type that is associated with perfect springs:

$$F = kx \quad \text{or} \quad \sigma = E\varepsilon \tag{9.1}$$

In this case, the mechanical analog will be in the form of a *spring element* assumed to obey Equation (9.1). In contrast, a fluid-like response as we would normally associate with thick, gummy, or viscous fluids can be described by a variety of stress, strain and time relationships. The simplest form of viscous fluid response is the *Newtonian fluid*, wherein:

$$\frac{d\varepsilon}{dt} = \frac{1}{\eta}\sigma \tag{9.2}$$

The material parameter, η is the *viscosity* and represents a measure of resistance to deformation with time, t. For this behavior model, it must be assumed that the rate of deformation is proportional to the stress with the mechanical analog taken to be a *dashpot element* or piston-in-cylinder mechanism; unlike the instantaneously responding spring, it takes time to move the piston as the viscous fluid must ooze past it for the device to deform. Many automatic door closure devices feature springs to pull the door shut and dashpots to control the rate of closure and prevent slamming.

Conceptually, the analog element are shown below in Figure 9.1:

Figure 9.1 Mechanical analog elements under stress with an elastic spring and time-dependent dashpot.

Starting with the well established stress/strain relationship in the linear regime:

$$\varepsilon = \frac{1}{E}\sigma \tag{9.3a}$$

Differentiation with respect for time, *t* under the assumption that strain and stress are functions of time:

$$\frac{d\varepsilon}{dt} = \frac{1}{E}\frac{d\sigma}{dt} \tag{9.3b}$$

and from Equation (9.2):

$$\frac{d\varepsilon}{dt} = \frac{1}{\eta}\sigma \tag{9.3c}$$

As mentioned earlier, the material parameter, η is the *Viscosity* and is a measure of the resistance to deformation with time (units of stress time). Generally, viscosity decreases with increasing temperature, and vice versa; just picture molasses on hot versus cold days. As long as viscosity is changing smoothly with temperature, and other events such as combustion are not evident, one may assume:

$$\eta^{-1} = \eta_0^{-1} e^{-\Delta H/kT} \tag{9.4}$$

Equation (9.4) is the usual *Arrhenius* type of expression governing nonquantum mechanical rate phenomena. In Equation (9.4), ΔH is the *activation potential energy* necessary for the process to occur such as macromolecules sliding by each other. In addition, k is the *Boltzman constant* ($k = 1.39 \times 10^{-23}$ J/ °K or 6.79×10^{-23} in-lb/ °R) and T the *absolute* temperature. While relatively simple in nature, the two elements (spring and dashpot) are sufficient to develop reasonably accurate viscoelastic models and response equations.

9.3 MAXWELL VISCOELASTIC MODEL

Because the spring and dashpot represent portions of the overall response, mechanical models may be constructed from these two elements. In general, the model must produce a specific constitutive equation of the form:

$$\varepsilon = f(\sigma, t) \tag{9.5}$$

Inherently, the functional form of Equation (9.5) will contain material variables and temperature dependence through Equation (9.4). Using this form of constitutive relationship, the simplest viscoelastic model and named after the prolific nineteenth-century scientist Maxwell, is the series arrangement of a single spring and dashpot shown in Figure 9.2 .

Figure 9.2 The Maxwell model for viscoelastic material response.

For the Maxwell model, the derivation starts with the usual (and still necessary) assumption of equilibrium. As such, the free body diagrams of the spring or the dashpot yields a statically determinate situation:

$$\sigma = \sigma_s = \sigma_d \tag{9.6a}$$

where the subscripts "s" and "d" refer to the spring and dashpot elements, respectively. In this case, the element strains are independent since the two are in series:

$$\varepsilon = \varepsilon_s + \varepsilon_d \tag{9.6b}$$

Differentiating with respect for time, the following expression results:

$$\frac{d\varepsilon}{dt} = \frac{d\varepsilon_s}{dt} + \frac{d\varepsilon_d}{dt} \tag{9.6c}$$

Substituting the constitutive equations (Equations (9.3)) into the compatibility expression, and employing the equivalence between the total applied stress and model element stresses, the complete response equation for a *Maxwellian solid* becomes:

$$\frac{d\varepsilon}{dt} = \frac{1}{E}\frac{d\sigma}{dt} + \frac{1}{\eta}\sigma \tag{9.7}$$

At this point, the only remaining question is whether Equation (9.7) accurately predicts real material behavior. To answer this question, it is first necessary to identify and model two specific viscoelastic behaviors, namely *stress relaxation* and *creep*. Once appropriate and hopefully realistic models are derived, the next task is to develop physical tests to probe each, and finally, compare prediction to physical reality.

9.4 VISCOELASTIC TESTING

In some materials, the stresses and underlying loads can actually relax over time in a process appropriately named *stress relaxation*. As the name implies, it is the stresses (and internal loads that generate them), and not the deformation that diminish over time; stress relaxation experiments require that a specimen be quickly deformed (the mode is usually tensile), the deformation be fixed or held constant, and the resulting load/stress be monitored over time as shown in Figure 9.3. In addition to being a common viscoelastic test procedure, stress relaxation may be encountered in load-bearing designs such as the bolt or rivet fasteners mentioned earlier. Depending on the design of the frame and materials used, the aforementioned eyeglasses may also be an example of stress relaxation.

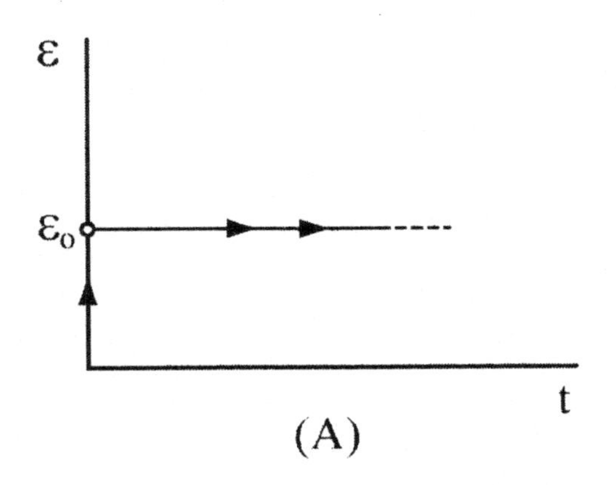

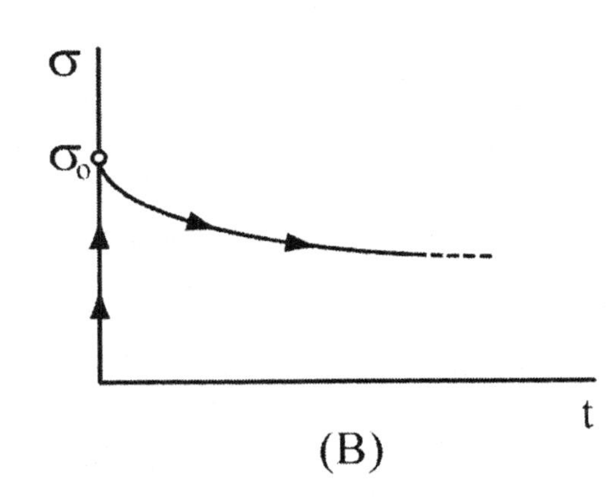

Figure 9.3 (A) Imposed strain history and (B) the resulting stress response observed during stress relaxation of a viscoelastic material.

The other component to viscoelastic behavior relates to the continuous deformation over time, even when the load/stress magnitude is held constant. In this instance, the continued deformation is denoted as *creep*. There is nothing scary or ghoulish about this behavior, the name simply relates to the very slow rate of deformations that just seems to "creep" along. In a creep experiment, a tensile specimen is rapidly loaded and held that way for some defined period. The strains that result from the loading, including those that accumulate with time are then measured and recorded. Figure 9.4 indicates a defined stress history (constant in this case) and the observed strain-response over time. Note that the strains accumulate positively with time for tensile loading as the specimen "creeps" forward in length.

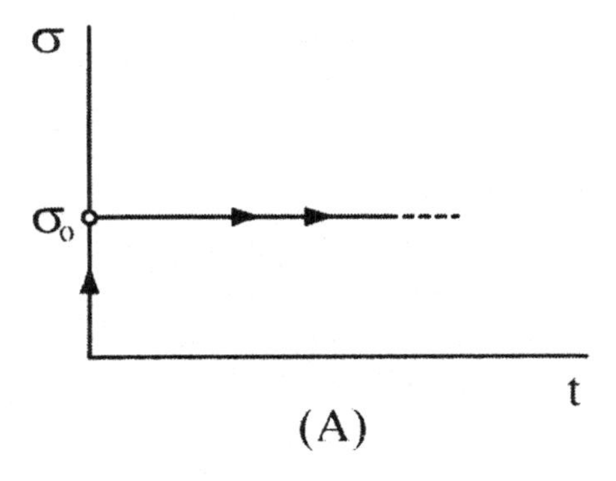

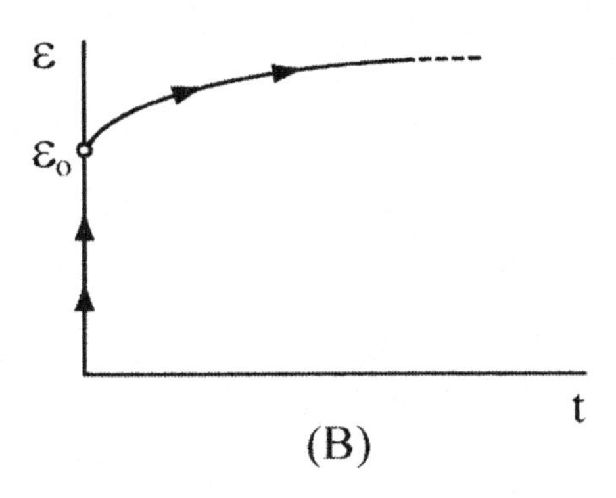

Figure 9.4 (A) Imposed stress history and the ensuing strain response (B) observed during the ensuing creep of a viscoelastic material.

Creep is actually a structural loading of great practical interest; many high-temperature devices such as those used to convert fossil or nuclear fuels into usable energy are often limited in process temperatures attainable due to the creep. In addition, the motion of roadbed asphalt due to vehicular loading is a more commonly observed manifestation of creep processes at work.

9.5 MAXWELL MATERIAL RESPONSE

For the situation of *stress relaxation*, the differential response defined by Equation (9.7) must be solved for the defined strain history ε = constant = ε_o such that $d\varepsilon/dt = 0$.

In this case:

$$\frac{d\varepsilon}{dt} \equiv 0 = \frac{1}{E}\frac{d\sigma}{dt} + \frac{1}{\eta}\sigma \tag{9.8a}$$

Stress and time terms can now be separated and the resulting solution can be assumed to be of the following form:

$$\sigma(t) = Ae^{Bt} \tag{9.8b}$$

In the above equation, A and B are constants to be particularized for the present case by using initial conditions; at time t = 0, the stress is known to be equal to σ_0, thus, A = σ_0. Substituting the solution $\sigma = \sigma_0 e^{Bt}$ into the general Maxwell response equation under constant strain conditions ($d\varepsilon/dt = 0$), one obtains:

$$0 = \frac{1}{E}\left(\sigma_0 Be^{Bt}\right) + \frac{1}{\eta}\sigma_0 e^{Bt} \tag{9.8c}$$

This results in $B = -E/\eta$ with units of inverse time. Finally, the equation governing the stress relaxation of a Maxwell material as a function of time is a decaying exponential of the form:

$$\sigma(t) = \sigma_0 e^{-\frac{Et}{\eta}} \tag{9.9a}$$

Exponentially decaying responses such as this are often characterized by a *relaxation time* or *time constant, τ*. Using this form of the exponent, the decaying stress can be alternately expressed as:

$$\sigma(t) = \sigma_0 e^{-t/\tau} \tag{9.9b}$$

where $\tau = \eta/E$. When time t = τ has elapsed, the stress has relaxed to 37 percent of its original value, σ_0. Taking the natural logarithm of both sides of Equation (9.9b), one obtains:

$$\ln\sigma = \ln\sigma_0 - t/\tau \tag{9.9c}$$

Logarithmic pleasures aside, Equation (9.9c) affords an easy means of deciding whether an actual stress-time decay record is exponential, and the material is therefore Maxwellian; if the material is indeed a Maxwell solid, the log of stress versus time should be a straight line with the slope $-1/\tau$ or $-E/\eta$.

Unless service temperatures are extremely high, say in excess of 80 to 90 percent of the melting temperature, most materials do not relax as rapidly or completely as predicted by Equation (9.9). Consequently, the Maxwell prediction will only give approximate predictions of actual stress relaxation magnitudes. Finally, the relaxation time, τ may be employed as a gage factor for the assignment of relevant terms such as "fast" or "instantaneous," as opposed to "slow" or "static." Times that are of the order of τ are such that a significant viscoelastic response can occur. Conversely, times that are orders of magnitude less than τ do not allow for viscoelasticity to manifest itself, so only elastic deformations occur in such "fast" time intervals.

When the structural loading is well-defined and held constant as in *creep*, then $\sigma = \sigma_0$ and $d\sigma/dt=0$ such that the Maxwell response can be written as:

$$\frac{d\varepsilon}{dt} = \frac{1}{\eta}\sigma_0 \qquad (9.10a)$$

In this case, increments of strain accumulate linearly with time such that:

$$d\varepsilon = \frac{1}{\eta}\sigma_0 dt \qquad (9.10b)$$

If ε_0 represents the measured strain produced under the instantaneous imposition of load or stress, σ_0, then:

$$\varepsilon(t) = \varepsilon_0 + \frac{\sigma_0}{\eta} t \qquad (9.11)$$

Interestingly, Equation (9.11) predicts that a Maxwell material accumulates creep strains linearly with time; while not necessarily a bad approximation to reality, most materials in their (moderate) service temperature ranges exhibit creep rates that are decidedly nonlinear with respect to stress. Hence, the unfortunate reality is that most materials are not observed to exhibit a Maxwell type responses in both stress relaxation and creep, unless temperatures are in excess in 90 percent of the melting point or glass transition temperature. Such temperatures are usually associated with drastic losses in static strength and are not common for intentional service. Despite the predictive shortcomings of the Maxwell model, it is still widely employed for approximate structural analysis, largely due to its mathematical simplicity.

Example Problem 9.1: A sample of 1100-H14 aluminum alloy used for a high-pressure fuel injector was rapidly strained to a stress of σ = 15,000 psi at a temperature of 131°F. Under this imposed stress and at a constant temperature, the following stress/time data were noted:

Tensile Stress (psi)	Time (minutes)
15,000	0
13,400	15
11,600	30
10,000	45
9,000	60

Can the material response be predicted adequately by the Maxwell stress relaxation relationship? If so, what is the time constant τ?

The first step is to plot the stress/time data semi-logarithmically for comparison to Equation (9.9) as shown in Figure 9.5. To the unaided eye, that data points clearly seem to lie on a straight-line as required for Equation (9.9c) to be valid. To calculate the time constant, employ Equation (9.9c) in the following form based on the data at t = 0 and t = 60 minutes:

$$\tau = -\frac{\Delta t}{\Delta(\ln\sigma)} = -\frac{60(\text{min})}{(9.10 - 9.62)} = 117 \text{ minutes}$$

If the modulus of elasticity of the aluminum is E = 10 x 10^6 psi at the relaxation temperature, then its viscosity is η = 11.7 x 10^8 psi-minute. Finally, it should be noted that the aluminum appears to possess a Maxwell response only so far as tested; it would be surprising if its true relaxation was as drastic as an extrapolation of the present data suggests.

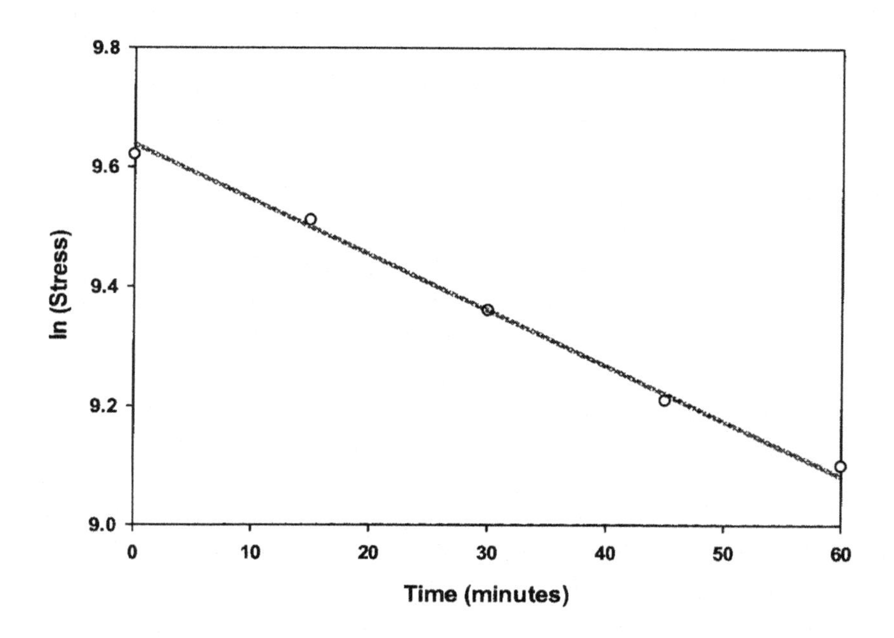

Figure 9.5 Natural log of the tensile-stress as a function of time for Example 9.1.

9.6 VOIGHT-KELVIN AND COMPOUND VISCOELASTIC MODELS

Given the limitations of the Maxwell model, alternative mechanical arrangements might yield improvements. Along these lines, the Voight-Kelvin two-element parallel arrangement of Figure 9.6 is one possibility.

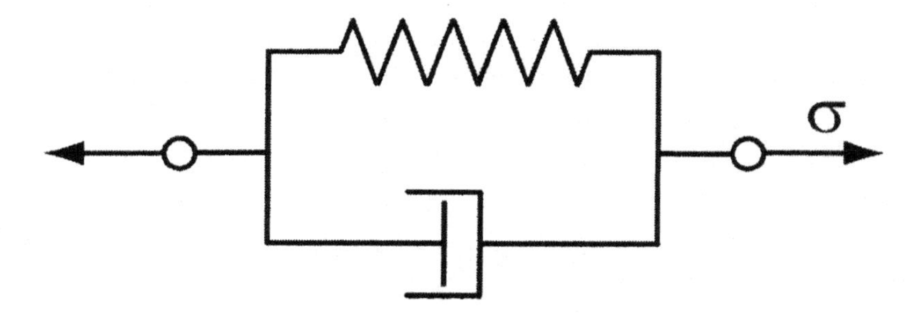

Figure 9.6 Voight-Kelvin viscoelastic response model.

In this configuration, a free body diagram of the spring or the dashpot in a parallel arrangement yields a statically indeterminate situation:

$$\sigma = \sigma_s + \sigma_d \tag{9.12a}$$

When the elements are parallel, the strains are no longer independent, so deformation compatibility between them must be maintained:

$$\varepsilon = \varepsilon_s = \varepsilon_d \tag{9.12b}$$

Differentiating with respect to time, the following expression is obtained:

$$\frac{d\varepsilon}{dt} = \frac{d\varepsilon_s}{dt} = \frac{d\varepsilon_d}{dt} \tag{9.12c}$$

Recalling the constitutive equations:

$$\varepsilon_s = \frac{1}{E}\sigma_s \quad \text{or} \quad \frac{d\varepsilon_s}{dt} = \frac{1}{E}\frac{d\sigma_s}{dt} \tag{9.12d}$$

and

$$\frac{d\varepsilon_d}{dt} = \frac{1}{\eta}\sigma_d \tag{9.12e}$$

Substituting the constitutive relations into the equilibrium statement, and noting that all strains are equivalent, one finally obtains the differential equation of the Voight-Kelvin material model:

$$\eta\frac{d\varepsilon}{dt} + E\varepsilon = \sigma \tag{9.13}$$

Although showing a simple (and certainly desirable) elegance, there are notable shortcomings to this model. For instance, the model can only deform with time, and not instantaneously; thus, the rapid elastic straining commonly observed in most materials is not possible. Secondly, if the level of strain is held constant, then $d\varepsilon/dt = 0$ and the viscous element is deactivated; in this case, $E\varepsilon = \sigma$ in Equation (9.13) and no relaxation is possible.

Example 9.2: While the Voight-Kelvin model just discussed finds its usage limited, the concept can be employed with more elements that can be designated as "compound." For instance, a three-element compound model shown in Figure 9.7 could be derived as follows.

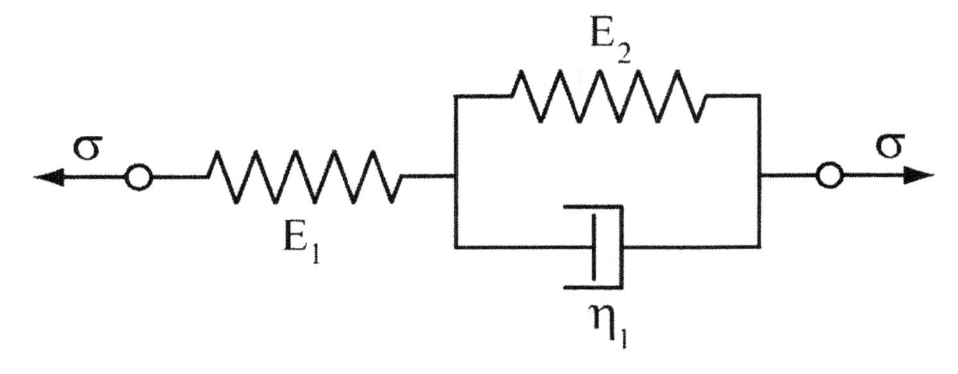

Figure 9.7 Three-element viscoelastic model.

We again begin with the all too familiar equilibrium for spring denoted as E_1:

$$\sigma = \sigma_{s1}$$

and for the Kelvin section for spring E_2

$$\sigma = \sigma_d + \sigma_{s2}$$

In terms of compatibility:

$$\varepsilon = \varepsilon_{s1} + \varepsilon_{kelvin} = \varepsilon_{s1} + \varepsilon_{s2} = \varepsilon_{s1} + \varepsilon_d$$

The constitutive relations take on the following form:

$$\sigma_{s1} = E_1 \varepsilon_{s1}$$

$$\sigma_{s2} = E_2 \varepsilon_{s2}$$

and

$$\frac{d\varepsilon_d}{dt} = \frac{1}{\eta}\sigma_d$$

Solving these equations simultaneously, one arrives at the constitutive relationship for the entire model:

$$\frac{d\varepsilon}{dt} + \frac{E_2}{\eta}\varepsilon = \frac{1}{E_1}\frac{d\sigma}{dt} + \frac{\sigma}{\eta}\left(1 + \frac{E_2}{E_1}\right)$$

Using this new and improved three-element formulation, the model can now respond elastically to instantaneous loadings (through spring E_1), as well as provide a reasonable stress relaxation response. Unfortunately, the new model reaches a terminal value of creep strain in an unreal fashion.

In reality, there is no end to the multiple-element models that might be invented, beginning with the four-element configuration of a Maxwell arrangement in series with a Voight-Kelvin assemblage shown in Figure 9.8; such a model would feature two elastic constants, E_1 and E_2, as well as two viscosities, denoted η_1 and η_2.

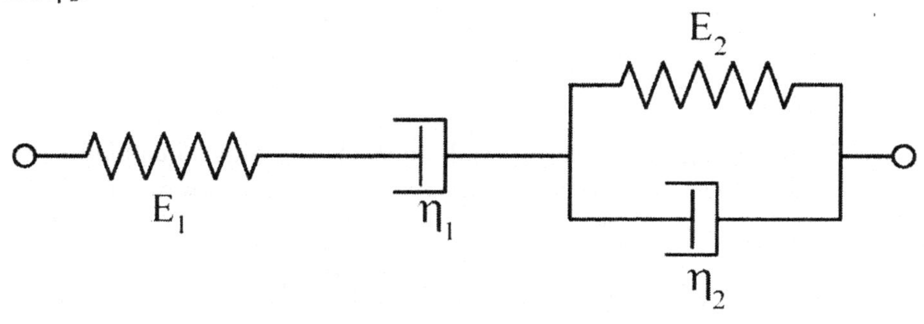

Figure 9.8 Four-element model for a viscoelastic response.

Although the differential response equations can be readily obtained via the analytical steps just described, the solutions become more and more intractable. Generally, numerical solutions become the only practical means of solving the response equations of such complex multi-element models.

9.7 GENERAL LINEAR VISCOELASTIC RESPONSE

A more formal approach to describing the viscoelastic response of a material would tend to dispense with mechanical analog models altogether, and utilize generalized linear differential response equations. In this generalized case, nonzero coefficients would be chosen to represent various elastic and viscous components of the actual response. Using this approach, one may then write for any component of stress, σ_{ij} and strain, ε_{ij}:

$$\sum_{m=0}^{a} A_m \frac{d^m \sigma_{ij}}{dt^m} = \sum_{n=0}^{b} B_n \frac{d^n \varepsilon_{ij}}{dt^n} \tag{9.14}$$

Nonzero values of the coefficients A_m and B_n are independent of stress, implying that strain is proportional to stress at any chosen time. Note that for the tensile response of a Maxwell material, B_1 = unity with all subsequent coefficients zero or $B_n = 0$. In addition, $A_0 = 1/\eta$, $A_1 = 1/E$, and all other values of A_m are zero. Finally, for large ranges in service stress, many polymeric materials display a more complex and nonlinear relation between stress and strain-accumulation.

9.8 REAL POLYMER VISCOELASTICITY

Owing to the complexities of each molecular structure, the proportionalities between stress, strain, time, and temperature vary greatly from one polymeric solid to another. Given these varieties, it is sometimes advisable to utilize responses that are experimentally determined, rather than attempting to invent and then force fit models. The procedures for doing this in terms of the expected behaviors are outlined below.

Creep Compliance: For conditions of constant stress, creep-strain accumulation maybe related to the applied stress by the following relationship:

$$\varepsilon(t) = D(t)\,\sigma \tag{9.15a}$$

where *D(t)* is the time-dependent *creep compliance* of the material. Although usually defined experimentally, the creep compliance may be related to specific viscoelastic models. For example, if one had a Maxwell solid, $D(t) = (1/E + t/\eta)$. However, Equation (9.15a) may also be applied to any material including nonlinear ones, because *D(t)* can be any time-dependent function that matches physical reality. For a three-dimensional systems of imposed stress, the tensile-derived creep compliance, *D(t)*, can be used to relate *equivalent strain,* $\bar{\varepsilon}$ and *equivalent stress,* $\bar{\sigma}$ as:

$$\bar{\varepsilon}(t) = D(t)\,\bar{\sigma} \tag{9.15b}$$

Finally, if the material is indeed a linear viscoelastic solid, rather than solving Equation (9.14), one can apply what is know as the *Boltzmann superposition principle*; that is, if a new value of stress $\sigma_{ij,p}$ is applied at time, t_p, then:

$$\varepsilon_{ij}(t) = D(t)\,\sigma_{ij,0} + D(t-t_1)\,\sigma_{ij,1} + D(t-t_2)\,\sigma_{ij,2} + \ldots \tag{9.16}$$

Relaxation Modulus: Under conditions of constant strain, the time-dependent stress may be conveniently expressed in the following form:

$$\sigma(t) = E(t)\,\varepsilon \tag{9.17}$$

Here, *E(t)* is the *relaxation modulus* of the viscoelastic solid that is explicitly time-dependent and usually experimentally defined. Although analytical models are not necessary to define *E(t)*, they can be used when available. For example, if the material were Maxwellian in its response, then $E(t) = E\exp(-Et/\eta)$. The utility of the relaxation modulus is in its ability to relate equivalent values of stress and strain:

$$\bar{\sigma}(t) = E(t)\,\bar{\varepsilon} \tag{9.18}$$

9.9 EXPERIMENTAL RESPONSE DETERMINATION

As stated in the previous section, the creep compliance *D(t)* and relaxation modulus *E(t)* maybe obtained from experimental creep and stress relaxation data. However, the creep and stress relaxation behavior of materials undergoing viscoelastic behavior are usually (and considerably) influenced by changes in temperature as aptly shown in Figure 9.9.

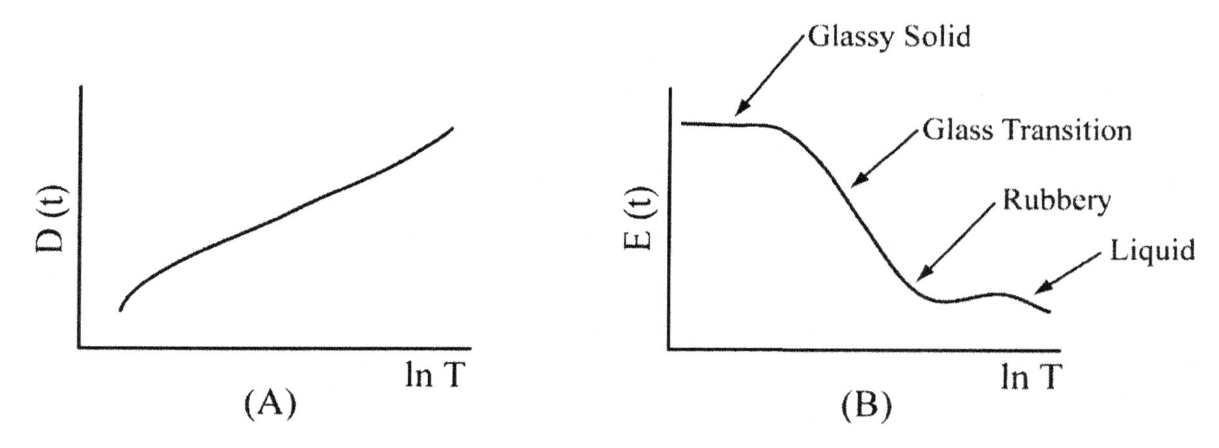

Figure 9.9 Anticipated mechanical response of noncrystalline materials as a function of temperature such as, (A) creep compliance, and (B) relaxation modulus.

As illustrated in the figure, both temperature and time plays a major rule in influencing the viscoelastic response of a material. Using the relaxation modulus *E(t)* as an example (Figure 9.9B), the behavior of noncrystalline or partially crystalline polymers is usually described by terms such as glassy or rubbery; in this case, glassy refers to the molecular configurations being more or less random as in the case of a super cooled liquid. As also shown in Figure 9.9, a rapid change in both creep compliance and relaxation modulus is seen during the transition region, indicating that the effect of temperature as well as time dependent effects must be considered if a viscoelastic design is to be successful.

Ideally, it would be most convenient to study the effects of time and temperature together and establish an all-encompassing relationship between them. Such a relationship would be decidedly advantageous, as it would allow tests at high temperatures over short intervals to assess longer-term behaviors. In turn, this would enable a master curve to be established from collected creep compliance and relaxation modulus data for any selected reference temperature. Fortunately, all of this is indeed possible by using the method of reduced variables, which affords a simple way of separating the two main variables of time and temperature in the phenomenological functions representing viscoelastic behavior.

For example, Figure 9.10a shows creep compliance curves for a linear viscoelastic material obtained at various temperatures. If the data of Figure 9.10A is cross-plotted against temperature for various selected times, the creep compliance curves have the distribution shown in Figure 9.9B.

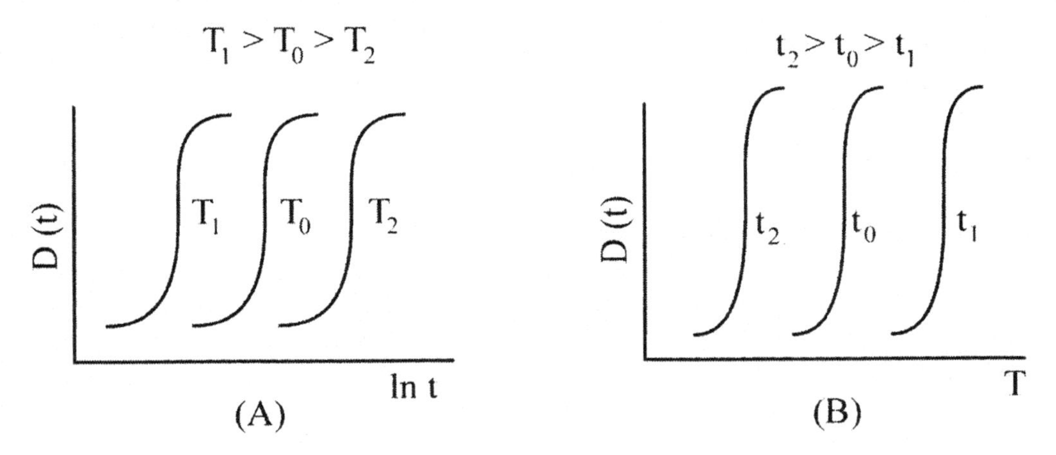

Figure 9.10 Influence of time and temperature on creep compliance.

Similar behaviors apply to the relaxation modulus as illustrated in Figure 9.11. Interestingly, both Figures 9.10 and 9.11 indicate that the dependence of behaviors (creep compliance and relaxation modulus) with temperature is similar with time, suggesting that some relationship between time and temperature behavior must exist in the all-important transition region for linear viscoelastic materials.

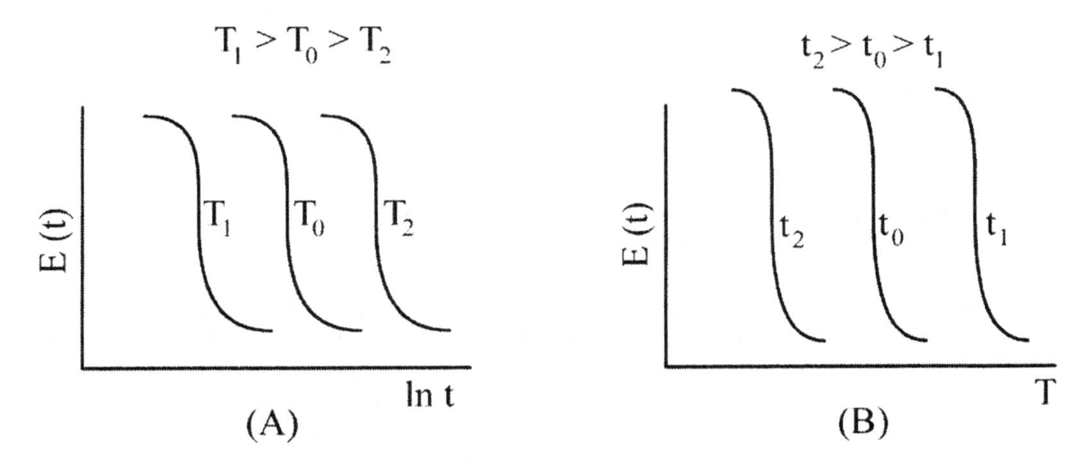

Figure 9.11 Influence of time and temperature on relaxation modulus.

Unfortunately, classical thermodynamics does not tend to apply in the transition region. Instead, consider the creep compliance *D(t)* and relaxation modulus *E(t)* in the transition region as a function of logarithmic time-lost at a constant temperature, T_0 as shown in Figures 9.10A and 9.11B. If the creep compliance or relaxation modulus is required at a higher temperature, it can be obtained by simply shifting the compliance or modulus curve for temperature T_0 to the shorter times. Similarly, if a compliance curve is desired for a lower temperature T_2, it can be obtained by shifting the compliance curve for temperatures T_0 toward longer times. This so-called *time-temperature equivalence* obtained through the method of reduce variables, indicates that temperature and logarithmic time are equivalent parameters and allows the experimental time scale to be extended. It is therefore a common practice to obtain viscoelastic properties at several temperatures to extend the experimentally accessible region.

A material whose creep compliance or relaxation-modulus curves only change their position on the logarithmic time scale as the temperature is varied is called a thermo-rheologically simple material. For these materials, the creep compliance D(ln(t), t) or relaxation modulus E(ln(t), T) can be shown to depend on a single combined quantity, ζ called the *reduced time* as:

$$D(\ln(t), T) = D(\zeta) \tag{9.19a}$$

and

$$E(\ln(t), T) = E(\zeta) \tag{9.19b}$$

where

$$\zeta = \ln(t) - f(T) \tag{9.19c}$$

The functions $D(\zeta)$ and $E(\zeta)$ are called *master curves* in creep compliance and relaxation modulus, and represent the shape common to all curves. On the other hand, $f(T)$ is a function of temperature only and fixes the position of the creep compliance or relaxation modulus curve on a logarithmic time scale. The following examples will illustrate their use with the method of reduced variables to obtain a master curve for creep compliance and relaxation modulus from experimental data.

Example 9.3 (Creep Compliance): A typical distribution of creep compliance with time and temperature is shown in Figure 9.12; this is simply an enlargement of the transition regions shown in Figure 9.11A. The application of the method of reduced variables requires the transformation of a function D(ln(t), T), where t is time and T is temperature, into a function D (ln(ζ)) for a selected reference temperature, T_0. In the expression, ζ = t/a_t and is the reduced time while a_t represents a shift factor that is a function of temperature only. In order to see whether a particular material obeys the time temperature superposition in terms in creep compliance, the following procedure is suggested:

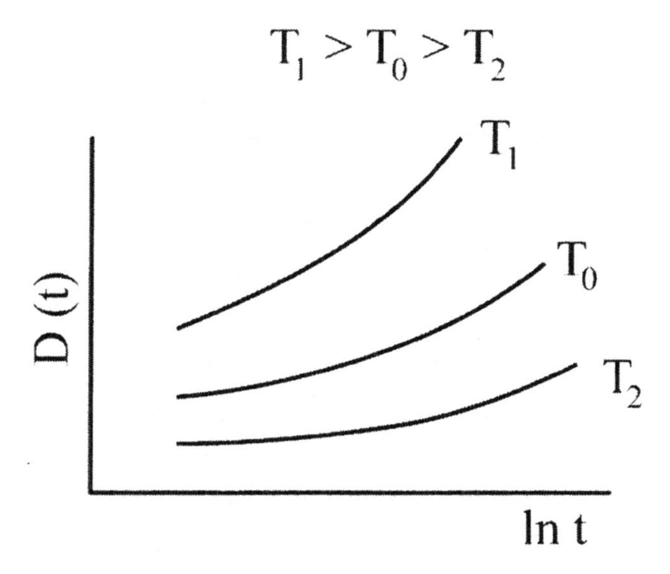

Figure 9.12 Typical distribution of creep compliance with temperature for a linear viscoelastic material.

First, conduct creep experiments at different temperatures for a selected time range holding temperatures constant for each test; curves such as the one shown in Figure 9.12 will be the result.

Next, obtain the creep compliance for a given temperature as:

$$D(t) = \frac{E(t)}{\sigma_0} \tag{9.20}$$

for various times and multiply these values by a factor $T\rho/T_0\rho_0$ where T is any arbitrary temperature, T_0 is a selected temperature (generally room temperature), ρ is the density at temperature T, and ρ_0 the density at temperature T_0.

Once this operation is completed, plot the corrected value of the creep compliance:

$$D(t)\frac{T\rho}{T_0\rho_0} \tag{9.21}$$

against the natural log of the reduced time, $\ln(\zeta)$ as shown in Figure 9.13. For this transformation:

$$\varepsilon = \frac{t}{a_t} \tag{9.22}$$

where

$$a_t = \frac{\eta}{\eta_0}\left(\frac{T_0\rho_0}{T_\rho}\right) \tag{9.23}$$

For uncrossed linked polymers, η_0 is the viscosity at temperature T_0 while η is the viscosity obtained at temperature T.

Finally, determine if the creep compliance curves at different temperatures can be made to coincide by shifting them parallel to the ln(t) axis in such a way that points at one value of creep compliance coincide for different temperatures (a smooth master creep compliance curve). Plot the *shift factor*, a_t, as a function of temperature as shown in Figure 9.14.

Example 9.4 (Relaxation Modulus): Again, the application of the method of reduced variables is used to construct a master curve for the relaxation modulus of a selected reference temperature. To begin, a typical distribution of relaxation modulus with time and temperature is shown in Figure 9.15. A transformation of the function $E(\ln(t), T)$ into a function $E(\ln(\zeta))$ for a reference temperature to where $\zeta = t/a_t$ enables the construction of a master curve for temperature T_0, where the shift factor a_t is again a function of temperature only. The procedure for deciding whether a particular material obeys the time-temperature superposition in terms of relaxation modulus similar to that for creep compliance.

At first, simply conduct stress relaxation experiments at different temperatures for a selected time range holding temperature constant for each test; curves such as those illustrated in Figure 9.15 will be obtained.

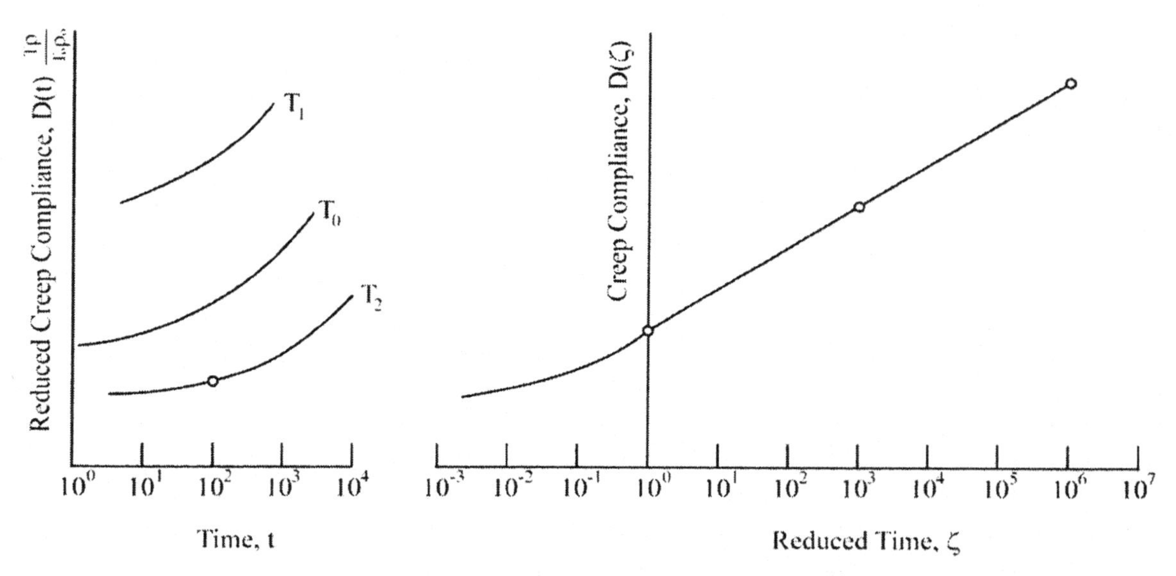

Figure 9.13 Master creep compliance curve.

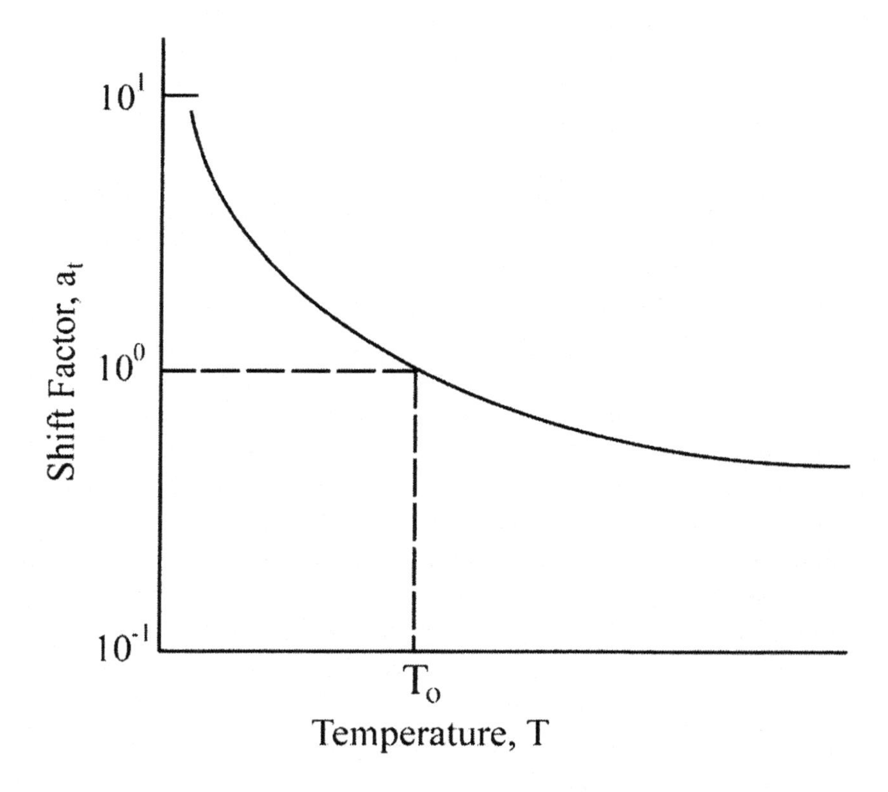

Figure 9.14 Variation of shift factor with temperatures.

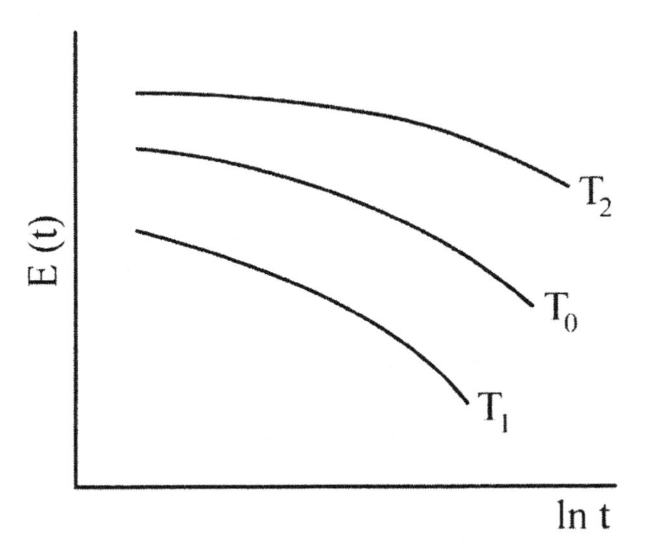

Figure 9.15 Typical distribution of relaxation modulus with time and temperature for a linear viscoelastic material.

Using this data, obtain the relaxation modulus for various times:

$$E(t) = \frac{\sigma(t)}{\varepsilon_0} \tag{9.24}$$

Next, multiply these values by a factor $T\rho/T_0\rho_0$ where T is still any arbitrary temperature, T_0 is a selected temperature (generally room temperature), ρ is the density at temperature T, and ρ_0 is the density at temperature T_0.

Plot the corrected value of the relaxation modulus:

$$E(t)\frac{T\rho}{T_0\rho_0} \tag{9.25}$$

against the natural log of the reduced time, ln(ζ) as shown in Figure 9.16 using Equations 9.22 and 9.23.

Next, determine if the relaxation modulus at different temperatures can be made to coincide by shifting them parallel to the ln(t) axis in such a way that points at one value of relaxation modulus coincide for different temperatures (a smooth master creep compliance curve). Plot the *shift factor*, a_t, as a function of temperature as already shown in Figure 9.14.

Example Problem 9-5: The variation of relaxation modulus of a viscoelastic polymer (for toys) as a function of time and temperature is shown in Figure 9.17. Construct a relaxation modulus master curve for a reference temperature $T_0 = 70^{\circ}F$ by the method of reduced variables.

For the analysis, assume that $\rho_0 = \rho$ (no density change) and that $(\eta/\eta_0)_{35F} = 2.0x10^5$ and $(\eta/\eta_0)_{105F} = 6.7x10^{-6}$. The corrected values of the relaxation modulus $E(\zeta) = E(t)\ T_0\rho_0/T\ \rho$ are first plotted versus time,

t as shown in Figure 9.17. The shift factor is computed as $a_t = \eta/\eta_0(T_0\rho_0/T\rho)$ and is then used to compute the reduced times as $\zeta = t/a_t$. The completed master curve is shown in Figure 9.18.

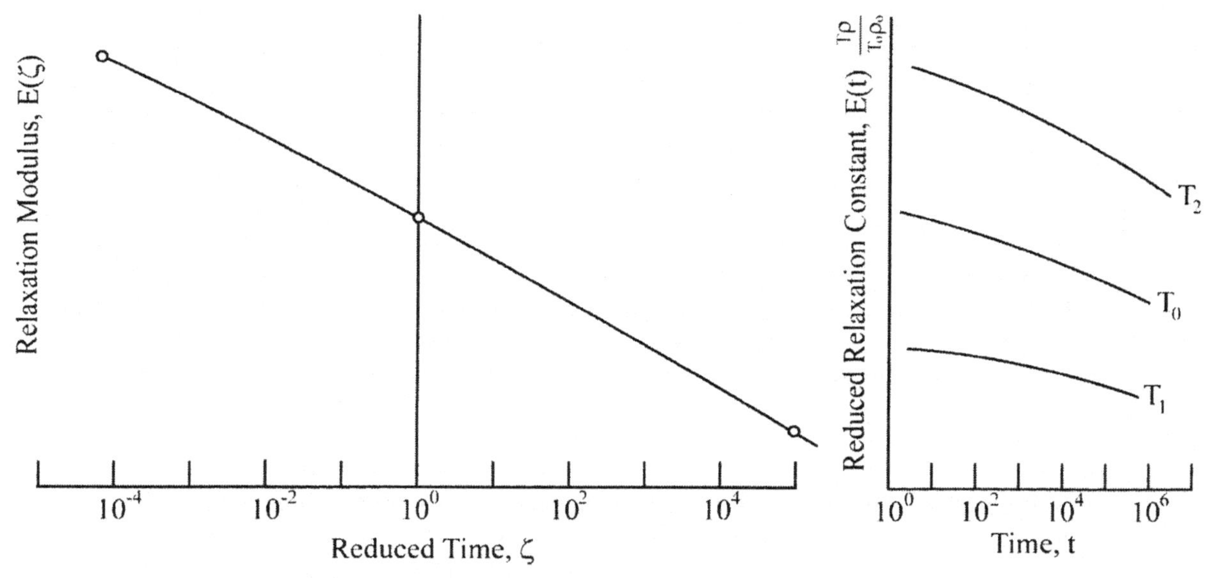

Figure 9.16 Master relaxation modulus curve.

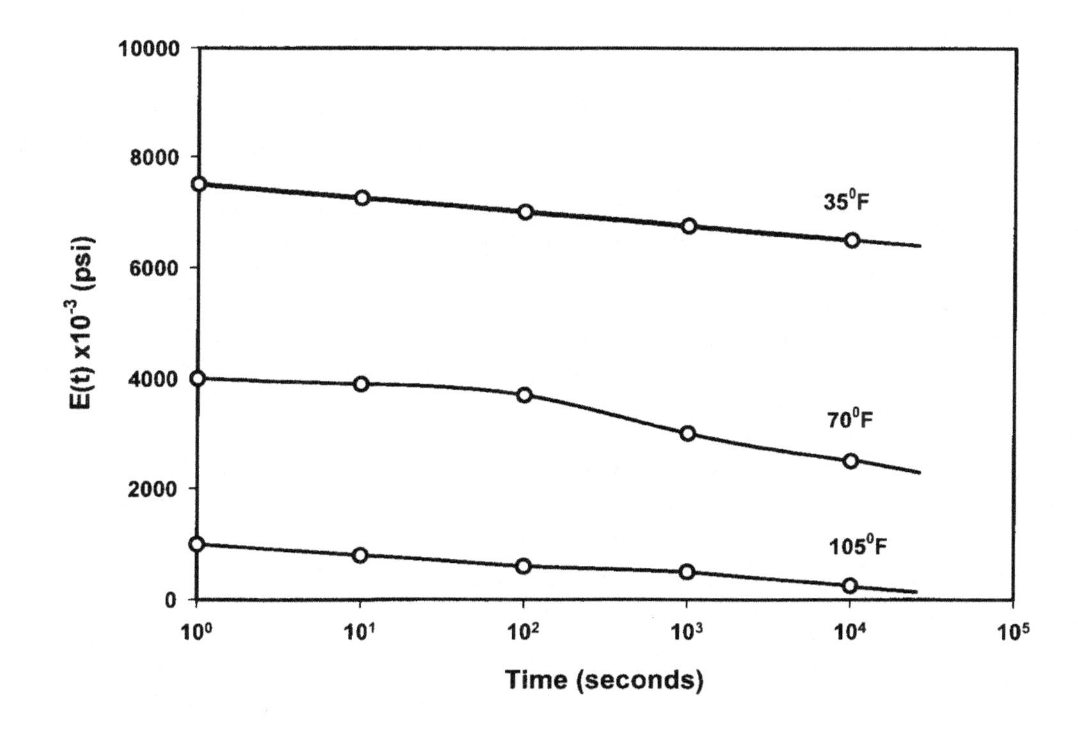

Figure 9.17 Relaxation modulus curve at temperature for Example Problem 9.5.

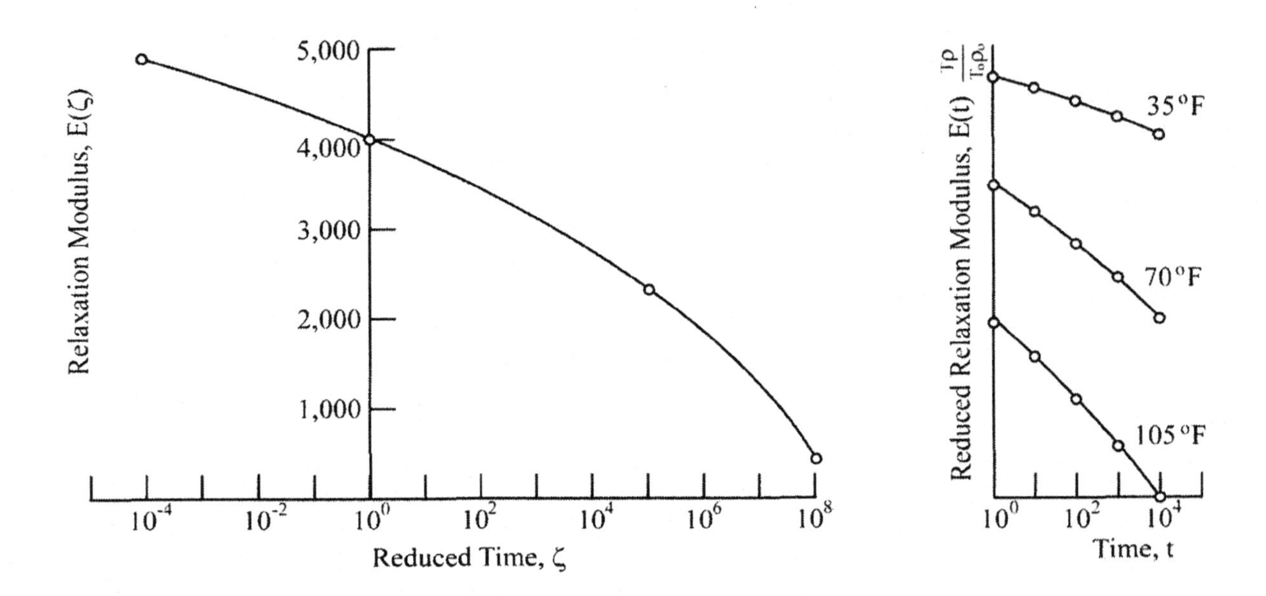

Figure 9.18 Master relaxation modulus curve for Example Problem 9.5.

PRACTICE EXERCISES

1. Much to his chagrin, the venerable and apparently eternal Gumby has been gripped between the loading heads of a tensile test machine 5 inches apart, and then loaded at a constant head speed of 1.5 in/minute with the following data noted:

σ (psi)	t (sec)	σ (psi)	t (sec)
0	0	2,100	7
300	1	3,000	10
900	3	3,900	13
1,500	5	5,000	16.67

At a Gumby stress of σ_G = 5,000 psi, the testing machine was shut off, thus fixing the corresponding elongation as a constant. Considering the instant the machine was shut off as t = 0, the following data were recorded:

σ (psi)	t (min)	σ_G (psi)	t (min)
5,000	0	1,070	20
4,630	1	730	25
4,230	2	490	30
3,970	3	230	40
3,670	4	100	50
3,400	5	50	60
2,310	10	23	70
1,570	15		

Using this data, determine: (a) if Gumby behaves as a Maxwell material, (b) his elastic constant, E, (c) his viscosity constant, η, and (d) Gumby's relaxation time constant, τ.

2. The unfortunate "specimen" analyzed in Problem 1 had a constant tensile strain of 0.12 in/in applied instantaneously. Find the tensile stress in our stretched pal after 30 min.

3. A polymeric material used for aircraft seats behaves as a Maxwell material with an elastic constant of E = 35,000 psi and a viscosity constant η = 400,000 psi-minute. If a constant stress of σ = 1400 psi is applied instantaneously to a tensile specimen of this material, what is the strain in the material after 40 minutes?

4. Plastics used for hardware mounts on computer "Motherboards" are known to be Maxwell materials with an elastic constant E = 40,000 psi and a viscosity constant η = 500,000 psi-minute. If a constant stress of σ = 2,000 psi is instantaneously applied to a tensile specimen of this material, held for 30 minutes and then released, what is the tensile strain in the material after 31 and 60 minutes?

5. The same Maxwell material used in Problem 4 with E = 40,000 psi and η = 500,000 psi-minute has a constant tensile strain of ε = 0.10 in/in applied instantaneously to a tensile specimen, held for 10 minutes, and then released. What is the permanent tensile strain in the material?

6. A Maxwell type viscoelastic material used for motorcycle seals has an elastic constant of E=20,000 psi and a viscous constant of η = 450,000 psi-minute. If the material experiences the stress history shown on the next page, what is the strain in the material at 30 min.?

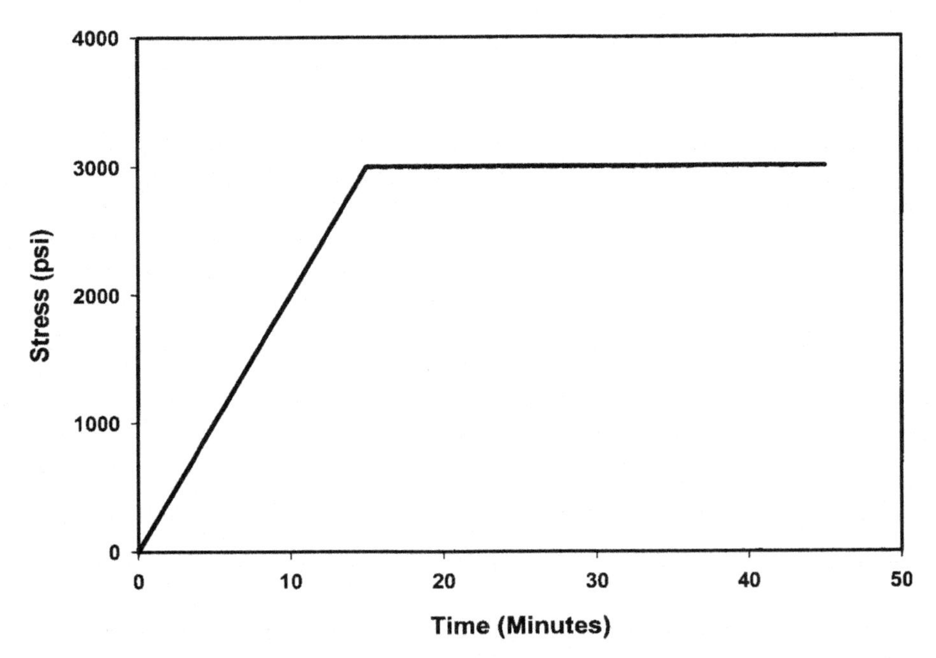

7. Shown below are the stress relaxation characteristics of annealed Electrolytic Tough Pitch copper wire at 70°C. (a) If an electrical probe is subjected to a uni-axially compressive contact stress of σ=15,000 psi at 70°C, determine how long will it take the contact stress to relax to 5,000 psi.

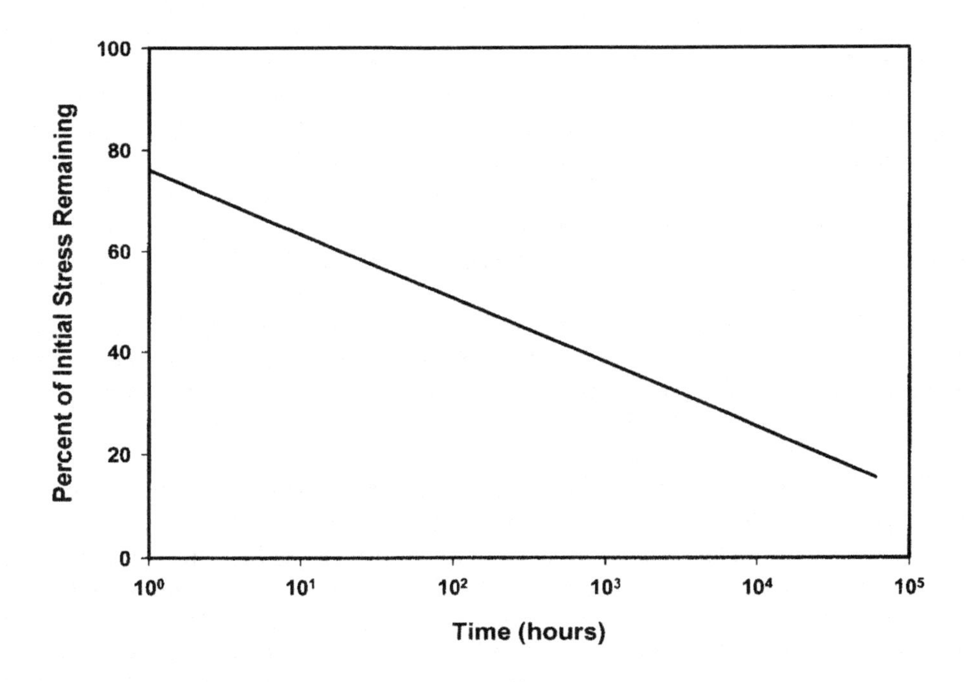

8. A sample of polyethylene for piping was loaded rapidly and found to have a linear tensile stress-strain behavior with an elastic constant of E = 30,000 psi. In a stress relaxation experiment, a sample of the same material is relaxed to a constant, non-zero value of tensile stress. Discuss why the viscoelastic model for this material might be a parallel spring and dashpot in series with a second spring.

9. The material discussed in Problem 8 was loaded instantaneously with a constant load which produced a tensile stress of σ = 1,000 psi. The specimen continued to elongate with time, approaching a constant tensile strain of 9.0 percent. What is the elastic constant of the spring which is in parallel with the dashpot?

10. The variation of creep compliance as a function of time and temperature is shown in the graph below. Construct a creep compliance matrix curve for a reference temperature T_0 = 70°F using the method of reduced variables.

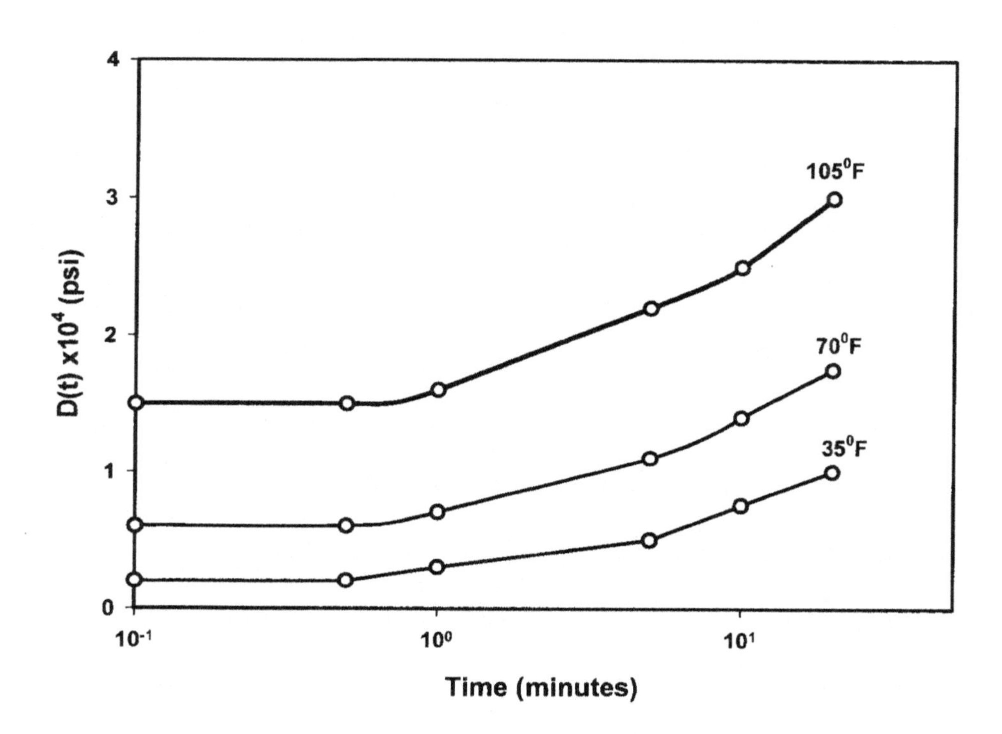

CHAPTER TEN

CREEP

10.1 INTRODUCTION

Clearly, a strong driving force exists for raising the operating temperatures of machines and structures involved in energy conversion, as the greater the difference between operating and ambient temperatures, so goes the thermodynamic efficiency. Hence, it should come as no surprise that many sectors of contemporary technology seek to employ materials at increasingly elevated temperatures. Unfortunately, the efficiency of energy conversion processes is usually material limited, where failure may be a "traditional breakage," and/or an intolerable accumulation of deformations over time via *creep*.

As discussed in the previous chapter, *creep* is that less than ghoulish mechanical response exhibited by materials wherein strains accumulate over time even under constant load or stress. For organics, this behavior usually occurs at lower temperatures including normal ambients. Conversely, higher temperatures are usually required for creep to occur for most alloys and structural ceramics. Creep strains are usually sufficient in magnitude to be of engineering concern when the service temperature (absolute) exceeds 30 percent of the melting temperature, T_m of the solid. For noncrystalline solids, the glass transition temperature, T_g serves the same scaling role.

10.2 CREEP BEHAVIOR

Assuming that both temperature and stress are sufficiently high that they generate measurable creep strains, a typical tensile strain response over time would look like the curve shown in Figure 10.1.

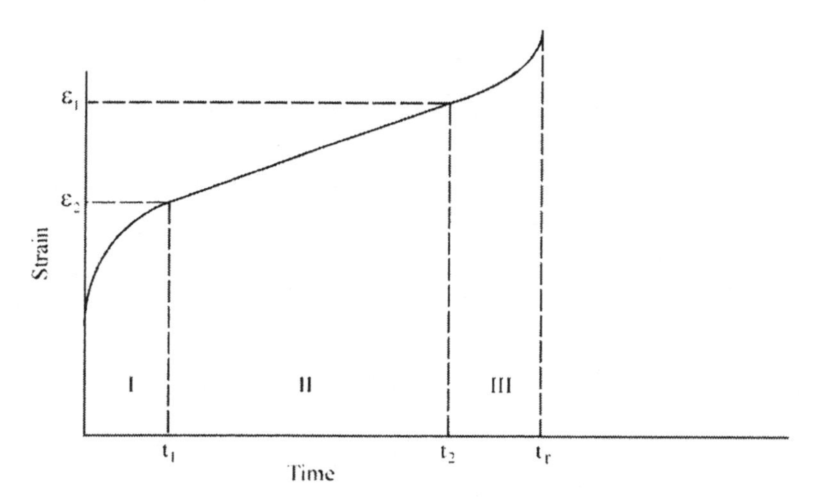

Figure 10.1 Typical creep deformation (over time) indicating three distinct regions of the response.

A variety of materials, including metals, polymers, and ceramics exhibit the type of response depicted in Figure 10.1. However, such creep in ceramics is relatively hard to achieve since the material will usually fail in fracture long before the stress levels are sufficient to generate creep strains. If creep does indeed occur for any of the materials mentioned, they all experience some level of *transient creep*, wherein the strain rate decreases with time as denoted by region I. The change in strain rate (decreasing) in the transient creep range is due to the competition of two opposing processes, strain hardening and annealing. When the rate is highest at the beginning, annealing or thermally activated softening tends to dominate. As the stored strain hardening is relieved by annealing, the overall effect is reduced. Eventually, continued strain hardening and any ongoing annealing processes reach a *steady-state* standoff, at which time the response of region II becomes established. Finally, region III or *tertiary creep* represents the onset of structural instability leading to fracture/rupture; this region terminates at what is known as the *time-to-rupture,* or t_r.

As far as creep is concerned, there are two design situations that typically occur in engineering practice. In the first, there are limits on the total amount of strain tolerated by the structural element. An example of a strain-limited design might be a turbine blade (engine or generator) that is designed to just clear the housing for reasons of aerodynamic efficiency; with creep strains accumulating at elevated temperatures, the blade will elongate and eventually contact the stationary housing. A second type of design situation may exist when strains are not of much importance, but the lifetime is critical. An illustrative example might be the tubes carrying heat transfer fluids for boilers or other fuel conversion systems. Although sagging and distortion of the tubes within the boiler may not be too critical, the tube bursting within the combustion zone would most certainly constitute a design failure.

10.3 CREEP STRAIN PREDICTION

Creep is a truly complicated matter as the accumulation of strain is dependent upon the stress-levels employed, service temperature(s), and the unique and apparently inherent resistance to creep deformations of any given materials. When considering these dependencies, many different predictive equations have been proposed, some purely empirical while others are derived from considerations of the most fundamental material considerations. While the methodology described herein correlates reasonably well with most experimental data, it can also be demonstrated to have basic theoretical underpinnings.

For our analysis, the total strain, ε_{tot} as a function of time can be expressed as the sum of the instantaneous, transient, and steady state accumulations:

$$\varepsilon_{tot} = \varepsilon_0 + \beta t^n + \dot{\varepsilon}_{ss} t \qquad (10.1)$$

Here, ε_0 is the elastic or elastic/plastic strain instantaneously generated at loading. In terms of time dependence, βt^n is the transient creep term, where β and n depend on stress, temperature, and the material; for most materials of engineering importance, the exponent will be in the range of $0 < n < 1.0$. Finally, the strain accumulated under steady state creep conditions is simply the product of the steady state creep rate and time or $\dot{\varepsilon}_{ss} t$. From Figure 10.1, the steady state or linear creep rate is defined as:

$$\dot{\varepsilon}_{ss} = \frac{\varepsilon_2 - \varepsilon_1}{t_2 - t_1} \qquad (10.2)$$

Creep strains associated with a region III response are usually not sought, as structural failure is imminent. Instead, the time factor in Equation (10.1) is typically limited to some fraction of the time-to-rupture or t_r. Finally, the transient contribution from βt^n is often ignored, as the strains associated with this region of creep may be insignificant in comparison to those encountered under steady state conditions. Under these assumptions and practical realities, the total strain as a function of time simplifies to:

$$\varepsilon_{tot} = \varepsilon_0 + \dot{\varepsilon}_{ss} t \qquad (10.3)$$

If the loads change frequently during the service life, the transient contribution cannot be simply wished away and nearly all of the creep encountered could be time dependent in nature; in this situation, Equation (10.3) would not necessarily give an accurate prediction.

10.4 CREEP STRAIN PREDICTION

As the name implies, the steady state rate, $\dot{\varepsilon}_{ss}$ is the only creep influence in Equation (10.3) when all transient effects are ignored. Unfortunately, this is where the simplicity ends (in case you thought things were too easy), as even the steady-state rate has a complicated dependency on the stress, σ, absolute temperature, T (units of Rankine or Kelvin), and the resistance to creep deformation of each material. Schematically, stress and temperature effects on creep strains are depicted in Figure 10.2. As shown by the figure, raising either the stress level or the temperature results in an increase in the creep rate.

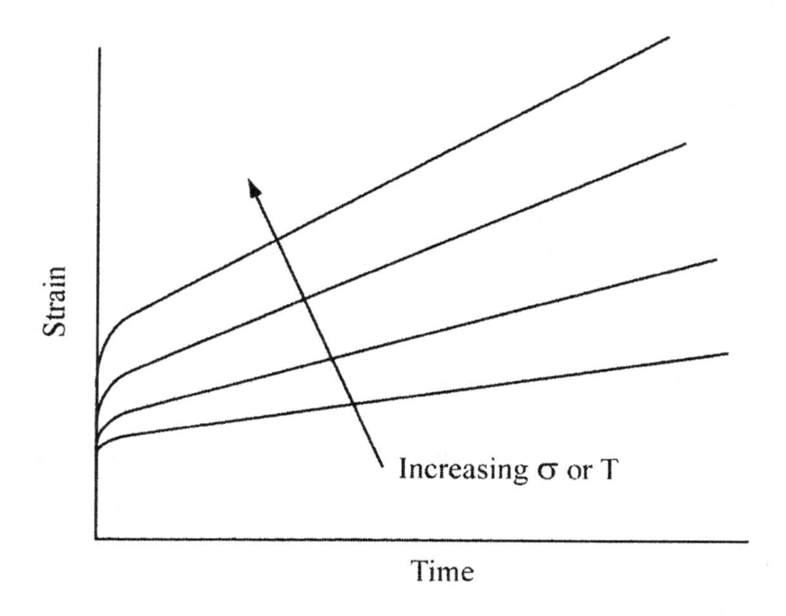

Figure 10.2 Creep strain versus time curves for various stresses and temperatures.

In order to formally determine the relationship between $\dot{\varepsilon}_{ss}$ and T alluded to by the figure, it is first necessary to select a constant stress level that is sufficient to produce measurable strains in a reasonable amount of time (say 10^2-10^3 hours range). The creep curves, $\dot{\varepsilon}_{ss}$ values, and temperatures are indicated in Figure 10.3.

Although creep is certainly complicated (it actually gets worse), the most successful correlation of the various steady state rates with temperature is the one that also describes many similar processes; to see the potential correlation, we must plot the strain rate versus the inverse of the absolute temperature as shown in Figure 10.4. It was not a simple stroke of luck that led to this graphical representation, as the resulting straight line (semi-logarithmic) correlation suggests that an Arrhenius type of rate equation governs steady state creep rates:

$$\dot{\varepsilon}_{ss} = Ae^{\frac{-\Delta H}{kT}} \tag{10.4}$$

In Equation (10.4), A is a material- and stress-dependent coefficient, ΔH, the potential energy necessary to drive the creep usually know as the activation energy, k, the Boltzmann gas constant that is employed for historical reasons, and T, the service temperature (in absolute units). Using this correlation, the slope of Figure 10.4 then equals the quantity $-\Delta H/k$.

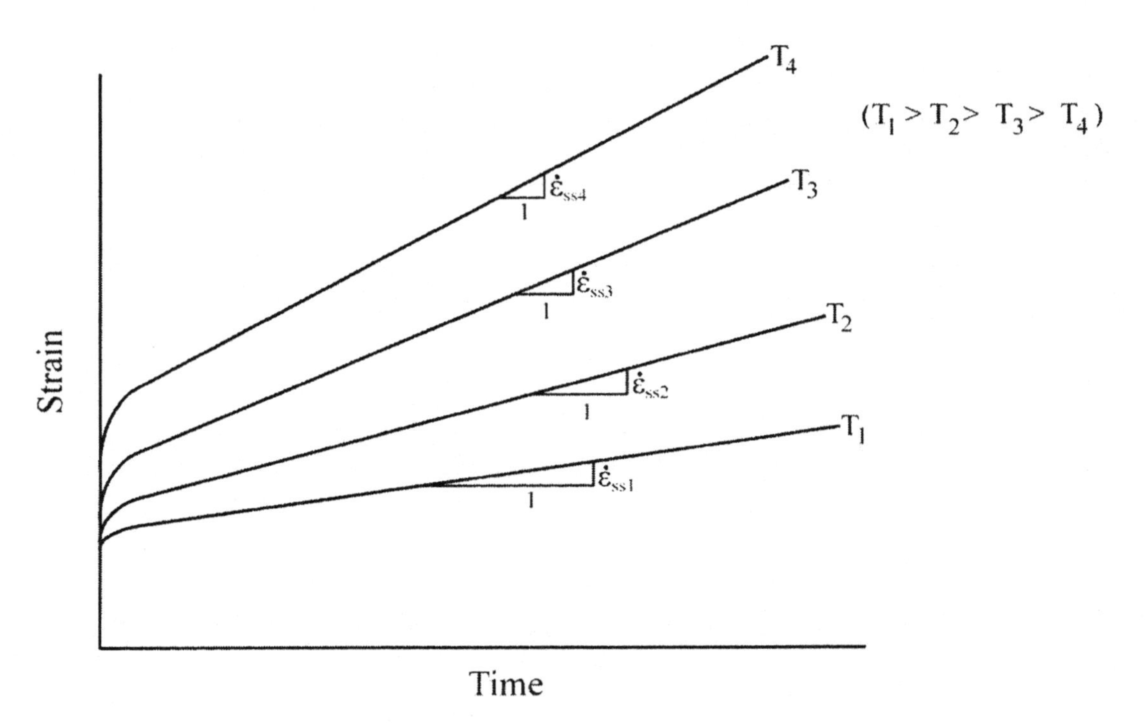

Figure 10.3 Creep curves for various temperatures at constant stress.

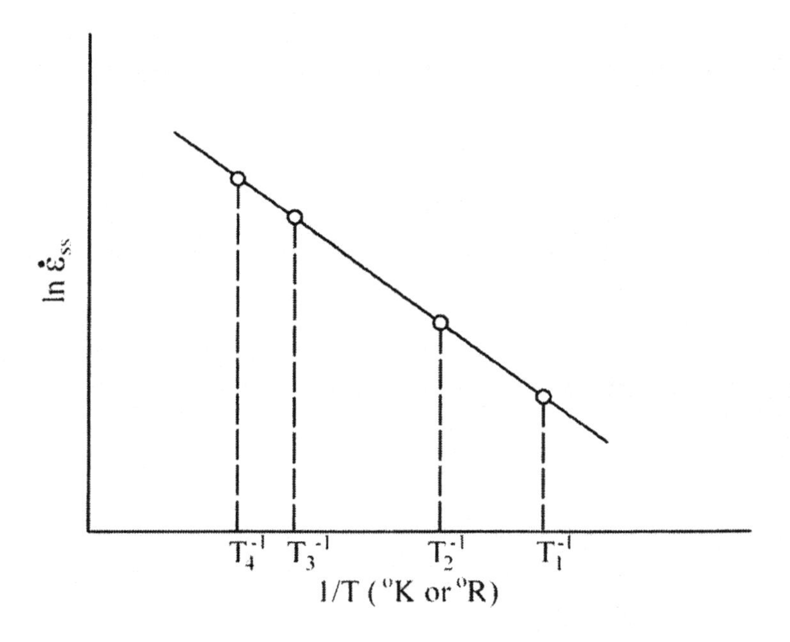

Figure 10.4 Steady state strain rate versus temperature at constant stress.

Figure 10.4 was derived from creep data at constant stress, so no stress-dependence can be observed. On the other hand, if two sets of data similar to that in Figure 10.4 were collected, each taken at a constant, but different, stress, two possibilities for the response exist as indicated in Figure 10.5. In case (A), the slopes of the two Arrhenius plots are equal and *ΔH* is clearly not a function of stress. Some materials may indeed exhibit this response, especially if the range in temperature and stress is not excessive. For case (B), the slope, or -ΔH/k, is different for the two stresses employed; in this type of response, the activation energy is clearly stress dependent or simply, $\Delta H = \Delta H(\sigma)$.

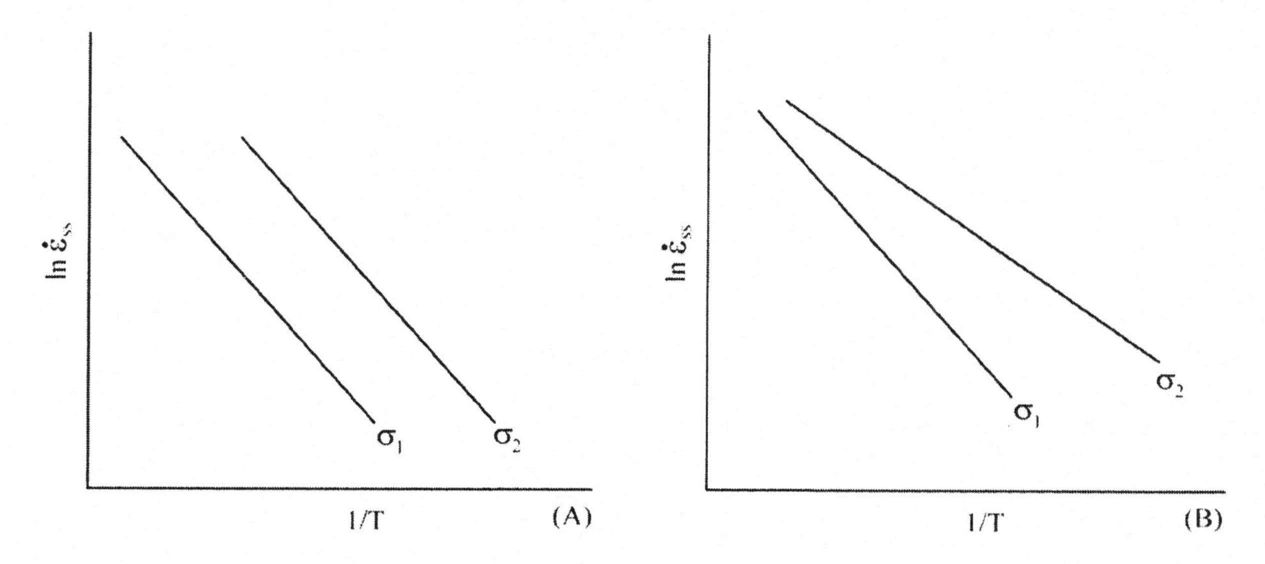

Figure 10.5 Arrhenius plots of $\dot{\varepsilon}_{ss}$ versus 1/T employing two stress levels, with (A), ΔH = constant and (B), $\Delta H = \Delta H(\sigma)$.

In general, Equation (10.4) must be rewritten such that the stress dependence of the activation energy is explicitly indicated:

$$\dot{\varepsilon}_{ss} = Ae^{\frac{-\Delta H(\sigma)}{kT}} \tag{10.5}$$

While not shown, it must be assumed that the coefficient, *A* is stress and material dependent as before. Moreover, the stress dependence of the steady-state creep rate must also be explored. As would be expected, such relationships do not present themselves very easily, so empirical observations are often used; the correlations between $\dot{\varepsilon}_{ss}$ and stress are based on the observation of a power law dependence of the following form: $\dot{\varepsilon}_{ss} \propto \sigma^m$.

In order to determine the full relationship, a series of creep curves must first be generated as already shown in Figure 10.3; however, in this case, a temperature of convenience will be selected and held constant for each creep test while stresses σ_1, σ_2, σ_3, σ_4 and so on will be varied, one for each creep curve. Once these tests are completed, the steady state rates as a function of stress at one fixed temperature will be known. The most general successful correlation of creep rate and stress of this type is illustrated in Figure 10.6.

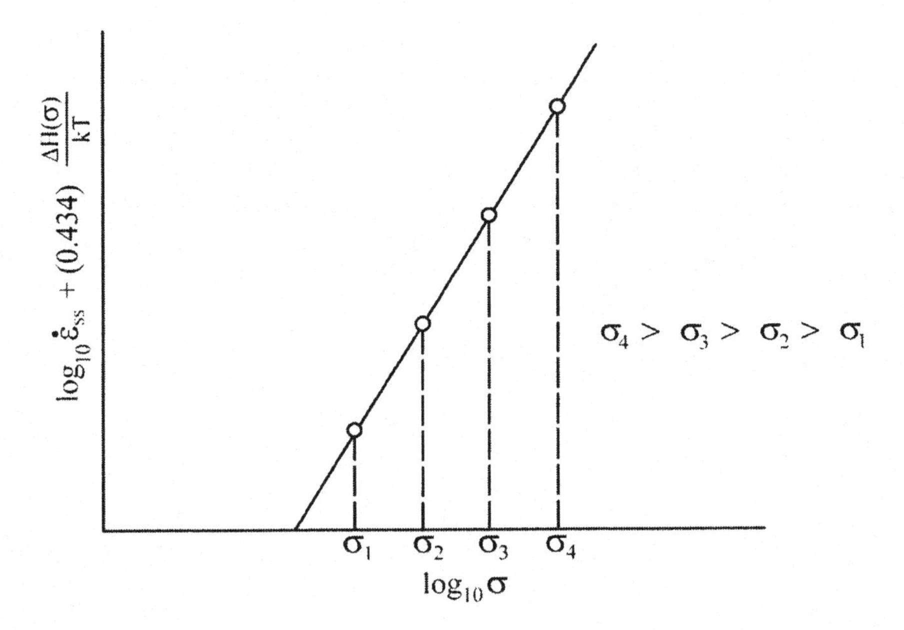

Figure 10.6 Steady state creep rate versus stress at constant temperature.

As indicated in the figure, the power relationship holds for the stress dependence of the coefficient, *A*. Taking the antilogarithm of the straight line, and noting that $\log_{10}{}^{e} = 0.434$, one obtains the Dorn-Miller equation for creep:

$$\dot{\varepsilon}_{ss} = \phi\sigma^{m} e^{\frac{-\Delta H(\sigma)}{kT}} \tag{10.6}$$

Here, *m* is the computed slope of the best fit straight line in Figure 10.6, or not so simply:

$$m = \frac{\partial}{\partial\left(\log_{10}\sigma\right)}\left[\log_{10}\dot{\varepsilon}_{ss} + (.434)\frac{\Delta H(\sigma)}{kT}\right] \tag{10.7}$$

Because no two materials are ever alike, there are many examples of heat-resistant materials whose activation energies are independent of stress over the range design interest. For these solids with ΔH = constant, the slope is given by the following relationship:

$$m = \frac{\partial\left[\log_{10}\dot{\varepsilon}_{ss}\right]}{\partial\left[\log_{10}\sigma\right]} \tag{10.8}$$

Regardless of whether Equation (10.7) or (10.8) is appropriate, *m* is usually in the range of $4.0 \leq m \leq 6.0$. Moreover, theoretical analyses of the role that micro-deformation mechanisms play in creep yield similar results. Finally, the Dorn-Miller relationship (Equation (10.6)) is customarily non-dimensionalized into the following form:

$$\frac{\dot{\varepsilon}_{ss}}{\dot{\varepsilon}_{0}} = \left(\frac{\sigma}{\sigma_0}\right)^{m} e^{-\frac{\Delta H(\sigma)}{kT}} \tag{10.9a}$$

In this form, new material parameters have been introduced, such that:

$$\phi = \frac{\dot{\varepsilon}_0}{\sigma_0^m} \tag{10.9b}$$

If we arbitrarily place $\varepsilon_0 = 1.0\ \text{hr}^{-1}$, the following form results:

$$\dot{\varepsilon}_{ss} = \left(\frac{\sigma}{\sigma_0}\right)^m e^{-\frac{\Delta H(\sigma)}{kT}} \tag{10.9c}$$

Example Problem 10.1: A chemical reactor chamber material of interest is an iron-based alloy, containing molybdenum and rhenium for high temperature strength. Results of steady state strain rate measurements for two stress levels as a function of temperature for this material are given in Figure 10.7. Determine the activation energy and if the material parameter shows a stress dependency.

For the data obtained at a stress level of $\sigma = 4{,}000$ psi:

$$-\frac{\Delta H}{k} = \frac{\delta(\ln \dot{\varepsilon}_{ss})}{\delta(1/T)} = \frac{-123\text{x}10^5}{1} = -123{,}000°\text{R}$$

$$\Delta H = -123{,}000 \cdot k = 8.35\text{x}10^{-18}\,\text{in} - \text{lb}$$

A similar calculation performed on the data for $\sigma = 6{,}000$ psi yields identical results, indicating that the activation energy is undeniably stress independent, at least for the two stress levels under scrutiny.

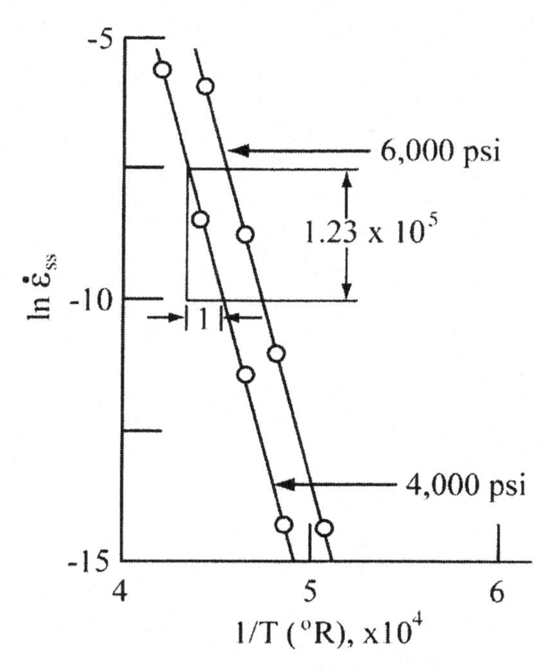

Figure 10.7 Steady state creep data for two stress levels.

Example Problem 10.2: The same alloy of the previous example exhibits a dependence of the steady-state strain rate on stress as indicated in Figure 10.8; based on the data available (including Example Problem 10.1), it was assumed that the activation energy was not stress dependent. Given this data and assumptions, determine the two remaining parameters in the Dorn-Miller relationship, namely σ_0 and m.

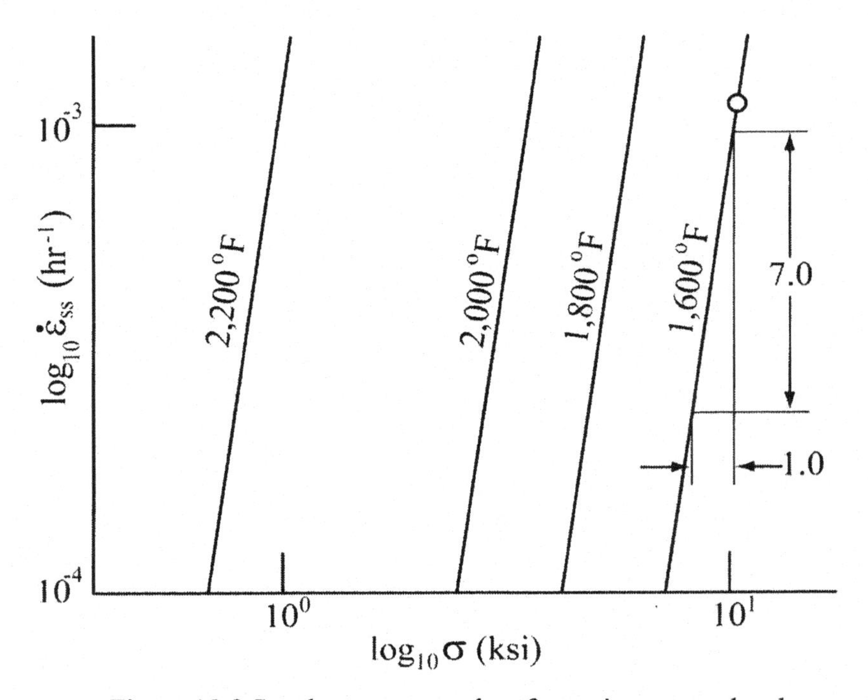

Figure 10.8 Steady state creep data for various stress levels.

An examination of the data and plots reveals that the lines generated at 1600°F and 1800°F appear to have the same slope, m that can be directly measured as:

$$m = 7.0$$

With m determined, σ_0 is calculated from known data points:

$$0.001 = \left(\frac{9,400}{\sigma_0}\right)^{7.0} e^{-\left(\frac{123,000}{2,060}\right)}$$

such that σ_0 = 5.01 psi. For stresses in units of psi and time in hours, the complete Dorn-Miller steady state creep rate equation for this material becomes:

$$\dot{\varepsilon}_{ss} = \left(\frac{\sigma}{5.01}\right)^{7.0} e^{-\left(\frac{123,000}{T}\right)}$$

Example Problem 10.3: A thin-walled Pebble Bed Reactor (PBR) tank made of the same material (see Example Problems 10.1 and 10.2) was designed to carry pressurized Helium at 800 psi and 1500°F. Determine the number of hours it took the vessel diameter to increase 1.0 percent and exceed design tolerances? For the doomed tubing, the mean radius-to-thickness ratio is 8.0.

At a diametral increase of 1 percent, the total circumferential strain is:

$$\frac{\Delta(\pi d)}{\pi d_0} = \frac{\Delta d}{d_0} = .01$$

Using Equation (10.3), the total accumulated strain can be related to the steady state creep rate and time:

$$0.01 = \varepsilon_0 + \dot{\varepsilon}_{ss} t$$

In order to determine the time for the specified diametral increase to occur, we must first look at the elastic/instantaneous strain, ε_0 that is easily calculated from the pressure induced stress and E.

$$\varepsilon_0 = \frac{\sigma}{E} = \frac{P r_m}{30x10^6 \, x \, \text{thickness}} = 0.000213$$

Clearly, the elastic strain is small when compared to the expected diametral increase and can therefore be ignored without jeopardizing much accuracy. Hence, the creep equation can be simplified to the following form:

$$0.01 \cong \dot{\varepsilon}_{ss} t$$

At this point, the time to the specified increase can be readily calculated as t = 3,240 hours or approximately 135 days.

10.5 CREEP RUPTURE

In many design situations involving creep, the constraint will not be the accumulated deformations, but complete failure or *creep rupture.* A prime example mentioned earlier is a boiler tube for any variety of heating or energy conversion system; while the deformations of the tube may be of little consequence, the eventual rupture could be messy to fatal, especially if toxic waste is the fluid. In this case, prediction of creep failure might be accomplished by establishing an upper limit to the amount of strain that can be accumulated.

For rupture to occur, creep strains must accumulate to the point where strain incompatibilities are generated at interfaces within the structural material, most notably the grain boundaries in polycrystalline solids such as metals and ceramics. In these situations, the strain-induced micro-voids grow, coalesce, and form macro-scale cracks. Eventually, there is insufficient material sustaining the load and rupture (failure) ensues. As described earlier, the rupture phenomenon is essentially a thermally activated rate process; if true, one might expect the time-to-rupture, t_r to be governed by an Arrhenius relationship such as:

$$t_r^{-1} = t_0^{-1} e^{\frac{-\Delta H}{kT}} \tag{10.10}$$

Interestingly, the activation energy, ΔH_r relevant to the total creep rupture process will not generally be the same as the activation energy for a steady-state strain accumulation. In addition, the new coefficient,

t_0 is a material parameter that may prove to be stress-dependent for some materials. Decomposing Equation (10.10) logarithmically, one obtains:

$$-\log_{10} t_r = -\log_{10} t_0 - \frac{\Delta H_r}{kT}\log_{10} e = -\log_{10} t_0 - (0.434)\frac{\Delta H_r}{kT} \tag{10.11a}$$

Rearranging terms, Equation (10.11a) becomes:

$$(0.434)\frac{\Delta H_r}{kT} = -\log_{10} t_0 + \log_{10} t_r \tag{10.11b}$$

For differing materials, two distinct types of rupture response have been observed. The first type, observed by Larson and Miller, indicates that ΔH_r is the only stress-dependent term. If so, then the Stress-dependent activation energies for the rupture process may be found by considering data sets for various rupture times and temperatures, a set of such points being collected for one stress level, then another, and so forth. In this case, t_0 will be invariant with respect to stress while ΔH_r will exhibit a dependency. Figure 10.9 illustrates the Larson-Miller creep rupture response; note that $\log_{10}t_0$ is the ordinate intercept as the inverse of temperature goes to zero or $1/T \rightarrow 0$. The range in inverse temperature $\Delta(1/T)$ as indicated in Figure 10.9 corresponds to temperatures for which creep rupture occurs in moderate amounts of testing time, perhaps 10^2 to 10^4 hours. In practice, the temperatures used to collect the data of Figure 10.9 may be higher than those encountered in actual service and are chosen to speed up the rupture process. However, care must always be exercised to ensure that the observed response is valid for actual service temperatures. The slopes of the lines shown in Figure 10.9 are equal to $0.434(\Delta H_r/k)$, and are clearly stress dependent.

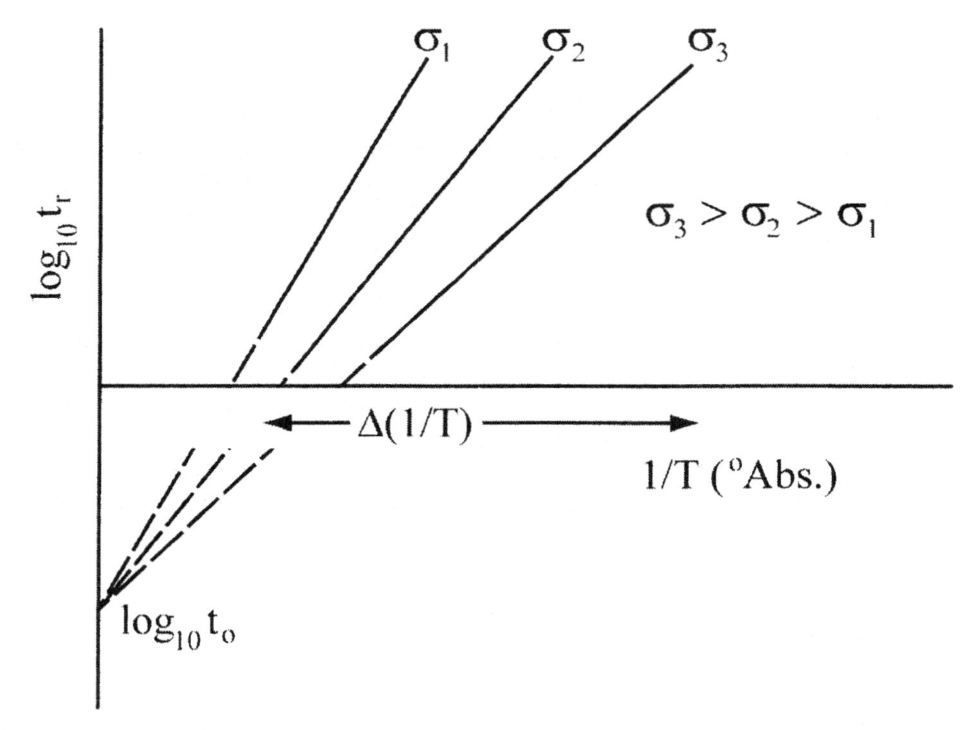

Figure 10.9 Larson-Miller creep rupture data correlation.

If the only stress-dependent parameter in Equation (10.11b) is ΔH_r, then the intercept point, $\log_{10}t_0$ will be common to all failure lines. Numerically speaking, t_0 is very small with rate process theory suggesting it should be of the order of 10^{-20} hours. Experimentally, it varies, depending on the material in question, from 10^{-15} to 10^{-30} hours. Of course, the infinite temperature corresponding to $t_r = t_0$ is physically meaningless. With these considerations in mind, it is customary to define the Larson-Miller constant, C, as:

$$-C \equiv \log_{10} t_0 \tag{10.12}$$

Thus, C = 20 theoretically, and ranges from 15 to 30 experimentally; as is usually the case, experimentally determined values are preferable when they exist. Using this constant, the parametric failure prediction method known as the Larson-Miller approach defines the Larson-Miller parameter, P, as:

$$P = T\left({}^{\circ}\text{Abs}\right)\left[C + \log_{10} t_r\right] \tag{10.13}$$

Since P is proportional to ΔH_r, it must be a function of stress only. If on the other hand, the functional dependence of P or ΔH_r are known, then a general predictive equation for rupture could be written to relate stress, temperature, and time-to-rupture in a manner commensurate with the resistance of each material to failure. However, because the underlying stress relationship is not known, a parametric correlation such as the Larson-Miller must suffice to relate the variables for any given material as is indicated Figure 10.10. The resulting correlation is referred to as a Master Life Curve whose shape depends on the material, and is the result of many experiments of the type used to generate the failure lines in Figure 10.7.

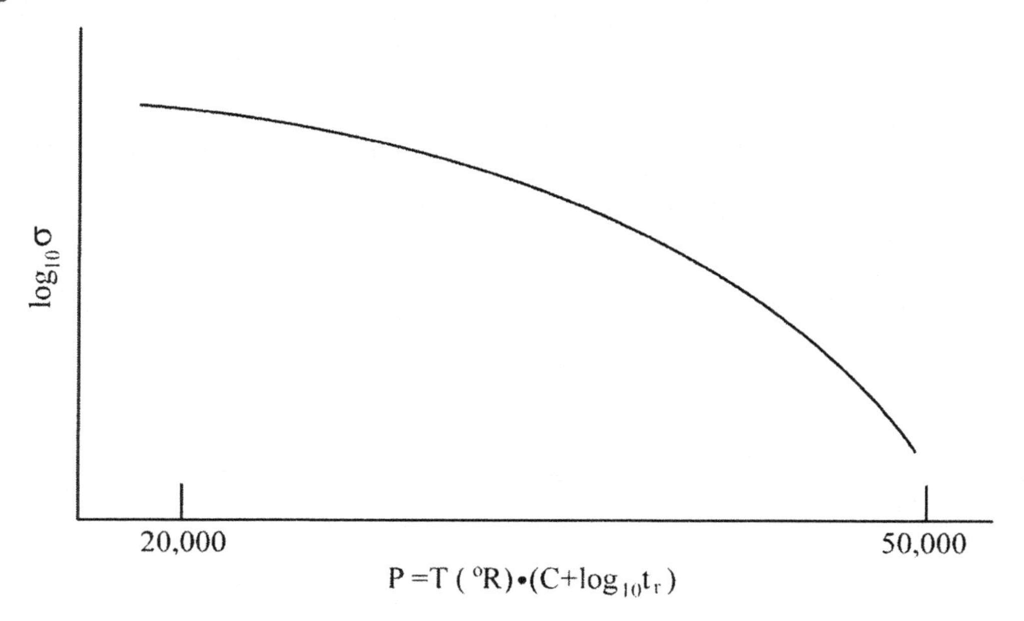

Figure 10.10 The Larson-Miller master life curve.

Although the Larson-Miller parametric approach to predicting creep rupture is the most commonly used in engineering practice, the same response has proven to be more precisely modeled by a second approach usually referred to as the *Dorn parametric method.* If the lines in Figure 10.9 do not converge to a common point, but instead appear to the parallel, then ΔH_r is not stress dependent. Because the intercept in this case, $\log_{10} t_{10}$, would depend on stress, t_0 becomes the stress-dependent quantity in this model. Equation (10.11a) is therefore rearranged to yield:

$$\log_{10} t_0 = \log_{10} t_r - (0.434)\frac{\Delta H_r}{kT} = f(\text{stress}) \qquad (10.14)$$

Taking the antilog form of Equation (10.14) yields:

$$t_0 = t_r e^{\frac{-\Delta H_r}{kT}} = P' \qquad (10.15)$$

where P' is the *Dorn parameter* that may vary from 10^{-15} to 10^{-30} hours based on practical levels of stress, temperature, and anticipated lifetimes. A hypothetical master life curve for the Dorn parametric correlation is shown in Figure 10.11.

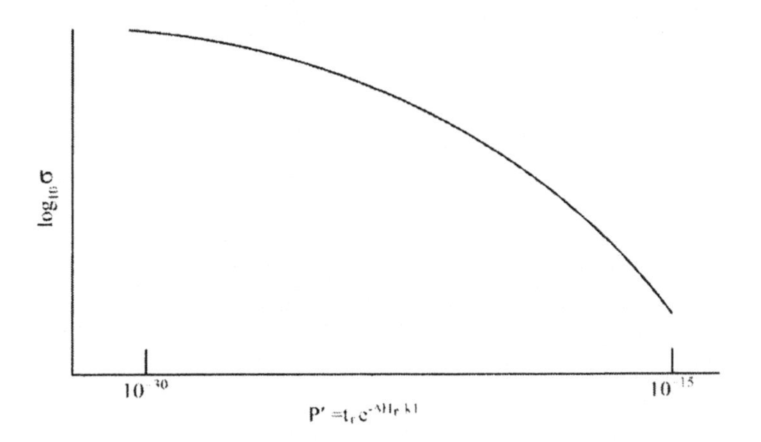

Figure 10.11 Dorn master life curve.

Example Problem 10.4: Shown below in Figure 10.12 is the Larson-Miller master life curve for the alloy described previously in Example Problems 10.1 through 10.3.

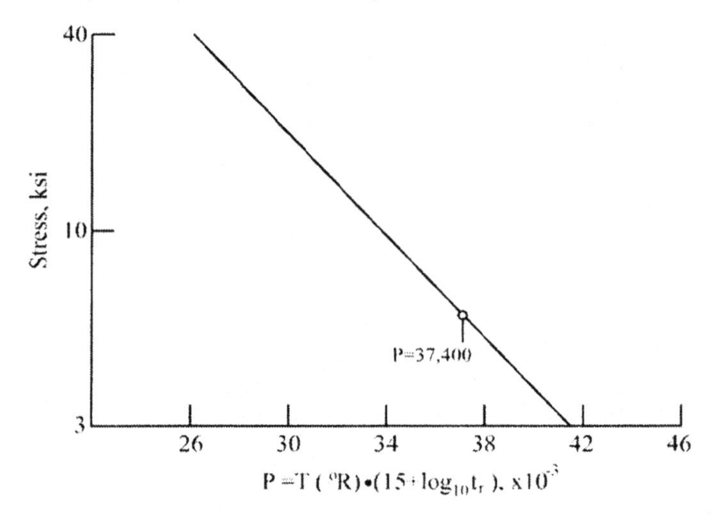

Figure 10.12 Larsen-Miller master life curve for the material of Example Problem 10.4.

As indicated in the figure, the Larson-Miller parameter corresponding to a stress of σ = 6000 psi is P = 37,400. At this stress level, what temperatures correspond to failure in: 1,000 hrs, 10,000 hrs, and 100,000 hrs.?

By inspection, it can easily be determined from the figure and definition of the Larsen-Miller parameter, *P* that:

$$T\,(15+3) = 37{,}400^{\circ}R \ \ (1{,}617^{\circ}F)$$

$$T\,(15+4) = 37{,}400^{\circ}R \ \ (1{,}508^{\circ}F)$$

$$T\,(15+15) = 37{,}400^{\circ}R \ \ (1{,}410^{\circ}F)$$

10.6 CREEP DAMAGE MECHANISMS

Creep deformations are essentially thermally aided plastic flow. In other words, and unlike conventional plastic flow which requires increasing stress increments for plastic strain accumulation, creep flow occurs under constant load and (true) stress conditions. In reality, the diffusional atomic motion that occurs during creep, but not low temperature straining, gives rise to damage mechanisms not observed at ordinary testing temperatures. Moreover, even standard (rapid) tensile tests performed at elevated temperatures indicate the same damage mechanisms as their room temperature counterpart, since there is insufficient time to allow for extensive diffusional motion.

In polycrystalline metals composed of many small single-crystals or grains, the boundaries provide sites for relatively enhanced diffusional motion of atoms, or ions. As grains elongate in the direction of maximum principal stress, their boundaries (interfaces of relatively high crystalline disorder) allow adjacent grains to "slide" relative to each other under the influence of local shear stresses. Such sliding while the grains themselves deform, translate, and/or rotate, will lead to deformation incompatibilities at the grain boundary. If the totality of motion cannot relieve the strain incompatibility at the grain boundary, cracks and voids must eventually form. Figure 10.13 illustrates this behavior for two common void formation mechanisms. The observation of grain boundary voids between adjacent grains and at boundary triple junctions is clear evidence that creep deformations have taken place. In contrast, static load failures occurring in "short" times will lead to porosity generation mainly within the grains. The appearance of this characteristic micro-damage in a failed member should lead the examiner to conclude that creep deformations were, at the least, present and may well have caused the observed failure.

10.7 HIGH-TEMPERATURE MATERIALS

In selecting materials for elevated temperature service, two principal concerns must be addressed. Obviously, one must be limiting the excessive accumulations of strains, and preventing premature failure by creep rupture. However, not all materials lose strength at the same rate as temperatures increase. In fact, tailored metallic alloys have been specifically developed to retain strength and creep resistance at elevated temperatures, even though their room temperature properties may be less than optimal. In addition, if elevated temperature exposure occurs in air or any other chemically active environment, chemical degradation will also be a significant factor. It should therefore come as no great surprise to see many an engineer running for the hills since the control of creep is not an easy or relished role.

Due to the dissociation of water, or simply because the environment is the usual atmospheric mix during combustion, oxygen is often present in high-temperature environments. In these instances, oxidation cannot be tolerated indefinitely, as the structure will literally burn up. On the other hand, iron, and many steel alloys will continuously oxidize since the surface layers are easily permeated by fresh oxygen, leading to the gradual bulk degradation of the member. Iron-based alloys that are heat resistant must contain elements that form stable, impermeable, and fracture resistant oxide layers; otherwise, cracked surfaces would expose the "fresh" alloy underneath. Alloys such as the AISI 300 series stainless steels contain large weight fractions of nickel and chromium, and can therefore develop reasonably durable surface oxide layers.

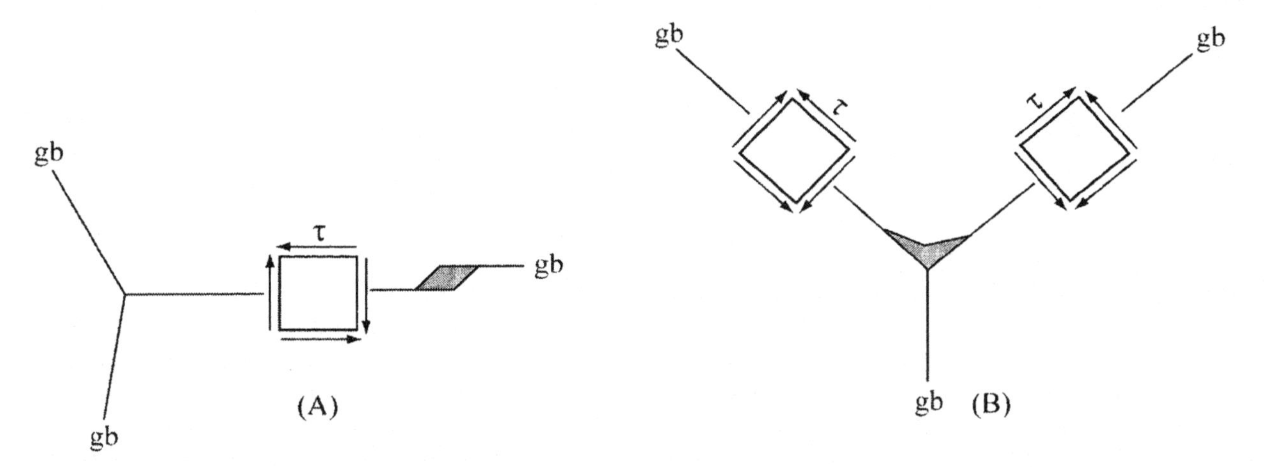

Figure 10.13 Void formation in creep for (A) straight grain boundary with jog, and (B) triple junction between three grains.

Not surprisingly, many nickel-bearing alloys have been specifically developed for elevated-temperature application and carry well-known proprietary names such as the Inconels, Monels, and Renes to name a few. Unfortunately, the price of nickel periodically becomes quite high, so the search for substitutes, preferably with a lower price tag is a contemporary endeavor of some urgency. One solution to the oxidation problem is to employ a material that is either already oxidized or extremely stable in the presence of oxide-rich environments. Metallic oxide ceramics are of course, unaffected by oxygen as they have already been burnt, so to speak. Many, but not necessarily all metal carbides and metal nitrides display notable resistance to oxidation. In this context, ceramic materials are potentially viable design alternatives. Generically speaking, ceramics still suffer from the basic drawback (they are brittle) that has historically led humanity to adopt metals in preference to cheaper and more abundant ceramics. Recent advances in toughening ceramics by creating composites may be changing this calculus.

As noted earlier, grain boundaries clearly play a dominant role in creep rupture. In general, the coarser the grain size, the lower the grain boundary area per unit volume ratio and the longer the rupture lifetime. Unfortunately, large grain sizes are usually associated with lowered static strength ratios. To help compensate, custom grain structures have been devised to increase rupture resistance. One approach uses directionally solidified grains (or even single crystals) to minimize grain boundary area perpendicular to the maximum principal stress direction. This approach is frequently used in specialized structural elements such as turbine blades. The emergence of nano-grained and amorphous or so call "liquid" metals (not sent from the future to kill rebel leaders) offer many advantages as well. Undoubtedly, the next decade or two will see major advances in tailoring the high temperature response of a material, especially as composite and "nano" technologies emerge.

PRACTICE EXERCISES

1. A critical part of any high performance auto exhaust system is the Muffler Bearing; the following strain time creep data (in/in) were obtained for a candidate low alloy steel at 900°F.

Time (hours)	3,000 (Psi)	5,000 (Psi)	8,000 (Psi)	11,000 (Psi)
50	0.02	0.045	0.099	0.135
200	.028	.053	.140	.238
400	.030	.060	.169	.305
600	.031	.069	.195	.368
800	.034	.076	.224	.433
1,000	.036	.084	.253	.500

The Dorn-Miller steady state creep rate equation is expected to be valid for this material:

$$\dot{\varepsilon}_{ss} = \left(\frac{\sigma}{\sigma_0}\right)^m e^{-\Delta H/kT}$$

At constant temperature (T = 1360°R), the rate equation becomes:

$$\dot{\varepsilon}_{ss} = B\sigma^m, \text{where } B = \left(\frac{1}{\sigma_0}\right)^m e^{-\Delta H/kT}$$

Find the steady state rate for each of the stresses employed above, and from a log-log plot of $\dot{\varepsilon}_{ss}$ versus σ, evaluate *m* and *B*

2. Steady state creep rates were determined for a nickel-cobalt-iron high temperature alloy designated for turbine blades for various stresses and temperatures, as noted below. Determine the activation energy ΔH (σ) at stresses of 20, 30 and 40 ksi. In addition, find the exponent, *m* and coefficient, σ_0 in the Dorn-Miller equation using a plot of $\log_{10} \dot{\varepsilon}_{ss} + (0.434)\ \Delta H(\sigma)/kT$ versus $\log_{10}\sigma$,

	Steady State Creep Rate (in./in./hr.) at Stresses of:		
Temperature (°F)	20 ksi	30 ksi	40 ksi
1,140	7.1×10^{-6}	1.3×10^{-4}	3.3×10^{-2}
1,040	1.6×10^{-6}	4.3×10^{-5}	1.0×10^{-2}
940	3.0×10^{-7}	8.7×10^{-6}	2.7×10^{-3}
840	-----------	1.5×10^{-6}	6.2×10^{-4}

3. The Starship Enterprise uses Dylithium Crystals in a matter/anti-mater, quantum intermix, 11th dimensional string-tuner/singularity-controller employing gravitron particles to heat water for coffee and tea. In order to keep the crew properly caffeinated, it operates at a stress of $\sigma = 16{,}000$ psi with the corresponding steady-state creep rates found to be:

Temperature (°F)	Steady State Creep Rate (in./in./hr.)
900	2.6×10^{-5}
1,000	1.27×10^{-3}
1,100	4.10×10^{-2}
1,200	8.5×10^{-1}

Find the activation energy, ΔH, for the crystals at 16,000 psi in order to keep the crew happy and avoid a reversal of the current expansion of the universe, thus leading to the "Big Crunch," and the end of all space-time (definitely very bad).

4. The high-pressure pipe fabricated from the steel of Problem 3 is to carry a moving and very toxic fluid at an internal pressure of 2,000 psi at an operating temperature of 800°F. The pipe has a mean radius of 5.0 inches, and a wall thickness of 0.5 inches. The stress exponent, m = 5.0, and it may be safely assumed that the activation energy previously determined is independent of stress. How many hours will it take for the diameter of the pipe to increase by 1.0 percent due to creep? Note: you may wish to use the data of Problem 3 to evaluate σ_0 and definitely avoid drinking the fluid (just say no!).

5. The following creep rupture data were secured for a high alloy steel (20 percent Cr, 20 percent Ni, and 20 percent Co) for aircraft engine mounts. The Larson-Miller constant C has been determined to be 17.5. Plot a master life curve for the alloy, $\log_{10} \sigma$ vs. $T(°\text{Abs})(C + \log_{10} t_r)$.

Temperature (°F)	Stress (ksi)	Rupture Time (hrs.)	Temperature (°F)	Stress (ksi)	Rupture Time (hrs.)
1,000	70	1,677	1,350	17.5	16,964
	80	433		20	9,529
	90	109		25	900
	100	22		28	342
				34	59
1,100	50	3,149	1,500	10	14,109
	60	264		12.5	5,052
	70	109		15	1,354
	80	25		20	267
				22.5	88
1,200	25	43,978		25	25
	30	11,937			
	35	2,243			
	40	756			
	50	93			
	60	26			

6. From the master life curve formed in Problem 5, determine the maximum operating stress for a tensile member designed to last 300,000 hrs. at 1,325°F. If the data for the point on the master life curve used in the first part were to be obtained from a 1,000 hr. creep rupture test, what would be the best temperature?

7. A boiler tube hanger is to be made of the heat-resisting alloy from Example Problems 10.1 through 10.4. The thermal environment is 1,400°F, and the hanger is to elongate only 10 percent in 5 years. If the hanger is to have a stress safety factor of two, what is the maximum allowable stress for the hanger? As an added confidence factor, one might wish to know how long it will be to rupture. For the stress and temperature in the first part, what is the predicted time to rupture?

8. The following creep data were obtained for polypropylene at room temperature and 11.2 MPa stress level. Further data, when analyzed gives:

$\dot{\varepsilon}_{ss} = 8.6 \times 10^{-5}$/minute for $\sigma = 9.78$MPa

$\dot{\varepsilon}_{ss} = 7.1 \times 10^{-5}$/minute for $\sigma = 8.82$MPa

Time (minutes)	**Strain**
0	0
1	3.67×10^{-3}
3	5.41×10^{-3}
5	6.82×10^{-3}
7	7.86×10^{-3}
9	8.24×10^{-3}
11	8.57×10^{-3}
13	8.99×10^{-3}
15	9.32×10^{-3}
25	10.8×10^{-3}
35	11.9×10^{-3}

For this material, what are the constants *B* and *m* in $\dot{\varepsilon}_{ss} = B\sigma^m$ at room temperature?

NUMERICAL ANSWERS TO PROBLEM ASSIGNMENTS

CHAPTER ONE

All strains are in units of micro-strain, (μ in./in. or μm/ m)

1: $\varepsilon_I = 922 \quad \varepsilon_{II} = -122 \quad \varepsilon_{III} = 0 \quad \gamma_{max} = 1044$

2: $\varepsilon_I = 847 \quad \varepsilon_{II} = -47 \quad \varepsilon_{III} = 0 \quad \gamma_{max} = 894$

3: $\varepsilon_I = 1023 \quad \varepsilon_{II} = -323 \quad \varepsilon_{III} = 200 \quad \gamma_{max} = 1346$

4: $\varepsilon_I = 1047 \quad \varepsilon_{II} = 153 \quad \varepsilon_{III} = 0 \quad \gamma_{max} = 894$ (in plane); 1047 (x, z-plane)

5. Note: answers (C.B.D.) refers to "Cannot Be Determined"

Case	ε_I	ε_{II}	ε_{III}
i	600	-200	C.B.D
ii	424	-524	C.B.D
iii	205	-705	C.B.D

CHAPTER TWO

1: σ_I = 31.2 ksi σ_{II} = -11.2 ksi θ=22.5^0

2: σ_I = 8 ksi σ_{II} = -2 ksi θ=26.6^0

3: σ_I = 15 ksi σ_{II} = -15 ksi θ=45^0

4 : σ_I = 12.2 ksi σ_{II} = 2.8 ksi θ=29^0

5: σ_I = 31.2 MPa σ_{II} = -11.2 MPa σ_{III} =5 MPa

6: σ_I = 31.2 MPa σ_{II} = -11.2 MPa σ_{III} =-15 MPa

7: σ_I = 31.2 MPa σ_{II} = -11.2 MPa σ_{III} =-15 MPa (same as #6 except σ_{xx} is now Principal)

8: σ_I =11.6 MPa σ_{II} = -9 MPa σ_{III} =-25.3 MPa

n_{x1}= 0.0266 n_{y1}= -0.8638 n_{z1}= -0.5031

n_{x2}= -0.6199 n_{y2}= 0.3806 n_{z2}= -0.6862

n_{x3}= 0.7833 n_{y3}= 0.3307 n_{z3}= -0.5264

9: a. σ_I = 15 ksi σ_{II} = 7.5 ksi b. σ_I = =σ_{II} = 7.5 ksi

10: σ_I =338 MPa σ_{II} = 0 MPa σ_{III} =-58 MPa θ=22.5^0

11: 0.5" Diameter Section: σ_I =22.2 ksi σ_{II} = 0 ksi σ_{III} =-3.76 ksi θ=22.5^0

0.75" Diameter Section: σ_I =36.2 ksi σ_{II} = 0 ksi σ_{III} =-3.85 ksi θ=18.1^0

12: a. Sorry, can't give it away! b. ρ = 0.01387"

13: P=55-58 kip (depends on how you define the stress and K_σ)

14: a. σ=735 psi b. σ=±1306 psi c. σ=-572 ksi (top) and σ=2041 ksi (bottom)

CHAPTER THREE

1: a. $\begin{pmatrix} 48.7 & 4.5 & 0 \\ 4.5 & 37.5 & 0 \\ 0 & 0 & 28.5 \end{pmatrix}$ ksi b. $\begin{pmatrix} 9 & 0 & 6.75 \\ 0 & -15.8 & 0 \\ 6.75 & 0 & 6.75 \end{pmatrix}$ ksi

2: a. $\varepsilon = 4.7 \times 10^{-3}$ b. σ =-739 MPa using stiffness at 800^0C (-832 MPa using average stiffness over temperature range)

3: a. σ_{st}=-22.8 ksi b. σ_{Al}=-11.4 ksi c. $|\delta|=0.06\text{x}10^{-3}$L (to left so steel contracts)

4: σ_{Al} =132 MPa σ_{SiC} = -506 MPa

5: T_f=67^0C

6: σ_{zr} = 557 MPa; 28% increase over the 1-D model

7: Effective Modulus (by ROM) = 50.1 GPa

8: Fiber volume fraction (by ROM) = 0.38

9: Volume fraction of fiber is 0.38

10: $\varepsilon_{ij} = \begin{pmatrix} 0.88 & 4.82 & 0 \\ 4.82 & 13.1 & 0 \\ 0 & 0 & ? \end{pmatrix} \text{x}10^{-3}$ With no out-of-plane data given, ε_{zz} cannot be calculated.

11: a. $E_{app} = \dfrac{E}{1-\nu^2}$ b. $E_{app} = E\left(\dfrac{1+\nu}{1+\nu-2\nu^2}\right)$ For a Poisson ration of ν = 0.3, the E_{app} values are 1.1E and 1.16E, respectively

CHAPTER FOUR

1. **a.** σ_p = 37,000 psi **b.** σ_y = 43,000 psi **c.** E = 10.3 x 10^6 psi **d.** u_r = 66.5 in.-lb./in.^{3}
   e. TS = 44,000 psi **f.** % Elong. = 0.9%
2. **a.** σ_p = 32,000 psi **b.** UYP = 42, 5000 psi **c.** LYP = 41,000 psi **d.** σ_y = 42,500 psi
   e. E = 30 x 10^6 psi **f.** u_r = 17 in.-lb./ in.^{3} **g.** TS = 68,500 psi **h.** % Elong. = 25%
   i. L_{franc} = 2.5 inches **j.** % Red Area = 40%
3. **a.** E = 12 x 10^6 psi **b.** σ_p = 38,000 psi **c.** σ_y =42,500 psi **d.** σ_{uts} = 57,000 psi
   e. u_r = 60 in.-lb./in.^{3} **f.** T = 22,450 in-lb/in^3, **g.** 40%
4. a. TS = 27,000 psi b. % Elong. = 21% c. T = 4790 in.-lb./in^3
5. a. ε_{tot} = 0.0039 b. ε_{el} = 0.0024; ε_{pl} = 0.0015 c. ε_{pl} = 0.0062
6. a. ΔL = 0.0176 inches b. ΔW = 0.246 x 10^{-3} inches, ΔT = 0.082 x 10^{-3} inch

c. $\Delta V = 0.00145$ in.^{3}

7. **b.** $E_{init} = 25 \times 10^6$ psi & $E_{tan} = 50$ psi **c.** $E_{sec} = 20 \times 10^6$ psi **e.** TS = 27,000 psi
   f. % Elong. = 0.25%
8. a. $E_{sec} = 71.4$ psi b. Res. = 414 in.-lb./in.^{3} c. Hyst. = 303 in.-lb./ in.^{3}
9. Using Modulus of toughness; a. 3000 in.-lb./in.^{3} b. 5720 in.-lb./in.^{3} c. 4740 in.-lb./in.^{3}
10. a. $A_{min} = 3.1$ in.^{2} b. $A_{min} = 3.4$ in.^{2} c. $A_{min,\,tot} = 3.4$ in.^{2}
11. a. $\sigma_p = -30{,}612$ psi b. $\sigma_y = -31{,}633$ psi c. $E = 30.6 \times 10^6$ psi
12. **a.** CS = 1193 psi **b.** % Cont. = 0.216 **c.** $E_{int} = 3.08 \times 10^6$ psi **d.** $E_{sec} = 1.81 \times 10^6$ psi
   e. $E_{tan} = 0.77 \times 10^6$ psi **f.** T = 1.95 in.-lb./in.^{3}
13. $\nu = 0.5$
14. P = 2,704 lbs.

CHAPTER FIVE

1. a. $\sigma_y = 42.4$ ksi (MSST) or 37.1 ksi (DET)
2. $\sigma_y = 255$ MPa (MSST) or 223 MPa (DET)
3. a. Elastic b. Just yields c. Elastic-plastic
4. Torque at yield = 220,831 in.-lb. (MSST) or 255,000 in.-lb. (DET)
5. Thickness = 3.75 mm (MSST) or 3.25 mm (DET)
6. Torque at yield = 143,600 in.-lb. using thin-walled solutions
7. a. $\sigma_{II} = 233$ MPa; $\sigma_I = 3\sigma_{II} = 700$ MPa and $\sigma_I = 3\sigma_{II}$ for load trajectory
   b. $\sigma_{II} = 450$ MPa and $\sigma_I = -250$ MPa
8. $\sigma_y = 55$ ksi
9. $\sigma_y = 47$ ksi

CHAPTER SIX

1. For brass: $\sigma_0 = 80{,}000$ psi and $n = 0.14$
   For stainless steel: $\sigma_0 = 158{,}000$ psi and $n = 0.106$
2. $\sigma_0 = \sim 200{,}000$ psi and $n = 0.565$
3. $\%R.A. = 100(1 - e^{-\varepsilon_t})$
4. P = 131,000 lbs. $\Delta L = 11.2$ inches
5. a. TS = 58,279 psi b. %R.A. = 22.1%

CHAPTER SEVEN

1. a. 6.68 GPa $\approx$ E/10 b. 2.14×10^{-5} m
2. a. 133 N-m b. 141 kN

3. a. 0.004 in. b. 0.01 in. c. 0.03 in.
4. a. σ_{design} = 60 ksi b. Structure is indeed Safe for σ_{design} = 60 ksi

   c. $a_{crit} \cong 0.11$ inches (by iteration) when applied stress is 60 ksi
5. a. 0.176 in. b. 0.028 in. c. 0.011 in.
6. a. P=1570 MPa b. P=8.8 MPa
7. a. 61,650 in.-lb. b. 0.00282 in.

CHAPTER EIGHT

1. a. Fatigue Strength = 17 ksi b. σ_a = 13.4 ksi
2. b. Endurance Limit = 90 ksi b. σ_m = 51.6 ksi
3. $n_3 = 1.8 \times 10^7$ cycles
4. Lifetime = 153 days
5. A=1.54 x 10^{-13} and m=4.2 Hence: $da/dN = 1.54 \times 10^{-13}(\Delta K_I)^{4.15}$
6. For the students pleasure ☺
7. $N_c = 3.87 \times 10^6$ cycles
8. a. σ_{ao} = 50 ksi b. σ_{ao} = 16.7 ksi

CHAPTER NINE

1. **a.** A log (σ) vs. time plot of the relaxation data is linear, so the material is Maxwellian.

   b. E = 60,000 psi **c.** η = 780,400 psi-minutes **d.** τ = 13 minutes
2. σ (at 30 minutes) = 716 psi
3. ε (at 40 minutes) = 0.18 in./in.
4. After release, the plastic strain = 0.12 in./in.; remains at that value to infinity and beyond..
5. After release, the plastic strain is 0.055 in./in.
6. ε_{tot} = 0.30 in./in.
7. a. t=10^4 hours
8. Have fun ☺
9. E = 17,600 psi

CHAPTER TEN

1. m = 2.69; B = 4.54 x 10^{-15}

	3 ksi	5 ksi	8 ksi	11 ksi
$\dot{\varepsilon}$	1.0 x 10^{-5}/ hr.	3.88 x 10^{-5}/ hr.	1.412 x 10^{-4}/ hr.	3.25 x 10^{-4}/hr.

2. m = 5.0 and σ_0 = 2834 psi

$\Delta H(20) = 2.63 \times 10^{-18}$ in.-lb. $\Delta H(30) = 2.37 \times 10^{-18}$ in.-lb. $\Delta H(40) = 2.-07 \times 10^{-18}$ in.-lb.

3. $\Delta H = 5.09 \times 10^{-18}$ in.-lb.
4. t = 8971 hr. = 373 days
5. You know the routine ☺
6. a. σ=13.7 ksi b. T=1540 °F
7. a. $\sigma_{allowable}$ = 4962 psi b. $t=5.61\times10^{6}$ hrs = 640 years
8. m =2.63 $B=2.31\times10^{-7}$

INDEX

A
Activation potential energy, 178
Adiabatic energy, 130
Alignment, 14
Alloying, 52
American Society for Testing and Materials
  (ASTM), 76, 144
Anelasticity, 51
Angle of Twist, 102
Angular deformation, 41
Anisotropy, 53, 63–65
Annealing, 85
Arbitrary stress states, 104–105
Arrhenius, 178, 201
ASTM. *See* American Society for Testing
  and Materials
Axis rotations, 24–32

B
Beams, 40
Bending
  four-point, 90
  moment, 155
  strength, 89–90
  stress, 89
Biaxial stresses, 39
Bi-material combinations, 57
Bi-metallic strips, 57
Boltzmann
  constant, 178
  superposition principle, 186
Brinell scale, 90
Brittle failure, 133–135, 140
Brittle fracture, 133
Brittle glass, 151
Brittle metal, 45
Brittle strength, 88–89

C
Cartesian coordinate system, 111
Cartridge brass, 127
Cast iron, 98
Central crack, 137
Ceramic coated turbine blades, 61
Chain rule, 7
Characteristic distance, 42
Circle periphery, 11
Circular bar, 101
Circumferentially crack, 138
Coefficient of Linear Thermal Expansion, 55
Cohesive strength, 132–133
Combined stress, 133–135
Complete displacement fields, 4
Compliances, 65
Component sheets, 63
Concrete, 79, 99
Constant stress, 167, 202
Constitutive equations, 53
Constrained thermal strains, 56–63
Contact stress, 73
Continuous displacement fields, 4
Cooling gases, 62
Correction factors, 138
Crack, 129. *See also* Fatigue crack; Fracture
  central, 137
  circumferentially, 138
  edge, geometries, 137, 139
  environment-aided, 135
  extension, 135–139
  initiation, 154
  mode I, 136
  penny-shaped, 131
  plastic zone, 141
  propagation, 168
  stability, 139, 140
Creep, 179
  behavior, 199–200
  compliance, 188
  damage mechanisms, 211
  diffusional, 211
  Larson-Miller, 208, 209
  master, 191
  rupture, 207–211
  steady-state, 199
  strain, 200–207
  tertiary, 200
  transient, 199
  void formation in, 212
Creep model
  Arrhenius, 178, 201
Critical flaw sizes, 145
Cross-sectional areas, 59
Cumulative damage concepts, 162–165
Cyclic amplitude, 163
Cyclic loading, 161
Cylindrical configuration, 76

D
Damage, 163
Dashpot element, 177
Deformation, 75, 122
  angular, 41
  exaggerated slip, 86
  normal, 1

plane stress, 143
relative displacement and, 3
shear, 1
DET. *See* Distortion Energy Theory
Deviatoric stress tensor, 109
Diametral interference, 56
Differentiation, 178
Dilation, 99
Dimples
elongated, 150
equiaxed, 148
fracture surface, 149
Direction cosines, 38
Directionality, 21
Discontinuous yielding behavior, 85
Dislocations, 129
Displacement fields, 2–5
Distortion, 1
Distortion Energy Theory (DET), 109–113, 116
Dorn parameter, 210
Dorn parametric method, 209
Dorn-Miller relationship, 204
Dual directionality, 19
Ductile, 45, 81
fracture, 148
overloads, 88
transgranular fracture, 148

E

Edge crack geometries, 137, 139
Elastic
constant transformations, 69–71
limit, 59, 85, 148
magnitudes, 54
modulus, 52, 74, 78, 79, 81, 83, 135, 151
response, 51–53, 77–80
strain energy per unit volume stored, 93
stress concentration factor, 43
Electrolytic tough pitch copper, 196
Electronic circuit boards, 61
Elliptical hole, 44
Elongation, 126
Endurance limit, 157
Endurance strength, 157
Energy
activation potential, 178
adiabatic, 130
DET, 109–113, 116
strain, 93, 110, 140
Environment-aided cracking, 135
Equal stress amplitude, 160
Equilibrium
force, 25, 32
sectioned structural member in, 18
stress, 24–32
Exaggerated slip deformation, 86
Experimental response determination, 187–194
Extension crack, 135–139

F

Failure. *See also* Fatigue failure
analysis, 146
brittle, 133–135, 140
intergranular, 147
transgranular, 148
Failure criterion, 134, 135, 162
distortion energy, 108–112
Goodman, 162
Fast fracture, 166
Fatigue
completely reversed load, 157
concentration factor, 165
degradation, 135
fractography, 170–172
limit, 157
strength, 165, 173
Fatigue crack
initiation, 163, 166
propagation, 166–170
striation lines, 171
Fatigue failure, 157–158, 166
physical nature of, 166
FBD. *See* Free-body diagram
Fiber composite, 53
First law (thermodynamics), 93
Fixed grip, 131
Flat ellipses, 134
Flaw detection, 169
Force equilibrium, 25, 32
Four-point bending, 90
Fractography, 147–150
Fracture
brittle, 133
cleavage, 148
dimples, 149
ductile, 148
intergranular, 147
material, resistance, 139–140
mechanics design concepts, 145–147
plane appearance, 143
resistance, 139–140
strain at, 83
surface dimples, 149
in tension, 86–88
toughness, 139, 142
Free surface, 12–15
Free-body diagram (FBD), 58

G

General linear viscoelastic response, 186
General stress history, 161, 163, 164
Glass, 151

reinforced polymer, 68
transition temperature, 199
Goodman failure criterion, 162
Grains, 147, 211
Griffith, A.A., 129
Griffith structural problem, 139
Grooved shaft, 146

H
Hardened steel, 164
Hardening exponent, 123, 124
Hardening strain, 84–85, 123–125
Hardness, 90–91
Heat treatment, 52
High strength, 140–144
High-temperature materials, 211–212
Histogram cycles, 159
Homogenous strains, 85–86
Hooke, Robert, 53
Hookean material, 93
Hooke's Law, 53–55, 63–65
Hydrostatic, 109
Hysteresis, 94, 98

I
IFT. *See* Internal friction theory
Ignorance factor, 45
Ill-matched collision, 162
Imposed shear stresses, 53
Imposed strain history, 180
Independent elastic constraints, 67
Independent shear stress, 23
Independent strain components, 4
Inelastic tensile response, 81–83
Inherent complexity, 138
Initiation crack, 154
Initiation site, 170
instability, 132
Instantaneous strain, 121–123
Instantaneous stress, 123
Interatomic spacings, 45
Interference fit, 56
Intergranular failure, 147
Intergranular way, 147
Internal friction theory (IFT), 113–117
Internal porosity, 88
Isostrain, 64
Isotropic
solid, 52–56
steel, 54

K
Kelvin, 201
Kelvin effect, 12. *See also* Voight-Kelvin model

L
Laminated composite material, 64
Larson-Miller
approach, 209
constant, 214
creep master life curve, 209
creep rupture data correlation, 208
parameter, 209
Linear elastic shear stress, 53
Linear polypropylene, 80
Load, 126
histories, 51
instability, 125–126
Loading, 75
Low-ductility, 87, 129
Lüders band, 86

M
Martensitic steels, 169
Master creep compliance curve, 191
Master Life Curve, 189, 209
Master relaxation modulus curve, 193
Material composition, 75–76
Material fracture resistance, 139–140
Material stiffness, 52
Maximum load point, 125
Maximum shear
planes, 30
stress, 30, 33, 105–109
values, 36
Maxwell material, 180–183, 195
Maxwell viscoelastic model, 178–179
Maxwellian solid, 179
Mean stress, 160–162
Mechanical properties, 77
Mechanical responses, 77
Medium carbon annealed steel, 87
Microvoid coalescence, 149
Miner's rule, 163
Mode I crack, 136
Modulus of Elasticity, 52, 74, 78, 79, 81, 83, 135, 151
Modulus of Resilience, 93
Modulus of Rigidity, 52
Modulus of Rupture (MOR), 90, 99
Mohr, Otto, 8
Mohr's Circle, 10
for designated regions, 106
stress orientation of, 31
three-dimensional, 34
Moiré Interferometry, 12
MOR. *See* Modulus of Rupture
Multi-material system, 60

N
Natural flaw populations, 133

NDE. *See* Nondestructive evaluation
Near-yield regime, 97
Necking, 87
Newtonian fluid, 177
Newtons, 20
Nominal measure, 121
Nominal strain, 76
Nondestructive evaluation (NDE), 169
Nonlinear elastic response, 80
Normal deformations, 1
Normal points, 28
Normal strains, 2, 8
Normal stresses, 19
Notch effects, 165

O
Oblique plane, 25
Octahedral shear stress, 109
Offset yield method, 82
Orientation, 14
Orthotropic composites, 66–69
Orthotropic filament, 66
Orthotropic moduli, 67
Orthotropic sheet material, 70
Orthotropic solids, 67

P
Parabolic strain hardening law, 126
Paris Crack Propagation Law, 168
Pascals, 20
Penny-shaped crack, 131
Percent elongation, 83, 96, 97
Perfectly brittle materials, 129–132
Photoelasticity, 12
Pitting, 129
Plane element, 1, 3, 4
Plane strain, 140, 144–145, 153
Plane stress, 21
  brittle failure envelope, 134, 140
  conditions, 141
  deformation, 143
  element, 33
Plastic deformation, 121
Plastic strain, 81, 103
Plastic zone crack, 141
Plasticity, 45, 140–143
Plate with a circular hole, 50
Point load, 42
Poisson's ratio, 52, 67, 78, 89
Polycrystalline ceramics, 147
Polyethylene, 196
Polymers, 88, 120, 199
Polypropylene type material, 95
Porosity, 87, 88, 147, 211
Positive thermal expansion, 60
Potential stress, 21
Potential yield-surfaces, 107
Pressure vessels, 39–40
Principal material directions, 69
Principal stress, 29, 31, 33, 36
Principle axes, 11, 29
Principle strain, 11
Proof tests, 146
Proportional limit, 79
Pure shear load path, 118

Q
Quasi brittle fashion, 140

R
Radial swelling, 89
Radius of curvature, 44–46, 165
Rankine, 201
Real polymer viscoelasticity, 186
Reduced time, 189
Reinforced-composite panel, 66
Relative displacement, 3
Relaxation modulus, 187, 188
Relaxation time, 181
Residual stress, 160
Restraint, 60
Resultant force vector, 17
Rigid walls, 59
Rigidly restrained bar, 58
River patterns, 148
Rockwell, 90, 91
Rods, 40
Rubber, 1, 51
Rule of mixtures, 64

S
Saint Venant's principle, 42
Scanning electron micrograph, 148, 149
Scratches, 129
Secant modulus, 80, 98
Self-equilibrating, 57
Shaft geometry, 47
Sharp ellipses, 134
Sharpness, 44
Shear
  deformation, 1
  imposed, stresses, 53
  independent, stress, 23
  linear elastic, stress, 53
  lip, 88
  maximum, planes, 30
  maximum, stress, 30
  maximum, values, 36
  modulus, 52, 102
  octahedral, stress, 109
  pure, load path, 118
  strain, 9

stress, 19
yielding, 101–104
Shift factor, 190
Shrink-fit assembly, 56
Simple beam theory, 42
Small strain, 55
S-N curves, 158
S-N diagrams, 158–159
Spring element, 177
Stainless steel, 127
Standard test specimen, 144
Static fatigue, 135
Statically equivalent loadings, 42
Steady state creep data, 205
Steady-state standoff, 199
Steels
hardened, 164
isotropic, 54
martensitic, 169
medium carbon annealed, 87
stainless, 127
Stiffnesses, 65
Strain, 93
constrained thermal, 56–63
creep, 200–207
displacement fields and, 2–5
energy, 93, 110, 140
at fracture, 83
hardening, 84–85, 123–125
homogenous, 85–86
imposed, 180
independent, 4
instantaneous, 121
measurement, 5, 12
nominal, 76
normal, 2
parabolic, 126
plane, 140, 144–145, 153
plane state of, 10
plastic, 81
principle, 11
response, 53
rosette, 13, 14
sensitivity, 12
shear, 9
small, 55
space points, 10
stress, 51, 126
tensile, curve, 80, 95
thermal, 55–56
transformations, 5–12
true, 124
Strength
brittle, 88–89
coefficient, 125
cohesive, 132–133
endurance, 157
fatigue, 165, 173
high, 140–144
tensile, 88, 125, 162
theoretical cohesive, 132
ultimate, 125, 129
yield, 82, 84, 115, 145
Stress, 136. *See also* Crack; Fracture
analysis, 17–50
arbitrary, 104–105
bending, distribution, 89
biaxial, 39
combined, 133–135
concentration, 42–46, 137
concept of, 19
constant, 167, 202
contact, 73
corrosion, cracking, 135
deviatoric, 109
elastic, 43
equal, 136
equilibrium, 24–32
fatigue, 165
general, history, 161, 163, 164
guessers, 21
imposed shear, 53
independent shear, 23
instantaneous, 123
intensity, 137, 151, 174
invariants, 37
linear elastic shear, 53
maximum shear, 30, 33, 105–109
mean, 160–162
Mohr's Circle and, 31
normal, 19
octahedral shear, 109
at a point, 17–20
potential, 21
principal, 29, 31, 33, 36
relaxation, 179, 180
residual, 160
resulting state, 108
riser, 43
shear, 19
signs, 20–21
state, 68
strain, 51, 126
structural, analysis, 39–42
tangential normal, 43
tensile, 76, 80, 95
three-dimensional, 21–22, 32–36, 37
transformations, 24–32
true, 123
two-dimensional, 21, 104
uniaxial, 158
units of, 20

variations, 23
volume element, 130
zero-valued, 69
Striations, 170
Structural stress analysis, 39–42
Suddenness, 44
Surface energy, 151
Symmetry, 60

T
Tangent modulus, 79–80, 98–99
Tangential normal stress, 43
Temperature, 199, 211–212
Tensile elastic constants, 79
Tensile plate, 44
Tensile response curve, 81
Tensile specimen, 77
Tensile strength, 88, 125, 162
Tensile stress, 76, 80, 95
Tensile test variables, 75–76
Tension, 92–95
Tensor, 22, 65
Theoretical cohesive strength, 132
Theory validity, 117–118
Thermal strains, 55–56
Thermoelastic properties, 72
Thin-walled pressure vessel, 39, 108
Thin-walled tubular specimen, 103
Three-dimensional analyses, 37–39
Three-dimensional loading, 22
Three-dimensional Mohr's Circles, 34
Three-dimensional stress, 21–22, 32–36, 37
Three-element viscoelastic model, 184
Three-point bend test, 144
Time, 75, 181
Torsion, 40, 41
Toughness, 93, 94
Transformation equations, 7
Transformational equivalences, 6
Transgranular failure, 148
Transgranular fashion, 147
Transmitted force increment, 18
Tresca yield condition, 104
True strain, 124
True stress, 123
Tungsten carbide, 91
Twisting torque, 101
Two-dimensional stress, 21, 104

U
Ultimate tensile strength, 125, 129
Undetected flaws, 145
Uni-directional composite material, 70
Unidirectional plate, *71*
Upper yield points, 85

V
Viscoelastic models, 178–179, 183–185
Viscoelastic response, 186
Viscoelasticity, 177–178, 186
Voight-Kelvin model, 183–186
Volume fraction, 74

W
Work hardening, 84

Y
Yield criterion, 141
Yield points, 85–86, 106
Yield strength, 82, 84, 115, 145
Yielding, 82, 85–86
for arbitrary stress states, 104–105
in pure shear, 101–104
Young's Modulus, 78

Z
Zero-valued stresses, 69
Zirconia, 62, 73

CPSIA information can be obtained at www.ICGtesting.com
Printed in the USA
BVOW07s1544100815

412619BV00006B/9/P